AF506286

Granular Materials

Fundamentals and Applications

Granular Materials
Fundamentals and Applications

Edited by

S. Joseph Antony
*School of Process, Environmental and Materials Engineering,
University of Leeds, UK*

W. Hoyle
Consultant, Stockport, UK

Yulong Ding
*School of Process, Environmental and Materials Engineering,
University of Leeds, UK*

ISBN 0–85404–586–4

A catalogue record for this book is available from the British Library

Published by The Royal Society of Chemistry,
Thomas Graham House, Science Park, Milton Road,
Cambridge CB4 0WF, UK

Registered Charity Number 207890

For further information see our web site at www.rsc.org

Typeset by RefineCatch Limited, Bungay, Suffolk, UK
Printed by Athenaeum Press Ltd, Gateshead, Tyne and Wear, UK

Preface

Granular materials are an important part of several engineering applications of technological importance. The industries that handle granular materials include food preparation, pharmaceutical, consumer products, agricultural products, metal powders, mechanical, geotechnical, chemical, nuclear and green industries. In addition, high rates of innovation in these industries means that there is a continuous need for achieving excellence in predicting the fundamental behaviour of granular materials under different working environments. Analysis of physical systems involving granular materials requires clear understanding of their behaviour not only at the single particle level, but should also consider the solutions of multi-physics problem involving multi-scale phenomena from molecular to macroscopic scale.

In this book, we present the recent advances made in theoretical, computational and experimental approaches in understanding the often mysterious behaviour of granular materials, including industrial applications. The approaches presented here are complementary and provide a collective understanding of the behaviour of granular materials. All the chapters, with the right blend of basics and advances, are presented by leading experts from around the world, who have many years experience in their fields.

Chapters one to seven deal with the recent advances made on the fundamental concepts of granular materials. The first chapter presents the quasi-static deformation characteristics of granular materials at both small and large strains. The particles have been considered as a discrete system in the analysis. The second chapter focuses on the effects of mechanical periodic excitation on a granular medium, dealing with several fundamental issues. The third chapter presents the advances made on the constitutive modelling of granular materials using a continuum approach. The behaviour of particle interactions at elevated temperatures, including adhesion forces and sintering phenomena is presented in chapter four. In this chapter a novel off-line technique has been presented for the determination of actual high temperature interactions. The next chapter examines the critical state behaviour of granular materials under several initial conditions using three-dimensional discrete element modelling. The sixth chapter presents the key features of granular plasticity, providing an insight into the macroscopic trends from the lowest level description of granular microstructure.

The concluding chapter of the first section on the fundamental concepts describes the advances made on the use of Atomic Forces Microscopy (AFM) as a measurement tool to investigate particle interactions, and in particular those in the presence of polymers.

Chapters eight to thirteen present the recent advances made in experimental measurement techniques, and several industrial applications dealing with granular materials. The first of these chapters focuses on the measurement of granular forces in dry systems using AFM. Aspects covered include humid air and conditions under which the particle surfaces may be modified by liquid films or coatings, since such conditions are of great technological importance.

A weakness in the ability to develop sophisticated granular models is the lack of suitable experimental validation methods. Ideally, these methods should be in-situ and not perturb the process modelled in any way. The ninth chapter presents several examples of such methods that can be applied to dry and wet particulate systems. The tenth chapter reviews the fluidisation characteristics of fine particles together with aids in fluidising fine powders. Effects of temperature on fine powder fluidisation are also covered. This chapter ends with the discussion of selected applications and potential applications. The eleventh chapter introduces and reviews various modelling approaches at different length scales for the prediction of granulation under high shear conditions.

Rotary kilns are widely used in the processing of granular solids in the chemical and metallurgical industries. Chapter twelve describes work done at Cambridge University on the dynamics of particle motion in rotary kilns, using cold laboratory scale equipment. The final chapter presents both the theoretical and experimental aspects of granular motion in the transverse plane of rotating drums. Experimental results presented in this chapter were obtained exclusively by using the Positron Emission Particle Tracking (PEPT) technique.

We hope that this book will serve as an excellent reference for scientists, engineers and students working across a wide spectrum of engineering disciplines dealing with granular materials.

We are very grateful to all the contributors for giving so generously of their valuable time and making it possible to produce this book. The support provided by the Royal Society of Chemistry, Cambridge, UK in publishing this book is gratefully acknowledged by the editors.

S. Joseph Antony
Bill Hoyle
Yulong Ding

Contents

Fundamentals

CHAPTER 1

Rates of Stress in Dense Unbonded Frictional Materials During Slow Loading

MATTHEW R. KUHN

University of Portland, 5000 N. Willamette Blvd, Portland, OR 97203 U.S.A.
Email: kuhn@up.edu

1 Introduction

This chapter concerns the transmission and evolution of stress within granular materials during slow, quasi-static deformation. Stress is a continuum concept, and its application to assemblies of discrete grains requires an appreciation of the marked nonuniformity of stress when measured at the scale of individual grains or grain clusters. As an example, numerous experiments and simulations have demonstrated that externally applied forces are borne disproportionately by certain grains that are arranged in irregular and ever-changing networks of *force chains*.[1,2,3] Although much attention has recently been given to the transmission of force at low strains, the current work focuses on the transmission of stress within granular materials at both small and large strains.

When a densely packed assembly of unbonded particles is loaded in either triaxial compression or shear, the behaviour at small strains is nearly elastic, and the volume is slightly reduced by the initial loading (an initial Poisson ratio less than 0.5, Figure 1). Plastic deformation ensues at moderate strains, at which an initially dense material becomes dilatant, and this trend of increasing volume continues during strain hardening, at the peak strength, and during strain softening. At very large strains, the material reaches a steady condition of flow, referred to as the "critical state" in geotechnical engineering practice, in which the material flows at a constant, albeit expanded, volume while sustaining a constant shearing or compressive effort.[4] Besides studying behaviour at the initial and peak states, we will also consider experimental results at this steady state condition and the manner in which the inter-granular forces are distributed and changed during steady state flow. These conditions are investigated with

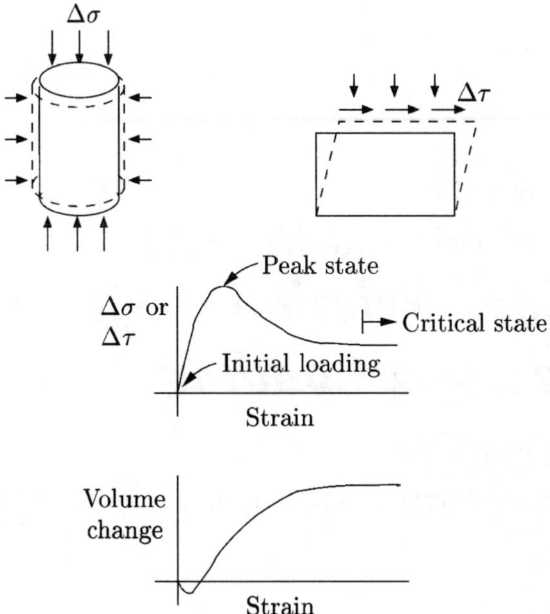

Figure 1   *Typical behaviour of a dense granular material during shearing or unconfined compression to large strains*

numerical simulations of an idealized assembly of circular disks. We will explore mechanisms that underlie the changing stress by separating the stress rate into various constituents and then study their relative influences during simulated loading.

Notation

In this chapter, vectors and tensors are represented in both indexed and unindexed forms with the use of upper and lower case glyphs: $\mathbf{A}$ or A_{ij} for tensors, and $\mathbf{a}$ or a_i for vectors. Inner products are computed as

$$\mathbf{a} \cdot \mathbf{b} = a_i b_i, \ \mathbf{A} \cdot \mathbf{B} = A_{ij} B_{ij}, \tag{1}$$

and tensors are often represented as the dyadic products of vectors:

$$\mathbf{a} \otimes \mathbf{b} = a_i b_j. \tag{2}$$

A juxtaposed tensor and vector will represent the conventional product

$$\mathbf{A}\mathbf{b} = A_{ij} b_j. \tag{3}$$

No contraction is implied with superscripts (e.g., $\mathbf{a}^c \mathbf{b}^c$). The *trace* of a tensor is defined as

$$\text{trace } (\mathbf{A}) = A_{11} + A_{22} + A_{33}. \tag{4}$$

Tensile stresses and extension strains are positive, although the pressure p will be positive for compressive conditions.

2   Partitioning the Stress Rate

The average Cauchy stress $\bar{\sigma}$ within a granular assembly can be computed as a weighted average of the contact forces between grains:[5,6]

$$\bar{\sigma} = \frac{1}{V} \sum_{c \in \mathcal{M}} \mathbf{f}^c \otimes \mathbf{l}^c, \tag{5}$$

where summation is applied to the set of $\mathcal{M}$ contacts within the assembly, and each contact c represents an ordered pair of contacting particles p and q, $c = (p, q)$. The sum is of the dyadic products $\mathbf{f}^c \otimes \mathbf{l}^c$, where $\mathbf{f}^c$ is the contact force exerted by particle q upon particle p, and *branch vector* $\mathbf{l}^c$ connects a reference (material) point on particle p to a reference point on particle q (Figure 2). For the numerical simulations in this study, these reference points are placed at the centres of circular disks. The current volume (area) of the three-dimensional (two-dimensional) region is represented by V. Equation 5 applies, of course, only under ideal conditions. The absence of kinetic terms limits Equation 5 to slow, quasi-static deformation, and the lack of body force terms implies a zero-gravity condition.[7] Equation 5 also avoids the complexities that are associated with peripheral particles in a finite region and with contacts that can transmit couples between particle pairs.[8] Although these complexities can be difficult to evaluate in physical experiments, they can be circumvented altogether in numerical simulations that exclude both gravity forces and contact moments, and in which the boundaries are periodic.

The formulation (5) for average stress has been employed in a number of ways.

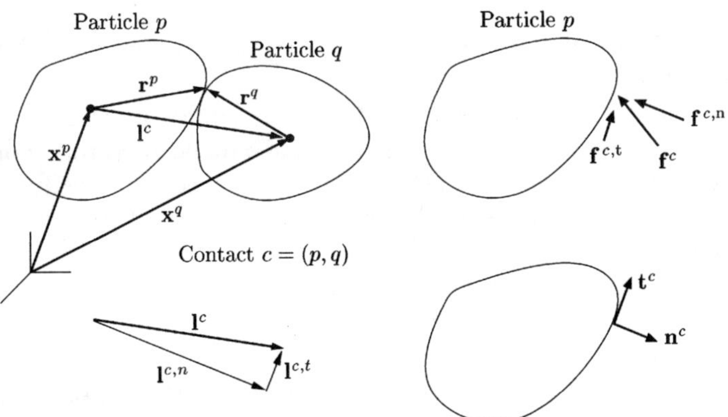

Figure 2   *Position and force for a pair of contacting particles*

Its use in numerical simulations allows the direct calculation of stress from the inter-particle contact forces, but without the supplemental operation of identifying boundary particles and computing the external forces on those particles, an advantage that is particularly appropriate when the boundaries are periodic. Equation 5 has also been the starting point for estimating a macro-scale stiffness from the micro-scale behaviour at particle contacts, and modest successes have been reported at small strains.[9–11] The equation has also led to important insights into the nature of stiffness and strength in granular materials. These insights have been primarily gained by partitioning the average stress $\bar{\sigma}$ into various contributions that arise from the distributions and directions of the contact forces $\mathbf{f}^c$. Several studies have partitioned the stress into two contributions: one from the normal components of the contact forces, and the other from the tangential components.[12–15] These studies have shown that the average stress is borne largely by the normal forces, and although the macroscopic strength is greatly reduced when contacts are rendered frictionless, the contribution of the tangential forces to $\bar{\sigma}$ is small, even when the contacts are frictional. Cundall[12] partitioned the average deviatoric stress into contributions from the tangential forces, from the anisotropic distribution of contact orientations, and from the anisotropic distribution of normal contact forces. The average deviatoric stress was shown to be supported primarily by the relative prevalence and intensity of the normal contact forces that are roughly aligned with the direction of the major principal compressive stress.

The current work concerns the manner in which the average stress $\bar{\sigma}$ *changes during loading*, and, in this regard, we consider a stress rate $d\bar{\sigma}/dt$ derived from Equation 5a. We will partition the stress in three different ways so that internal processes can be investigated within a model granular material. The separate participation of V, $\mathbf{f}^c$, and $\mathbf{l}^c$ in the stress rate will be investigated with the differential form

$$d\bar{\sigma} = -\frac{dV}{V}\bar{\sigma} + \sum_{c \in \mathcal{M}} d\mathbf{f}^c \otimes \mathbf{l}^c + \sum_{c \in \mathcal{M}} \mathbf{f}^c \otimes d\mathbf{l}^c, \tag{6}$$

by measuring the small changes dV, $d\mathbf{f}^c$, and $d\mathbf{l}^c$ that occur in a numerical simulation of biaxial compression.

We note that the Cauchy stress rate $d\bar{\sigma}/dt$ at a material point does not satisfy the objectivity condition and, as such, should not be used in a constitutive formulation. The increment $d\bar{\sigma}$ in (6) is clearly not objective, as neither the force increments $d\mathbf{f}^c$ nor the relative movements $d\mathbf{l}^c$ are objective. We could instead use an objective rate such as

$$\overset{\circ}{\mathbf{f}}{}^c = \dot{\mathbf{f}}^c - \mathbf{W}\mathbf{f}^c, \tag{7}$$

where $\mathbf{W}$ is a reference rate of material rotation. As another example, the Jaumann stress flux also includes the spin $\mathbf{W}$ of the material point at which the stress rate is being measured. The spin $\mathbf{W}$ could, however, be taken from *any other* material point, however distant, or as a weighting of the spins at numerous distant points, provided that the weights sum to one. By including distant spins in the computation of an objective rate, we must obviously forfeit the usual assumption of local action. In the current context, however, we compute an objective *average stress rate*, which is local only to an entire material region, and the principal of local action at a material point has no meaning in this setting. In the current simulations, we use the average spin $\overline{\mathbf{W}}$ among all particles in an assembly to compute an objective stress rate. Because this average spin is nearly zero for the conditions of biaxial compression that are being considered, the spin $\overline{\mathbf{W}}$ is entirely neglected in the stress increment of equation 6.

The expression for $d\overline{\sigma}$ in equation 6 does not explicitly account for changes in the set $\mathcal{M}$ of particle contacts, even though few contacts are likely to be persistent throughout a period of sustained deformation. We note, however, that the set $\mathcal{M}$ could also be the set of *all particle pairs*, not just those that are in contact (in this case, only the subset of contacting pairs would contribute non-zero forces $\mathbf{f}^c$). The numerical simulations in this study use a time-stepping procedure, and we account for the contacts that are newly formed (or disengaged) within the span of each time step, and we also account for the fractions of a time step over which such contacts are engaged.

2.1 Partition A of the Stress Increment $d\overline{\sigma}$

Form (6) will be used to explore the evolution of stress in granular materials, and it will serve as the basis for even finer partitions of the stress rate. The terms in equation 6 will be represented as

$$d\overline{\sigma} = d\overline{\sigma}^{dv} + d\overline{\sigma}^{df} + d\overline{\sigma}^{d\ell},\tag{8}$$

and are also listed in Table 1. At low strain, we would expect the second term, $d\overline{\sigma}^{df}$, to be dominant. Changes in the contact forces $d\mathbf{f}^c$ depend upon changes in the contact indentations, but even large changes in indentation will produce only

Table 1 *Partition A of the average cauchy stress increment $d\overline{\sigma}$*

No.	Symbol	Value[a]
A1	$d\overline{\sigma}^{dv}$	$-\dfrac{dV}{V}\overline{\sigma}$
A2	$d\overline{\sigma}^{df}$	$\dfrac{1}{V}\sum d\mathbf{f}^c \otimes \mathbf{l}^c$
A3	$d\overline{\sigma}^{d\ell}$	$\dfrac{1}{V}\sum \mathbf{f}^c \otimes d\mathbf{l}^c$
$\sum =$	$d\overline{\sigma}$	$d(\dfrac{1}{V}\sum \mathbf{f}^c \otimes \mathbf{l}^c)$

[a] Sums Σ are for the set of particle contacts $c \in \mathcal{M}$.

small increments $d\mathbf{l}^c$ in the branch vectors between the centres of contacting particles (Figure 2). (In the current simulations, the average contact indentation is only about 5×10^{-4} of the average particle size.) The volume contribution $d\bar{\sigma}^{dv}$ will also be negligible for dense packings at small strains, since the bulk modulus will be much larger than the stress.

A primary concern of the current study is the behaviour at large strains, where the material reaches a critical, steady state condition of zero stress change and zero volume change. For this situation, both $d\bar{\sigma}$ and $d\bar{\sigma}^{dv}$ in equation 8 will be zero, and the two contributions $d\bar{\sigma}^{df}$ and $d\bar{\sigma}^{d\ell}$ will be of comparable magnitude, although, as will be seen, neither term can be zero. We will investigate both contributions to the stress rate $d\bar{\sigma}$ during critical state deformation. These contributions, $d\bar{\sigma}^{df}$ and $d\bar{\sigma}^{d\ell}$, can be more fully understood by two further decompositions of these stress increments, as in the remaining Partitions B and C.

2.2 Partition B of the Stress Increments $d\bar{\sigma}^{df}$ and $d\bar{\sigma}^{d\ell}$

2.2.1 *Partition B of Stress Increment $d\bar{\sigma}^{df}$*

In this section, we derive four contributions to the changes in the contact forces $\{d\mathbf{f}^c: c \in \mathcal{M}\}$ and to their cumulative effect on the stress increment, $d\bar{\sigma}^{df}$. The stress increment $d\bar{\sigma}^{df}$, the second term in Partition A (Equations 6 and 8), can be expressed as the sum of these four contributions:

1. $d\bar{\sigma}^{df;\,\text{unif.-elast.}}$: This increment is the change in stress that would be produced by the force increments $d\mathbf{f}^c$ if the particles did not rotate and if the motions of their centres conformed to the mean deformation of the entire assembly. With this contribution, the contact mechanism is also assumed to be elastic. The particles will, of course, rotate; their movements will not conform to a uniform, affine deformation condition; and the contact interactions will likely produce frictional slipping and plastic deformation. By conducting realistic numerical simulations, we can determine the actual movements, rotations, and contact interactions of a model assembly. This data will allow the calculation of the following three corrections to $d\bar{\sigma}^{df;\,\text{unif.-elast.}}$.

2. $d\bar{\sigma}^{df;\,\text{fluct.-elast.}}$: This increment is the change in stress due solely to fluctuations of the particle motions from a uniform deformation field, but with elastic contacts and no particle rotations.

3. $d\bar{\sigma}^{df;\,\text{rotate-elast.}}$: This increment is the change in stress due solely to the measured particle rotations, but neglecting the effects of any inelastic contact behaviour.

4. $d\bar{\sigma}^{df;\,\text{slide}}$: This increment is the change in stress that can be attributed to inelastic contact behaviour.

Taken together, the stress increment $d\bar{\sigma}^{df}$ in equation 8 is the sum of the four contributions,

$$d\overline{\boldsymbol{\sigma}}^{df} = d\overline{\boldsymbol{\sigma}}^{df;\,\text{unif.-elast.}} + d\overline{\boldsymbol{\sigma}}^{df;\,\text{fluct.-elast.}} + d\overline{\boldsymbol{\sigma}}^{df;\,\text{rotate-elast.}} + d\overline{\boldsymbol{\sigma}}^{df;\,\text{slide}}, \tag{9}$$

and these contributions are detailed below.

The change in a contact force $d\mathbf{f}^c$ depends upon the relative movement $d\mathbf{v}^c$ of the particles p and q at the contact point c, with

$$d\mathbf{v}^c = d\mathbf{u}^q - d\mathbf{u}^p + d\Omega^q \mathbf{r}^{c,q} - d\Omega^p \mathbf{r}^{c,p}. \tag{10}$$

In this equation, movements $d\mathbf{u}^{(\cdot)}$ are the incremental particle translations, $d\Omega^{(\cdot)}$ are the incremental particle rotation tensors, and $\mathbf{r}^{c,(\cdot)}$ are the vectors joining the centres of particles p and q with the contact point c (Figure 2). The particle translations $d\mathbf{u}^{(\cdot)}$ can be expressed as fluctuations $d\Delta^{(\cdot)}$ from the uniform, average deformation field, or

$$d\mathbf{u}^{(\cdot)} = \overline{\mathbf{L}}\mathbf{x}^{(\cdot)} + d\Delta^{(\cdot)}, \tag{11}$$

where $\overline{\mathbf{L}}$ is the average Eulerian velocity gradient (the sum of the average rate of deformation $\overline{\mathbf{D}}$ and the average assembly spin $\overline{\mathbf{W}}$), and $\mathbf{x}^{(\cdot)}$ is the particle position (Figure 2).

For non-linear or inelastic contacts, the force increment will depend upon the history of movements at a contact. With the simple spring-slider mechanism that is used in the current simulations, the increment $d\mathbf{f}^c$ can be written as

$$d\mathbf{f}^c = \mathbf{g}^c\,(d\mathbf{v}^c, \mathbf{f}^c), \tag{12}$$

where $\mathbf{f}^c$ is the current contact force. The force increment in equation 12 can be expressed as

$$d\mathbf{f}^c = \mathbf{g}^{c,\text{elastic}}\,(d\mathbf{v}^c, \mathbf{f}^c) + d\mathbf{f}^{c,\text{slide}}, \tag{13}$$

where the first term gives the change in force that would occur in the absence of inelastic contact deformation. The second term in equation 13 is an adjustment that accounts for any inelastic contact behaviour, such as that due to frictional slipping, contact crushing, or other plastic deformations. The simulations in this study employ a simple linear spring at the contact between two particles, and this spring is placed in series with a frictional slider. For this linear spring-slider mechanism, the elastic force increment in Equation 13 is simply

$$\mathbf{g}^{c,\text{elastic}}\,(d\mathbf{v}^c, \mathbf{f}^c) = \mathbf{K}^c d\mathbf{v}^c \tag{14}$$

$$\mathbf{K}^c = k^{\mathrm{n}}\,\mathbf{n}^c \otimes \mathbf{n}^c + k^t\,(\boldsymbol{\delta} - \mathbf{n}^c \otimes \mathbf{n}^c), \tag{15}$$

where k^{n} and k^t are the normal and tangential spring stiffnesses, and δ is the Kronecker identity tensor.[16]

By gathering Equations 10, 11, and 14, the force increment $d\mathbf{f}^c$ in Equation 13 is

$$d\mathbf{f}^c = d\mathbf{f}^{c;\,\text{unif.-elast.}} + d\mathbf{f}^{c;\,\text{fluct.-elast.}} + d\mathbf{f}^{c;\,\text{rotate-elast.}} + d\mathbf{f}^{c;\,\text{slide}}, \tag{16}$$

with

$$d\mathbf{f}^{c;\,\text{unif.-elast.}} = \mathbf{K}^c(\overline{\mathbf{D}}dt)\mathbf{1}^c \tag{17}$$

$$d\mathbf{f}^{c;\,\text{fluct.-elast.}} = \mathbf{K}^c(d\Delta^q - d\Delta^p) \tag{18}$$

$$d\mathbf{f}^{c;\,\text{rotate.-elast.}} = \mathbf{K}^c\left[(d\Omega^q - \overline{\mathbf{W}}dt)\mathbf{r}^q - (d\Omega^p - \overline{\mathbf{W}}dt)\mathbf{r}^p\right] \tag{19}$$

$$d\mathbf{f}^{c;\,\text{slide}} = d\mathbf{f}^c - (d\mathbf{f}^{c;\,\text{unif.-elast.}} + d\mathbf{f}^{c;\,\text{fluct.-elast.}} + d\mathbf{f}^{c;\,\text{rotate-elast.}}). \tag{20}$$

The branch vector $\mathbf{1}^c$ in equation 17 is

$$\mathbf{1}^c = \mathbf{r}^p - \mathbf{r}^q = \mathbf{x}^q - \mathbf{x}^p \tag{21}$$

as shown in Figure 2. Because numerical simulations can track the particle motions and contact forces, each of the increments (17)–(20) can be computed for each contact within a granular assembly and then used to find the corresponding stress contributions in Equation 9:

$$d\overline{\sigma}^{df;\,\text{unif.-elast.}} = \frac{1}{V}\sum_{c\in\mathcal{M}} \mathbf{K}^c(\overline{\mathbf{D}}\mathbf{1}^c dt) \otimes \mathbf{1}^c \tag{22}$$

$$d\overline{\sigma}^{df;\,\text{fluct.-elast.}} = \frac{1}{V}\sum_{c\in\mathcal{M}} \mathbf{K}^c(d\Delta^q - d\Delta^p) \otimes \mathbf{1}^c \tag{23}$$

$$d\overline{\sigma}^{df;\,\text{rotate-elast.}} = \frac{1}{V}\sum_{c\in\mathcal{M}} \mathbf{K}^c[(d\Omega^q - \mathbf{W}dt)\mathbf{r}^q) - (d\Omega^p - \mathbf{W}dt)\mathbf{r}^p)] \otimes \mathbf{1}^c \tag{24}$$

$$d\overline{\sigma}^{df;\,\text{slide}} = \frac{1}{V}\sum_{c\in\mathcal{M}} (d\mathbf{f}^c - d\mathbf{f}^{c,\,\text{elast.}}) \otimes \mathbf{1}^c, \tag{25}$$

as is summarized in rows B1–B4 of Table 2. The presence of a linear contact stiffness tensor $\mathbf{K}^c$ in these equations applies only with linearly elastic contacts; but for non-linear contacts, stiffness $\mathbf{K}^c$ would be replaced by a non-linear function $\mathbf{g}^{c,\,\text{elastic}}$ as in Equation 14. If all contacts share the same linear stiffness $\mathbf{K}^c$, then the stress increment in Equation 22 is

$$d\overline{\sigma}^{df;\,\text{unif.-elast.}} = \frac{1}{V}\mathbf{K}^c\overline{\mathbf{D}}\,dt\sum_{c\in\mathcal{M}}\mathbf{1}^c \otimes \mathbf{1}^c, \tag{26}$$

so that the stress increment $d\overline{\sigma}^{df;\,\text{unif.-elast.}}$ depends only on the arrangement of particles (i.e., the assembly *fabric*) as expressed by the set of branch vectors $\{\mathbf{1}^c\colon c$

Table 2 *Partition B of the average cauchy stress increment $d\bar{\sigma}$*

No.	Symbol	Value[a]
A1	$d\bar{\sigma}^{dv}$	$-\dfrac{dV}{V}\bar{\sigma}$
B1	$d\bar{\sigma}^{df;\,\text{unif.-elast.}}$	$\dfrac{1}{V}\sum \mathbf{K}^c(\overline{\mathbf{D}}\,dt\,\mathbf{l}^c)\otimes\mathbf{l}^c = \dfrac{1}{V}\mathbf{K}^c\overline{\mathbf{D}}dt\,\Sigma\,\mathbf{l}^c\otimes\mathbf{l}^c$
B2	$d\bar{\sigma}^{df;\,\text{fluct-elast.}}$	$\dfrac{1}{V}\sum \mathbf{K}^c(d\mathbf{\Delta}^q - d\mathbf{\Delta}^p)\otimes\mathbf{l}^c$
B3	$d\bar{\sigma}^{df;\,\text{rotate-elast.}}$	$\dfrac{1}{V}\sum \mathbf{K}^c[(d\mathbf{\Omega}^q - \overline{\mathbf{W}}dt)\mathbf{r}^q) - (d\mathbf{\Omega}^p - \overline{\mathbf{W}}dt)\mathbf{r}^p)]\otimes\mathbf{l}^c$
B4	$d\bar{\sigma}^{df;\,\text{slide}}$	$\dfrac{1}{V}\sum (d\mathbf{f}^c - d\mathbf{f}^{c,\text{elast.}})\otimes\mathbf{l}^c$
B5	$d\bar{\sigma}^{d\ell;\,\text{elast.}}$	$\dfrac{1}{V}\sum \mathbf{f}^c\otimes d\mathbf{v}^{c,\text{elast.}}$
B6	$d\bar{\sigma}^{d\ell;\,\text{slide}}$	$\dfrac{1}{V}\sum \mathbf{f}^c\otimes d\mathbf{v}^{c,\text{slide}}$
B7	$d\bar{\sigma}^{d\ell;\,\text{rotate}}$	$-\dfrac{1}{V}\sum \mathbf{f}^c\otimes(d\boldsymbol{\theta}^q\times\mathbf{r}^{c,q} - d\boldsymbol{\theta}^p\times\mathbf{r}^{c,p})$
$\sum =$	$d\bar{\sigma}$	$d(\dfrac{1}{V}\sum \mathbf{f}^c\otimes\mathbf{l}^c)$

[a] Sums Σ are for the set of particle contacts $c \in \mathcal{M}$.

$\in \mathcal{M}\}$. The other increments (23)–(25) will depend upon the complex interactions of particles throughout the assembly, but these interactions can be directly measured in numerical simulations.

2.2.2 *Partition B of Stress Increment $d\bar{\sigma}^{d\ell}$*

As with Partition B of the contact force increments $d\mathbf{f}^c$, we now consider various sources of the inter-particle movements $d\mathbf{l}^c$ and their contributions to the corresponding stress increment $d\bar{\sigma}^{d\ell}$ in Equation 8. We choose to express $d\bar{\sigma}^{d\ell}$ as the sum of three contributions:

1. $d\bar{\sigma}^{d\ell;\,\text{elast.}}$: The change in stress attributed to the relative particle movements $d\mathbf{l}^c$ and the elastic *contact* deformations that would result from these movements. We note that the relative movements $d\mathbf{l}^c$ in Equation 8 are referenced to the motions at particle centres, rather than at the contacts themselves (see Equation 10). The stress contribution $d\bar{\sigma}^{d\ell;\,\text{elast.}}$ (and $d\bar{\sigma}^{d\ell;\,\text{slide}}$ as well) results from the relative motions of particle pairs *at their contact points*.

2. $d\bar{\sigma}^{d\ell;\,\text{elast.}}$: The change in stress that results from the relative particle movements $d\mathbf{l}^c$ and, specifically, the inelastic, sliding motions *at the contacts*.

3. $d\bar{\sigma}^{d\ell;\,\text{elast.}}$: The change in stress due to the effects of particle rotations on $d\mathbf{l}^c$.

The contact movement dv^c is the relative motion between particles p and q at their contact point c. This movement is given in Equation 10 or with the alternative form

$$dv^c = d\mathbf{l}^c + (d\boldsymbol{\theta}^q \times \mathbf{r}^{c,q} - d\boldsymbol{\theta}^p \times \mathbf{r}^{c,p}), \tag{27}$$

where $d\boldsymbol{\theta}^{(\cdot)}$ are incremental particle rotation vectors, and $\mathbf{r}^{c,(\cdot)}$ are vectors joining the centres of particles p and q with the contact point c (Figure 2). If we define the motion $-d\mathbf{l}^{c;\text{rotate}}$ as the expression in parentheses in (27), the equation can be rearranged as

$$d\mathbf{l}^c = dv^c + d\mathbf{l}^{c;\text{rotate}} \tag{28}$$

$$d\mathbf{l}^{c;\text{rotate}} = -(d\boldsymbol{\theta}^q \times \mathbf{r}^{c,q} - d\boldsymbol{\theta}^p \times \mathbf{r}^{c,p}). \tag{29}$$

The contact deformation dv^c can then be separated into elastic and inelastic (sliding) movements, so that

$$d\mathbf{l}^c = dv^{c;\text{elast.}} + dv^{c;\text{slide}} + d\mathbf{l}^{c;\text{rotate}}. \tag{30}$$

The three terms on the right yield three contributions to the stress increment $d\overline{\boldsymbol{\sigma}}^{d\ell}$ in Equations 6 and 8:

$$d\overline{\boldsymbol{\sigma}}^{d\ell} = d\overline{\boldsymbol{\sigma}}^{d\ell;\text{elast.}} + d\overline{\boldsymbol{\sigma}}^{d\ell;\text{slide}} + d\overline{\boldsymbol{\sigma}}^{d\ell;\text{rotate}}, \tag{31}$$

where

$$d\overline{\boldsymbol{\sigma}}^{d\ell;\text{elast.}} = \frac{1}{V} \sum_{c \in \mathcal{M}} \mathbf{f}^c \otimes dv^{c,\text{elast.}} \tag{32}$$

$$d\overline{\boldsymbol{\sigma}}^{d\ell;\text{slide}} = \frac{1}{V} \sum_{c \in \mathcal{M}} \mathbf{f}^c \otimes dv^{c,\text{slide}} \tag{33}$$

$$d\overline{\boldsymbol{\sigma}}^{d\ell;\text{rotate}} = \frac{1}{V} \sum_{c \in \mathcal{M}} \mathbf{f}^c \otimes (d\boldsymbol{\theta}^q \times \mathbf{r}^{c,q} - d\boldsymbol{\theta}^p \times \mathbf{r}^{c,p}). \tag{34}$$

When a deformation $\mathbf{D}(\mathbf{x})\,dt$ occurs within a continuum region V, the internal work done by the stress $\boldsymbol{\sigma}(\mathbf{x})$ is

$$dW = \int_v \boldsymbol{\sigma}(\mathbf{x}) \cdot (\mathbf{D}(\mathbf{x})dt)\,dV, \tag{35}$$

where the operator "$\cdot$" signifies the inner product $\sigma_{ij}D_{ij}$. In a discrete setting, Equation 35 can be approximated as the inner product

$$dW \approx V\left(\overline{\boldsymbol{\sigma}} \cdot \overline{\mathbf{D}}\right) dt \tag{36}$$

of the average stress $\overline{\boldsymbol{\sigma}}$ and the average rate of deformation $\overline{\mathbf{D}}$. For the numerical experiments of the current study, which employ flat periodic boundaries, the average stress can be computed with Equation 5, and the average deformation $\overline{\mathbf{D}}$ can be computed unambiguously from the boundary motions.

The work dW is invested in elastic deformation at the contacts, in frictional sliding at the contacts, and in other energetic investments. The latter, which include changes in kinetic energy, plastic deformations within grains, and viscous energy dissipation, are excluded in the current discussion as they were minimized in the slow, quasi-static simulations of this study. In the absence of these other investments, the work in Equation 35 is simply

$$dW = \sum_{c \in \mathcal{M}} \mathbf{f}^c \cdot \left(d\boldsymbol{v}^{c,\,\text{elast.}} + d\boldsymbol{v}^{c,\,\text{slide}}\right), \tag{37}$$

or, after substituting Equations 32 and 33,

$$dW = V \operatorname{trace}(d\overline{\boldsymbol{\sigma}}^{d\ell;\,\text{elast.}}) + V \operatorname{trace}(d\overline{\boldsymbol{\sigma}}^{d\ell;\,\text{slide}}), \tag{38}$$

where the trace of a tensor is defined in Equation 4. The sliding terms in Equations 37 and 38 are always positive, since work is only dissipated (and not produced) in frictional sliding. For steady, critical state deformation at large strains, we would expect no net change in elastic energy – the elastic terms in Equations 37 and 38 – as the external work is entirely invested in frictional sliding. Moreover, the stress increment trace ($d\overline{\boldsymbol{\sigma}}^{d\ell;\,\text{rotate}}$) should also be zero for the biaxial loading conditions of the current simulations. These presumptions of a zero trace ($d\overline{\boldsymbol{\sigma}}^{d\ell;\,\text{elast.}}$) and a zero trace ($d\overline{\boldsymbol{\sigma}}^{d\ell;\,\text{rotate}}$) will be verified by the numerical simulations. If true, we can summarize the conditions at the critical state as

$$\operatorname{trace}(d\overline{\boldsymbol{\sigma}}^{d\ell;\,\text{elast.}}) = \operatorname{trace}(d\overline{\boldsymbol{\sigma}}^{d\ell;\,\text{rotate}}) = 0 \tag{39}$$

$$\operatorname{trace}(d\overline{\boldsymbol{\sigma}}^{d\ell}) = \operatorname{trace}(d\overline{\boldsymbol{\sigma}}^{d\ell;\,\text{slide}}) > 0. \tag{40}$$

Since deformation proceeds at constant volume during critical state shearing, we can also conclude the following from Equations 6, 8, and 40:

$$d\overline{\boldsymbol{\sigma}}^{dv} = 0 \tag{41}$$

$$d\overline{\boldsymbol{\sigma}}^{df} = - d\overline{\boldsymbol{\sigma}}^{d\ell} < 0 \tag{42}$$

for deformation at the critical state.

2.3 Partition C of the Stress Increments $d\overline{\boldsymbol{\sigma}}^{df}$ and $d\overline{\boldsymbol{\sigma}}^{d\ell}$

Changes in the contact forces $d\mathbf{f}^c$ and relative particle positions $d\mathbf{l}^c$ are caused by changes in either the sizes or directions of the vectors $\mathbf{f}^c$ and $\mathbf{l}^c$. With Partition C, we split the vectors $\mathbf{f}^c$ and $\mathbf{l}^c$ into components that are normal and tangential to the contact surfaces, and then we track the changes in the sizes and directions of these components. A contact force $\mathbf{f}^c$ is the sum of two components (Figure 2),

$$\mathbf{f}^c = \mathbf{f}^{c,\mathrm{n}} + \mathbf{f}^{c,\mathrm{t}}, \tag{43}$$

and each component is the product of a scalar (size) and a unit direction vector,

$$\mathbf{f}^c = f^{c,\mathrm{n}}\mathbf{n}^c + f^{c,\mathrm{t}}\mathbf{t}^c, \tag{44}$$

where $\mathbf{n}^c$ is directed outward from particle p and is normal to the contact surface of the ordered pair $c = (p, q)$. In an unbonded material, the normal forces $f^{c,\mathrm{n}}$ are compressive (negative). Because the simulations in this study are two-dimensional, we adopt an unambiguous tangential direction vector $\mathbf{t}^c$ – a unit vector directed counterclockwise around particle p (Figure 3).
The force increment $d\mathbf{f}^c$ can be expressed as

$$d\mathbf{f}^c = df^{c,\mathrm{n}}\mathbf{n}^c + f^{c,\mathrm{n}}d\mathbf{n}^c + df^{c,\mathrm{t}}\mathbf{t}^c + f^{c,\mathrm{t}}d\mathbf{t}^c, \tag{45}$$

and the incremental change in the branch vector $\mathbf{l}^c$ can be expanded in a similar manner:

$$d\mathbf{l}^c = dl^{c,\mathrm{n}}\mathbf{n}^c + l^{c,\mathrm{n}}d\mathbf{n}^c + dl^{c,\mathrm{t}}\mathbf{t}^c + l^{c,\mathrm{t}}d\mathbf{t}^c. \tag{46}$$

The eight quantities on the right of Equations 45 and 46 can be substituted in Equation 6 to differentiate eight new contributions to the stress increment $d\overline{\sigma}$, which we write as

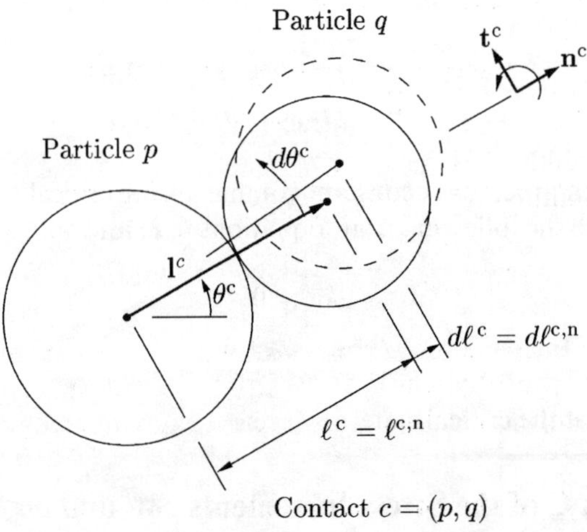

Figure 3   *Rotation $d\theta^c$ of the direction vectors $\mathbf{n}^c$ and $\mathbf{t}^c$ and change $d l^c$ in the branch vector length*

$$d\overline{\sigma} = d\overline{\sigma}^{dv} + d\overline{\sigma}^{df^n} + d\overline{\sigma}^{dn^f} + d\overline{\sigma}^{df^t} + d\overline{\sigma}^{dt^f} + d\overline{\sigma}^{d\ell^n} + d\overline{\sigma}^{dn^\ell} + d\overline{\sigma}^{d\ell^t} + d\overline{\sigma}^{dt^\ell}. \qquad (47)$$

The superscripts df^n, df^t, $d\ell^n$, and $d\ell^t$ denote the effects of changes in the scalar sizes of f^n, f^t, ℓ^n, and ℓ^t; whereas the superscripts dn^f, dt^f, dn^ℓ, and dt^ℓ denote the corresponding changes in the directions of $\mathbf{n}$ and $\mathbf{t}$ and their effects on $\mathbf{f}^c$ and $\mathbf{l}^c$.

For the two-dimensional setting of circular disk particles, the terms in Equations 45–47 can be simplified by noting that direction $\mathbf{n}^c$ is aligned with the branch vector $\mathbf{l}^c$ and that $d\mathbf{n}^c$ and $d\mathbf{t}^c$ depend upon the counterclockwise movement $d\theta^c$ of particle q as it revolves around particle p (Figure 3):

$$\mathbf{l}^c = \mathbf{l}^{c,\mathrm{n}} = \ell^c \mathbf{n}^c \qquad (48)$$

$$d\ell^{c,\mathrm{n}} = d\ell^c \qquad (49)$$

$$\ell^{c,\mathrm{t}} = dl^{c,\mathrm{t}} = 0 \qquad (50)$$

$$d\theta^c = (d\mathbf{l}^c/\ell^c) \cdot \mathbf{t}^c \qquad (51)$$

$$d\mathbf{n}^c = d\theta^c \mathbf{t}^c \qquad (52)$$

$$d\mathbf{t}^c = -\,d\theta^c \mathbf{n}^c, \qquad (53)$$

where l^c is the length of the branch vector and $d\ell^c$ is the incremental change of its length. For circular particles, the tangential component of a branch vector, $\mathbf{l}^c \cdot \mathbf{t}^c$, is zero and its increment $d\ell^{c,\mathrm{t}}$ is also zero (as in Equation 50), so that the stress contributions $d\overline{\sigma}^{d\ell t^t}$ and $d\overline{\sigma}^{dt^\ell}$ in Equation 47 are zero. Although the removal of these two contributions simplifies Equation 47, we will further expand this equation by separating the contributions of the normal and tangential components of forces $\mathbf{f}^c$ inside of the two terms $d\overline{\sigma}^{d\ell^n}$ and $d\overline{\sigma}^{dn^\ell}$. The Partition C is summarized in Table 3, which gives expressions for circular disks as well as for particles of general shape.

The contributions in Table 3 can be investigated for a material at the critical state, for which deformation proceeds both at constant volume ($dV = d\overline{\sigma}^{dv} = 0$) and at constant stress ($d\overline{\sigma} = 0$). The dominant stress contributions will be identified in Section 4 using the results of numerical simulations. We note, however, that for assemblies of circular disks at the critical state, the trace of the sum of the two contributions $d\overline{\sigma}^{df^n}$ and $d\overline{\sigma}^{d\ell^n;f^n}$ must be zero:

$$\mathrm{trace}\,(d\overline{\sigma}^{df^n} + d\overline{\sigma}^{d\ell^n;f^n}) = 0, \qquad (54)$$

since the trace of the remaining terms in Table 3 are zero. (Contribution $d\overline{\sigma}^{dv}$ is zero; the traces of contributions $d\overline{\sigma}^{dn^f}$, $d\overline{\sigma}^{df^t}$, $d\overline{\sigma}^{d\ell^n;f^t}$, and $d\overline{\sigma}^{dn^\ell;f^n}$ are zero; and the trace of the sum of $d\overline{\sigma}^{dt^f}$ and $d\overline{\sigma}^{dn^\ell;f^t}$ is zero.) Equation 54 can, therefore, be written as

$$\frac{1}{V}\sum_{c \in \mathcal{M}} (df^{c,\mathrm{n}}\ell^c + f^{c,\mathrm{n}}d\ell^c) = 0 \qquad (55)$$

Table 3　*Partition C of the average Cauchy stress increment $d\bar{\sigma}$*

No.	Symbol	Value[a]	Circular disks[a]
A1	$d\bar{\sigma}^{dv}$	$-\dfrac{dV}{V}\bar{\sigma}$	$-\dfrac{dV}{V}\bar{\sigma}$
C1	$d\bar{\sigma}^{df^n}$	$\dfrac{1}{V}\sum df^{c,n}\,\mathbf{n}^c \otimes \mathbf{l}^c$	$\dfrac{1}{V}\sum df^{c,n}\,\ell^c n_i^c n_j^c$
C2	$d\bar{\sigma}^{dn^f}$	$\dfrac{1}{V}\sum f^{c,n}\,d\mathbf{n}^c \otimes \mathbf{l}^c$	$\dfrac{1}{V}\sum f^{c,n}\,\ell^c d\theta^c t_i^c n_j^c$
C3	$d\bar{\sigma}^{df^t}$	$\dfrac{1}{V}\sum df^{c,t}\,\mathbf{t}^c \otimes \mathbf{l}^c$	$\dfrac{1}{V}\sum df^{c,t}\ell^c t_i^c n_j^c$
C4	$d\bar{\sigma}^{dt^f}$	$\dfrac{1}{V}\sum f^{c,t}\,d\mathbf{t}^c \otimes \mathbf{l}^c$	$-\dfrac{1}{V}\sum f^{c,t}\ell^c d\theta^c n_i^c n_j^c$
C5	$d\bar{\sigma}^{d\ell^n;f^n}$	$\dfrac{1}{V}\sum f^{c,n}\,d\ell^{c,n}\mathbf{n}^c \otimes \mathbf{n}^c$	$\dfrac{1}{V}\sum f^{c,n}d\ell^c n_i^c n_j^c$
C6	$d\bar{\sigma}^{d\ell^n;f^t}$	$\dfrac{1}{V}\sum f^{c,t}\,d\ell^{c,n}\mathbf{t}^c \otimes \mathbf{n}^c$	$\dfrac{1}{V}\sum f^{c,t}\,d\ell^c t_i^c n_j^c$
C7	$d\bar{\sigma}^{dn^l;f^n}$	$\dfrac{1}{V}\sum f^{c,n}\,\ell^{c,n}\mathbf{n}^c \otimes d\mathbf{n}^c$	$\dfrac{1}{V}\sum f^{c,n}\ell^c d\theta^c n_i^c t_j^c$
C8	$d\bar{\sigma}^{dn^l;f^t}$	$\dfrac{1}{V}\sum f^{c,t}\,\ell^{c,n}\mathbf{t}^c \otimes d\mathbf{n}^c$	$\dfrac{1}{V}\sum f^{c,t}\ell^c d\theta^c t_i^c t_j^c$
C9	$d\bar{\sigma}^{d\ell^t}$	$\dfrac{1}{V}\sum \mathbf{f}^c \otimes (d\ell^{c,t}\mathbf{t}^c)$	0
C10	$d\bar{\sigma}^{dt^\ell}$	$\dfrac{1}{V}\sum \mathbf{f}^c \otimes (\ell^{c,t}d\mathbf{t}^c)$	0
$\sum =$	$d\bar{\sigma}$	$d\left(\dfrac{1}{V}\sum \mathbf{f}^c \otimes \mathbf{l}^c\right)$	$d\left(\dfrac{1}{V}\sum f_i^c(\ell^c n_j^c)\right)$

[a] Sums Σ are for the set of particle contacts $c \in \mathcal{M}$.

for circular disks at the critical state. If the contacts are unbonded and linear, with $df^{c,n} = k^n\, d\ell^c$, then

$$\sum_{c \in \mathcal{M}} (k^n \ell^c + f^{c,n})d\ell^c = 0, \tag{56}$$

where the scalar normal contact forces, $f^{c,n}$, will be less than zero (compressive). The negative forces $f^{c,n}$ are equal to the product of k^n and the effective contact overlaps (i.e., twice the contact indentations). These overlaps will, of course, be much smaller than the particle diameters and the branch vector lengths ℓ^c, so that

$$k^n \ell^c + f^{c,n} \gg 0. \tag{57}$$

This observation and Equation 56 suggest that there will be zero change in the average length of branch vectors at the critical state:

$$\sum_{c \in \mathcal{M}} d\ell^c \approx 0. \tag{58}$$

That is, deformation at the critical state can be almost entirely attributed to the tangential motions of contacting particle pairs.

3 Simulation Methods

A conventional implementation of the numerical Discrete Element Method (DEM) was used to simulate the quasi-static behaviour of a large 2D granular assembly and to study Partitions A, B, and C of the stress rate $d\overline{\sigma}/dt$ at small, moderate, and large strains. The square assembly contained 10,816 unbonded circular disks of multiple diameters (Table 4).

The disk sizes were randomly distributed over a fairly narrow range of between $0.56\overline{D}$ and $1.7\overline{D}$, where $\overline{D}$ is the mean particle diameter. The material was created by slowly and isotropically compacting a sparse arrangement of particles, during which friction between particle pairs was disallowed (friction was later restored for the biaxial compression test). This compaction technique produced a material that was dense, random, and isotropic, at least when viewed at a macro-scale. The average initial void ratio was 0.1727 (solid fraction of 0.853), the average coordination number was 3.82, and the average overlap between neighboring particles was about 5×10^{-4} of $\overline{D}$. The square assembly was surrounded by periodic boundaries, a choice that eliminates any non-uniformities that might otherwise occur in the vicinity of rigid platens or assembly corners. The use of periodic boundaries also eliminates the ambiguous tabulation of average stress[8] that can result from inter-particle contacts along the assembly's boundary. The initial height and width of the square assembly were each about $102\overline{D}$.

A single loading test was conducted. The height of the assembly was reduced

Table 4 *Initial particle assembly*

Characteristic	Description
Particle shape	2D, circular disks
Number of particles	10,816
Assembly shape and size	Square, $102 \times 102\overline{D}$
Boundaries	Periodic
Diameter range	0.56 to $1.6\overline{D}$
Contact springs	Linear, k^n and k^t
Stiffness ratio, k^t/k^n	1
Friction coefficient	0.50 ($\phi_\mu = 26.6°$)
Avg. overlap	$4.5 \times 10^{-4}\overline{D}$
Void ratio	0.1727
Solid fraction	0.8527
Avg. coordination no.	3.82
Pressure, $\overline{p}_o/k^n$	4.7×10^{-4}

at a constant rate of compressive strain ($\dot{\varepsilon}_{22} < 0$) until its final height was less than half of the original height, while maintaining a constant average horizontal stress ($\dot{\sigma}_{12} = 0$), as in Figure 4. About 11×10^6 numerical time steps were required to reach the final vertical strain, $\bar{\varepsilon} = -0.59$, and at this rate of loading, the average imbalance of force on a particle was typically less than 4×10^{-4} times the average contact force.

During biaxial compression, a simple force mechanism was employed between contacting particles. The particles were unbonded, so that no inter-particle tensile forces could develop (Table 4). All contact forces were short-range – only particles that were touching could develop a repulsive contact force, and this force depended on the numerical overlap between a contacting particle pair. Linear normal and tangential contact springs were assigned equal stiffnesses ($k^n = k^t$), and slipping between particles would occur whenever the contact friction coefficient of 0.50 was attained – an inter-particle friction angle of 26.6°.

4   Results and Analysis

4.1   Average Mechanical Behavior

The average, macro-scale behaviour during simulated biaxial compression is illustrated in Figure 5, which gives the evolution of the dimensionless vertical compressive stress $\bar{\sigma}_{22}/\bar{p}_o$, where $\bar{p}_o$ is the (positive) initial pressure, $\bar{p}_o = -\frac{1}{2}(\bar{\sigma}_{11} + \bar{\sigma}_{22})$, at strain $\bar{\varepsilon}_{22} = 0$. The strain $-\bar{\varepsilon}_{22}$ in this figure is a Lagrangian *engineering strain*, computed as the change in the assembly's current height H divided by its

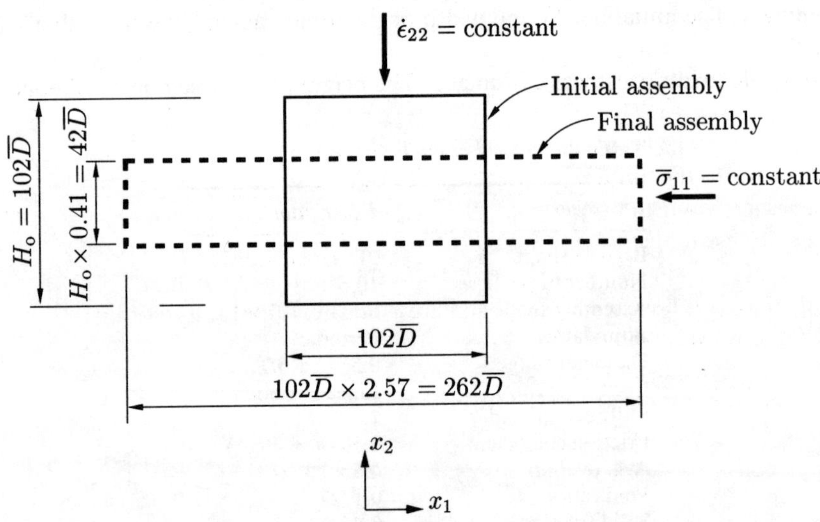

Figure 4   *Biaxial loading conditions for the simulated compression of a 10,816 disk assembly. The initial and final sizes are shown to scale. All boundaries are periodic*

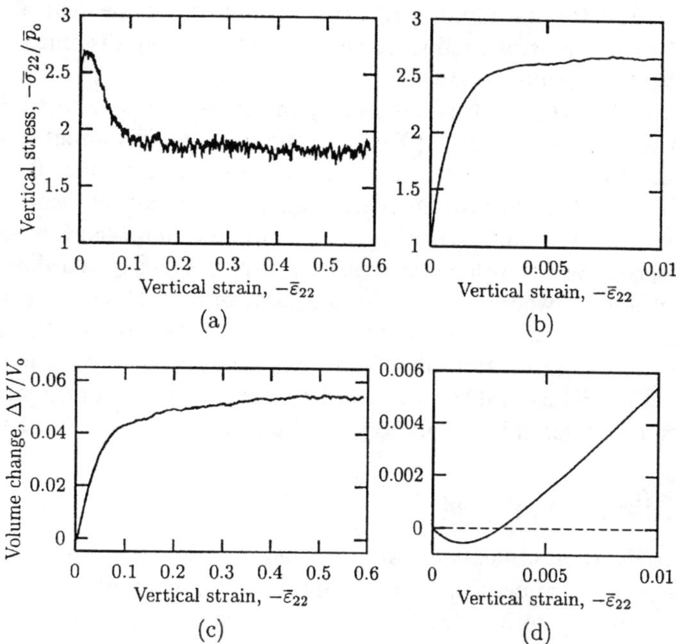

Figure 5 *Results from a numerical DEM simulation of the biaxial compression of 10,816 circular disks. Figures (b) and (d) show the behaviour at small strains*

initial height H_o (Figure 4). Elsewhere in the chapter, we employ Eulerian strain rates, usually the average vertical velocity gradient $\overline{L}_{22}$ – the (negative) rate of change of height divided by the current height, $(dH/dt)/H$. We also note that the average stress $\overline{\sigma}$ that is reported in Figure 5 and throughout this chapter is the *Cauchy stress*, a force per unit of *current area*. A Piola stress is often employed in laboratory practice, in which the applied forces are divided by an original, undeformed area. The vertical Cauchy stress and stress rates that are reported in this chapter will be smaller (less compressive) than their corresponding Piola values due to the quite large expansion of the assembly's width (Figure 4).

The behaviour in biaxial compression is as would be expected. The assembly has a large initial stiffness (Young's modulus), and the peak mobilised strength occurs at a strain $-\overline{\varepsilon}_{22} = 0.014$, much smaller than the final strain of 0.59. The peak state is followed by a period of vigorous softening, ending in an eventual steady, critical state mobilised strength. Because the shearing stresses $\overline{\sigma}_{12}$ and $\overline{\sigma}_{21}$ are negligible during this biaxial compression test, the mobilised friction angle ϕ_{mob} can be computed from the Coulomb equation,

$$\sin\phi_{\mathrm{mob}} = \frac{\overline{\sigma}_{22}/\overline{\sigma}_{11} - 1}{\overline{\sigma}_{22}/\overline{\sigma}_{11} + 1}, \tag{59}$$

where the constant horizontal stress $\overline{\sigma}_{11}$ is simply the initial mean pressure, $\overline{\sigma}_{11} = -\overline{p}_o$. The peak mobilised friction angle was 27.3°, slightly larger than the angle of inter-particle friction (26.6°, see Tables 4 and 5). At the critical state, the mobilised angle was greatly reduced, with $\phi_{mob} = 16.9°$, and was much lower than the inter-particle friction angle.

The assembly's volume was reduced by the initial loading, and this volume reduction corresponds to a (2D) Poisson ratio of 0.13. This initial compression was followed by vigorous dilation until the strain $-\bar{\varepsilon}_{22}$ had reached 0.45 (Figures 5c and 5d). Beyond this strain, deformation proceeded at nearly constant volume, with an eventual (2D) critical state void ratio of 0.236 (Table 5).

The Eulerian rate of volume change, $\dot{v}$, can be expressed as a dimensionless dilation rate $-\dot{v}/\overline{L}_{22}$, where $\overline{L}_{22}$ is the negative rate of vertical compression. At the peak state, dilation was quite vigorous, with $-\dot{v}/\overline{L}_{22} = 0.83$, so that the horizontal extension was occurring at a rate of 1.83 times the rate of vertical compression (Table 5). The dilation rate was reduced to zero at the advanced strain $-\bar{\varepsilon}_{22} = 0.45$, when the material had reached the critical state.

4.2   Partitions A, B, and C

We now analyze the simulation results in regard to the three partitions of stress rate at three states of deformation: during the initial loading, at the peak state, and at the critical state. The results are summarized in Table 6 through 11, whose entries correspond to their definitions in Tables 1–3. The tables are arranged in pairs that present the deviatoric and volumetric behaviours during biaxial com-

Table 5   *Strength and dilation*

	Initial loading	Peak state	Critical state
Engineering strain, $-\bar{\varepsilon}_{22}$	0	0.0135–0.0145	0.45[+]
Strength, $-\bar{\sigma}_{22}/\bar{p}_0$	1	2.69	1.82
Mobilised friction, ϕ_{mob}	0°	27.3°	16.9°
Stress rate, $d\bar{\sigma}_{22}/(\overline{L}_{22}dt\bar{p}_0)$	1764	≈ 0	0
Dilation rate, $-\dot{v}/\overline{L}_{22}$	−0.87	0.83	0
Void ratio	0.173	0.183	0.236
Avg. coordination no.	3.82	3.12	3.09
Fabric ratio	1	1.45	1.28

Table 6   *Partition A, rates of the deviator stress*

No.	Symbol	Deviator rates[a,b]		
		Initial loading	Peak state	Critical state
A1	$d\bar{\sigma}_q^{dv}$	0	−1.4	0
A2	$d\bar{\sigma}_q^{df}$	1766	5.3	2.8
A3	$d\bar{\sigma}_q^{d\ell}$	−1	−4.5	−2.8
$\Sigma=$	$d\bar{\sigma}_q$	1764	−0.6	0

[a] The deviator of each (·) contribution is the difference $d\bar{\sigma}_{22}^{(\cdot)} - d\bar{\sigma}_{11}^{(\cdot)}$.
[b] The rate of each (·) contribution is given in the dimensionless form $d\bar{\sigma}_q^{(\cdot)}/(\overline{L}_{22}dtp_0)$.

Table 7   *Partition A, rates of the average cauchy trace stress*

No.	Symbol	Trace rates[a,b]		
		Initial loading	Peak state	Critical state
A1	$d\overline{\sigma}_t^{dv}$	2	−3.1	0
A2	$d\overline{\sigma}_t^{df}$	1764	3.3	0.77
A3	$d\overline{\sigma}_t^{d\ell}$	−1	−0.8	−0.77
$\Sigma =$	$d\overline{\sigma}_q$	1764	−0.6	0

[a] The trace of each ($\cdot$) contribution is the sum $d\overline{\sigma}_{22}^{(\cdot)} + d\overline{\sigma}_{11}^{(\cdot)}$.
[b] The rate of each ($\cdot$) contribution is given in the dimensionless form $d\overline{\sigma}_t^{(\cdot)}/(\overline{L}_{22}dtp_o)$, where $\overline{L}_{22}dt$ is negative for these biaxial compression tests (Figure 4).

Table 8   *Partition B, rates of the average Cauchy deviator stress*

No.	Symbol	Deviator rates[a,b]		
		Initial loading	Peak state	Critical state
A1	$d\overline{\sigma}_q^{dv}$	0	−1.4	0
B1	$d\overline{\sigma}_q^{df; unif.-elast.}$	2607.7	4957.5	3503.7
B2	$d\overline{\sigma}_q^{df; fluct-elast.}$	−291.8	−829.3	−471.4
B3	$d\overline{\sigma}_q^{df; rotate-elast.}$	−533.3	−2546.9	−2048.7
B4	$d\overline{\sigma}_q^{df; slide}$	−16.5	−1575.0	−980.8
B5	$d\overline{\sigma}_q^{d\ell; elast.}$	−0.8	0	0
B6	$d\overline{\sigma}_q^{d\ell; slide}$	0	−0.7	−0.3
B7	$d\overline{\sigma}_q^{d\ell; rotate}$	−0.3	−3.8	−2.5
$\Sigma =$	$d\overline{\sigma}_q$	1764	−0.6	0

[a] The deviator of each ($\cdot$) contribution is the difference $d\overline{\sigma}_{22}^{(\cdot)} - d\overline{\sigma}_{11}^{(\cdot)}$.
[b] The rate of each ($\cdot$) contribution is given in the dimensionless form $d\overline{\sigma}_q^{(\cdot)}/(\overline{L}_{22}dtp_o)$.

Table 9   *Partition B, rates of the average Cauchy trace stress*

No.	Symbol	Trace rates[a,b]		
		Initial loading	Peak state	Critical state
A1	$d\overline{\sigma}_t^{dv}$	2	−3	0
B1	$d\overline{\sigma}_t^{df; unif.-elast.}$	1997	−551	472
B2	$d\overline{\sigma}_t^{df; fluct-elast.}$	−233	554	−472
B3	$d\overline{\sigma}_t^{df; rotate-elast.}$	0	0	0
B4	$d\overline{\sigma}_t^{df; slide}$	0	0	1
B5	$d\overline{\sigma}_t^{d\ell; elast.}$	−1	0	0
B6	$d\overline{\sigma}_t^{d\ell; slide}$	0	−1	−1
B7	$d\overline{\sigma}_t^{d\ell; rotate}$	0	0	0
$\Sigma =$	$d\overline{\sigma}_t$	1764	−1	0

[a] The trace of each ($\cdot$) contribution is the sum $d\overline{\sigma}_{22}^{(\cdot)} + d\overline{\sigma}_{11}^{(\cdot)}$.
[b] The rate of each ($\cdot$) contribution is given in the dimensionless form $d\overline{\sigma}_t^{(\cdot)}/(\overline{L}_{22}dtp_o)$.

Table 10   *Partition C, rates of the average Cauchy deviator stress*

No.	Symbol	Deviator rates[a,b]		
		Initial loading	Peak state	Critical state
A1	$d\overline{\sigma}_q^{dv}$	0	−1.4	0
C1	$d\overline{\sigma}_q^{df^n}$	1081	14.0	5.6
C2	$d\overline{\sigma}_q^{dn^f}$	−1	−4.5	−2.8
C3	$d\overline{\sigma}_q^{df^t}$	685	−4.1	0
C4	$d\overline{\sigma}_q^{dt^f}$	0	0	0
C5	$d\overline{\sigma}_q^{d\ell^n;f^n}$	−1	0	0
C6	$d\overline{\sigma}_q^{d\ell^n;f^t}$	0	0	0
C7	$d\overline{\sigma}_q^{dn^l;f^n}$	−1	−4.5	−2.8
C8	$d\overline{\sigma}_q^{dn^l;f^t}$	0	0	0
$\Sigma =$	$d\overline{\sigma}_q$	1764	−0.6	0

[a] The deviator of each ($\cdot$) contribution is the difference $d\overline{\sigma}_{22}^{(\cdot)} - d\overline{\sigma}_{11}^{(\cdot)}$.
[b] The rate of each ($\cdot$) contribution is given in the dimensionless form $d\overline{\sigma}_q^{(\cdot)}/(\overline{L}_{22}dtp_o)$.

Table 11   *Partition C, rates of the average Cauchy trace stress*

No.	Symbol	Trace rates[a,b,c]		
		Initial loading	Peak state	Critical state
A1	$d\overline{\sigma}_t^{dv}$	2	−3.1	0
C1	$d\overline{\sigma}_t^{df^n}$	1764	2.5	0
C2	$d\overline{\sigma}_t^{dn^f}$	—	—	—
C3	$d\overline{\sigma}_t^{df^t}$	—	—	—
C4	$d\overline{\sigma}_t^{dt^f}$	0	0.8	0.8
C5	$d\overline{\sigma}_t^{d\ell^n;f^n}$	−1	0	0
C6	$d\overline{\sigma}_t^{d\ell^n;f^t}$	—	—	—
C7	$d\overline{\sigma}_t^{dn^l;f^n}$	—	—	—
C8	$d\overline{\sigma}_t^{dn^l;f^t}$	0	−0.8	−0.8
$\Sigma =$	$d\overline{\sigma}_t$	1764	−0.6	0

[a] The trace of each ($\cdot$) contribution is the sum $d\overline{\sigma}_{22}^{(\cdot)} + d\overline{\sigma}_{11}^{(\cdot)}$.
[b] The rate of each ($\cdot$) contribution is given in the dimensionless form $d\overline{\sigma}_t^{(\cdot)}/(\overline{L}_{22}dtp_o)$.
[c] Excluded values (—) are for trace rates that must equal zero for circular disks.

pression loading (Figure 4). The former is expressed in a series of deviator stress rates (Tables 6, 8, and 10), where increments of deviator stress,

$$d\overline{\sigma}_q^{(\cdot)} = d\overline{\sigma}_{22}^{(\cdot)} - d\overline{\sigma}_{11}^{(\cdot)}, \tag{60}$$

are reported with a "q" subscript for each ($\cdot$) contribution to the three partitions. The volumetric behaviour is evaluated with the *trace* of the stress increments,

$$\text{trace}\,(d\overline{\sigma}^{(\cdot)}) = d\overline{\sigma}_t^{(\cdot)} = d\overline{\sigma}_{11}^{(\cdot)} + d\overline{\sigma}_{22}^{(\cdot)}, \tag{61}$$

and appear with a "t" subscript. These values will be referred to as simply "trace rates" or "trace increments" (Tables 7, 9, and 11). We recall that the horizontal stress $\overline{\sigma}_{11}$ was maintained constant throughout the loading, by allowing the width of the assembly to expand while the material was vertically compressed. Although the net increments of horizontal stress, $d\overline{\sigma}_{11}$, are zero, the various contributory increments $d\overline{\sigma}_{11}^{(\cdot)}$ might not be zero, and the deviator increments $d\overline{\sigma}_q^{(\cdot)}$ might, therefore, not equal their trace counterparts, $d\overline{\sigma}_t^{(\cdot)}$.

The increments in Tables 6–11 are expressed as *Eulerian dimensionless rates* of the following forms:

$$\frac{d\overline{\sigma}_q^{(\cdot)}}{(\overline{L}_{22}dt)\overline{p}_0} \text{ and } \frac{d\overline{\sigma}_t^{(\cdot)}}{(\overline{L}_{22}dt)\overline{p}_o}, \tag{62}$$

where $d\overline{\sigma}_q^{(\cdot)}$ and $d\overline{\sigma}_t^{(\cdot)}$ are incremental contributions to the average Cauchy stress, $\overline{L}_{22}dt$ is a (negative) increment of the Eulerian vertical strain, and $\overline{p}_o$ is the (positive) initial pressure, $\overline{p}_o = -\frac{1}{2}(\overline{\sigma}_{11} + \overline{\sigma}_{22})$. Because the values in Tables 6–9 are given in the form (62), we will refer to an increment $d\overline{\sigma}_{q\,\text{or}\,t}^{(\cdot)}$ as either an "increment" or a "rate".

The signs of numbers in the "deviator" and "trace" tables can be interpreted at follows:

- Deviator rates: Positive deviator rates, $d\overline{\sigma}_q^{(\cdot)}/(\overline{L}_{22}dt\overline{p}_o)$, contribute to material stiffness. Negative increments connote a softening influence.
- Trace rates: During initial loading, positive trace rates contribute to material compression, since their signs are aligned with the overall (positive) rate $d\overline{\sigma}_t/(\overline{L}_{22}dt\overline{p}_o)$. At the peak state, the overall rate is negative, and negative trace rates contribute to material dilation.

Changes in the contact forces $\mathbf{f}^c$ will produce changes in the average stress by way of the various contributions $d\overline{\sigma}^{df;(\cdot)}$. These effects can be interpreted as follows:

- Normal contact forces: positive rates $d\overline{\sigma}^{df^n}/(\overline{L}_{22}dt\overline{p}_o)$ connote a positive effect of the normal (compressive) forces between particles (Table 3, row C1). Positive contributory deviator rates are produced by predominant increases in the normal (compressive) forces among particle pairs that have a more vertical alignment. Positive contributory trace rates are produced by increases in the average normal (compressive) forces among particle pairs.
- Tangential contact forces: positive deviator rates $d\overline{\sigma}^{df^t}/(\overline{L}_{22}dt\overline{p}_o)$ are produced by increases of the tangential contact forces in the directions shown in Fig. 6 (Table 3, row C6).

4.2.1 *Results for Partitions of the Deviator Rate*

The following eleven results are extracted from information on the deviator rates in Table 6, 8, and 10.

1. During the initial loading ($\overline{\varepsilon}_{22} = 0$), the change in deviatoric stress is almost entirely due to changes in the contact forces, instead of changes in the

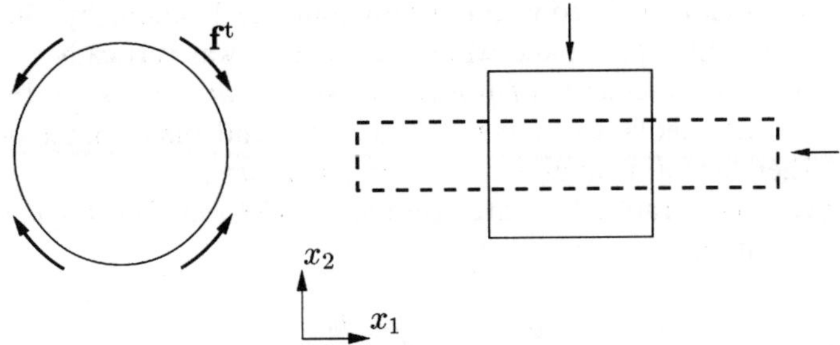

Figure 6   *Positive contributions to $d\bar{\sigma}_q^{df^t}$ for vertical biaxial compression*

branch vector lengths or the assembly volume (Table 6, row A2, Initial loading). The exclusive dominance of $d\bar{\sigma}_q^{df}$ is the basis of most small-strain estimates of the material behaviour that are derived from contact stiffnesses and micro-scale particle arrangements.

2. At initial loading, the change in deviatoric stress is due to changes in both the normal contact forces $d\bar{\sigma}_q^{df^n}$ and the tangential contact forces $d\bar{\sigma}_q^{df^t}$, as shown in Table 10, rows C1 and C3. The contact stiffness ratio k^t/k^n was 1.0 in the simulation, and the resulting ratio of tangential-to-normal contributions, $d\bar{\sigma}_q^{df^t}/d\bar{\sigma}_q^{df^n}$, was about 0.63 at the start of biaxial compression.

3. At initial loading, the measured Young's modulus $(= d\bar{\sigma}_{22}/d\bar{\varepsilon}_{22})$ is about 68% of the value that would be estimated on the basis of a uniform strain assumption and elastic contact behavior (68% is the ratio $d\bar{\sigma}_q/d\bar{\sigma}_q^{df;\,unif.\text{-}elast.}$, Table 8, rows B1 and Σ). The reduced modulus is due to the three other effects that were discussed as part of Partition B in Section 2: fluctuations in the particle translations, particle rotations, and contact sliding, which are also discussed in result 6 below.

4. The four Partition B rates of deviator stress, $d\bar{\sigma}_q^{df;(\cdot)}$, are summarized in rows B1–B4 of Table 9 and in Figure 7. As has already been stated, the contribution $d\bar{\sigma}_q^{df;\,unif.\text{-}elast.}$ depends only on the averaged orientations of the contacting pairs of particles and on the number of contacts (see eq. 26). That is, changes in this single stress contribution can be ascribed either to changes in fabric anisotropy or to changes in the average coordination number. The rate $d\bar{\sigma}_q^{df;\,unif.\text{-}elast.}$ increases between the initial loading and the peak state, and this increase is due to an increasing fabric anisotropy, as expressed in the following fabric ratio, an average ratio of the n_2 and n_1 components of the contact normal vectors:

$$\sum_{c\in\mathcal{M}} n_2^2 \Big/ \sum_{c\in\mathcal{M}} n_1^2, \tag{63}$$

the final row in Table 5.[17] The contacts become preferentially aligned in the (vertical) direction of loading, and this anisotropy would, by itself, contribute to a stiffening of the material between the initial and peak states. This increase in anisotropy is more important than the counteracting

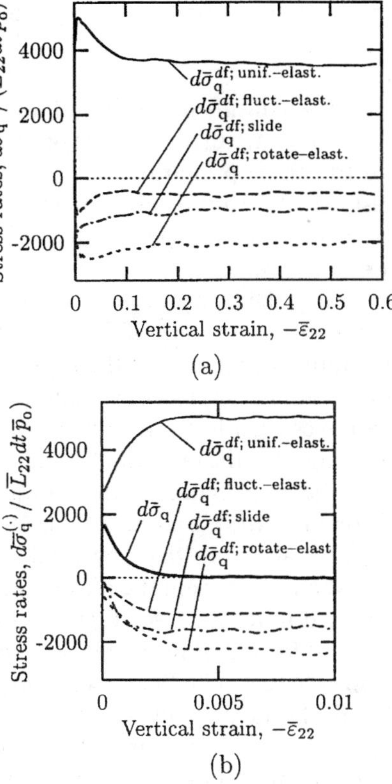

(a)

(b)

Figure 7 *Contributions in Partition B to the increment $d\bar{\sigma}_q^{df}$ of deviator stress, $d\bar{\sigma}_{22}^{df} - d\bar{\sigma}_{11}^{df}$, as discussed in results 4 and 6. The four contributions are expressed as dimensionless rates $d\bar{\sigma}_q^{(\cdot)}/(\bar{L}_{22}dt\bar{p}_o)$. The combined stress increment $d\bar{\sigma}_q$ is also shown as a rate in Figure 7b*

effect of a diminishing coordination number, which also occurs over the same range of strains (Table 5), and the combined effect is a 90% increase in $d\bar{\sigma}_q^{df;\,unif.-elast}$.

5. The rate $d\bar{\sigma}_q^{df;\,unif.-elast.}$ is reduced at strains above the peak state, and this reduction can be attributed to a paradoxical reduction in fabric anisotropy at these large strains, as is indicated in the final row of Table 5. The average coordination number, however, is stationary at strains beyond the peak state.

6. Throughout the loading, the initial deviatoric stiffness is continually reduced until the stiffness becomes negative with the softening that occurs beyond the peak state (Figure 5). Prior to the peak state, the reduction in stiffness is the result of three mechanisms: fluctuations in the particle translations, particle rotations, and contact sliding (Figure 7 and Table 8, rows B1–B4). These three mechanisms counteract the stiffening effect of an increasing fabric anisotropy, as in $d\bar{\sigma}_q^{df;\,unif.-elast.}$. Particle rotations are the dominant softening mechanism at all strains (Table 8, row B3). The soften-

ing effect of contact sliding, $d\bar{\sigma}_q^{df;\,\mathrm{slide}}$, is negligible at initial loading, is greatest at the peak state, and is substantially reduced at the critical state (Table 8, row B4).

7. Although the three mechanisms in result 6 all act to reduce the material stiffness, the intense softening that occurs beyond the peak state and until the strain $-\bar{\varepsilon}_{22} = 0.10$ can be primarily attributed to a reduction in fabric anisotropy (see result 5). This anisotropy is manifested in the declining rate $d\bar{\sigma}_q^{df;\,\mathrm{unif.\text{-}elast.}}$ that occurs over this range of strain (Figure 7; Table 8, row B1; and the final row of Table 5).

8. Throughout the loading, the contact forces change in a manner that tends to increase the deviator stress ($d\bar{\sigma}_q^{df}$, Table 6, row A2). These averaged changes in contact forces occur even at the peak and critical states, where the average stress is stationary. That is, a stationary stress does not imply stationary contact forces – neither individually nor in the mean.

9. At the peak state, changes in the tangential contact forces tend to diminish the deviator stress; whereas, changes in the normal contact forces tend to increase the deviator stress ($d\bar{\sigma}_q^{df^{n}}$ and $d\bar{\sigma}_q^{df^{t}}$ Table 10, rows C1 and C3, peak state). This unusual behaviour of the tangential forces is opposite that at the initial loading, where changes in the tangential forces have a stiffening effect, as was discussed in result 2 (see Figure 6 for an illustration of the directions of tangential force that promote stiffening). At the critical state, only the normal forces have a net influence on changes in the average stress.

10. At the peak and critical states, a tendency for the normal contact forces to increase the deviator stress (results 8 and 9) is counteracted by changes in the directions of the contact normal vectors, a result of the contacting particle pairs revolving around each other (Figure 3 and eq. 47). This effect is manifested in the two terms $d\bar{\sigma}_q^{dn^{f}}$ and $d\bar{\sigma}_q^{dn^{\ell};f^{n}}$ (Table 10, rows C2 and C7). Both effects are of second-order, as they both arise from the averaged changes $d\mathbf{n}^{c}$ in the directions of the contact normals, and their prominence in maintaining a stationary stress at large strains is due only because the first-order stiffnesses are reduced to such small values.

11. The appearance of non-zero stress contributions at the critical state in each of the three partitions demonstrates that, although stress and volume are constant, the internal state of the material is not stationary. Even in an assembly of only 10,816 particles, the particle motions are continually readjusting during deformation at the critical state. These adjustments produce averaged contributions that are non-zero, and the stationary stress is a result of the counteracting influences of the various averaged contributions.

4.2.2   *Results for Partitions of the Trace Rate*

The following four results are extracted from information on the trace rates in Tables 6, 8, and 10.

12. During initial loading, the material compresses (with a 2D Poisson ratio of

0.13, Table 5), but the fluctuations of particle translations from those of uniform, affine deformation have a net dilatant influence on the material. Chang and Liao[11] have shown that the assumptions of deformation and non-rotating particles lead to a Poisson ratio of zero (i.e., isochoric deformation) for particle assemblies with isotropic fabric and a contact stiffness ratio, k^t/k^n, of unity. The larger, more dilatant Poisson ratio that was measured in the current simulations can be ascribed to the negative rate $d\overline{\sigma}_q^{df;\,\text{fluct-elast.}}$ due to fluctuations of the particle motions from a uniform deformation field (Table 9, row B2, initial loading).

13. During initial loading, the particle rotations, by themselves, have no net effect on the material's volumetric behaviour (Table 9, row B3, initial loading), although rotations have a substantial softening effect on the initial deviatoric stiffness (see result 6).

14. At the peak state, the vigorous material dilation can be attributed to significant fluctuations of particle translations from those of a uniform deformation field. The positive volumetric stiffness $d\overline{\sigma}_t^{df;\,\text{fluct-elast.}}$ indicates a dilative influence of these fluctuations (Table 9, row B2, peak state), which counteracts the compression that would otherwise occur if the particle motions were constrained to a uniform deformation field.

15. As expected, the trace rate $d\overline{\sigma}_t^{d\ell;\,\text{slide}}/(\overline{L}_{22}dtp_o)$ is negative at the critical state (i.e., a positive increment $d\overline{\sigma}_t^{d\ell;\,\text{slide}}$) due to its association with frictional energy dissipation (inequality 40 and Table 9, row B6). The numeric value of this dissipation is quite small, however, when compared with the other contributory rates that were measured at the critical state.

References

1. G. de Josselin de Jong and A. Verruijt, *Cahiers du Groupe Français de Rhéologie*, 1969, **7**, 73.
2. A. Drescher and G. de Josselin de Jong, *J. Mech. Phys. Solids*, 1972, **20**, 337.
3. M. Oda, J. Konishi, and S. Nemat-Nasser, *Géotechnique*, 1980, **30**, 479.
4. D. Muir Wood, 'Soil Behaviour and Critical State Soil Mechanics', Cambridge University Press, Cambridge, U.K., 1990.
5. L. Rothenburg and A. P. S. Selvadurai, in 'Mechanics of Structured Media, Part B', ed. A. P. S. Selvadurai, Elsevier, Amsterdam, The Netherlands, 1981, 469.
6. J. Christoffersen, M. M. Mehrabadi, and S. Nemat-Nasser, *J. Appl. Mech.*, 1981, **48**, 339.
7. K. Bagi, *J. Appl. Mech.*, 1999, **66**, 934.
8. J. P. Bardet and I. Vardoulakis, *Int. J. Solids and Structures*, 2001, **38**, 353.
9. M. A. Koenders, *Acta Mechanica*, 1987, **70**, 31.
10. B. Cambou, in 'Powders & Grains 93', ed. C. Thornton, A.A. Balkema, Rotterdam, The Netherlands, 1993, 73.
11. C. S. Chang and C. L. Liao, *Appl. Mech. Rev.*, 1994, **47**, Part 2, S197.
12. P. A. Cundall and O. D. L. Strack, in 'Mechanics of Granular Materials: New Models and Constitutive Relations', eds. J. Jenkins and M. Satake, Elsevier Science Pub. B. V., Amsterdam, The Netherlands, 1983, 137.
13. C. Thornton and D. J. Barnes, *Acta Mechanica*, 1986, **64**, 45.

14. C. Thornton, *Géotechnique*, 2000, **50**, 43.
15. Y.-C. Chen and H.-Y. Hung, *Soils and Found.*, 1991, **31**, 148.
16. N. Gaspar, 'Structures and Heterogeneity in Deforming, Densely Packed Granular Materials', PhD Thesis, Kingston University, Kingston, U.K., 2002.
17. M. Satake, in 'Proc. IUTAM Symp. on Deformation and Failure of Granular Materials', eds. P. A. Vermeer and H. J. Luger, A.A. Balkema, Rotterdam, 1982, 63.

CHAPTER 2

Snapshots on Some States of Granular Matter: Billiard, Gas, Clustering, Liquid, Plastic, Solid?

P. EVESQUE

Lab MSSMat, UMR 8579 CNRS, Ecole Centrale Paris,
92295 Châtenay-Malabry, France
E-mail: evesque@mssmat.ecp.fr;

1   Introduction

The dynamics of granular media presents many distinct facets;[1] and it is currently a very active field of research in the various scientific fields: physics, mechanics, geophysics, chemistry, and pharmacy. The objectives of this paper are to primarily describe a few distinct problems, which illustrate some of the current trends.[2] The paper focuses on the effects of mechanical periodic excitation on a granular medium; but this does not really limit the scope, which starts with the behaviour of dynamic billiards, when only few grains are in a box, and ends with the propagation of an elastic wave in an elastic inhomogeneous medium when the medium is so dense that contacts cannot slide and change. Of course, one finds inbetween the gas$\leftrightarrow$"liquid" transition, the "liquid"$\leftrightarrow$plastic one and the "plastic"$\leftrightarrow$elastic-solid one, depending on the amplitude of vibration, its frequency and the density of the material, ... One can ask also if a "liquid" phase really exists or if a "jamming" transition is involved in the gas-plastic or liquid-plastic or plastic-elastic solid transition? Or, does it involve a new type of phase transition, as it is often argued? Is it possible to define some constitutive equations to these states of matter (phases)? If so, are they valid in the bulk? Do they depend on the boundary conditions? Are these states homogeneous; are they responding homogeneously to homogeneous mechanical perturbation? So many questions, so little time and so few complete answers. Anyway, it seems to me that starting with well defined states is better than studying their transitions. This is the goal of this chapter, to give some snapshots about these states.

So, the chapter starts with the puzzling behaviour of a granular gas in space and shows its natural tendency to form clusters, so that the gas survives only in the Knudsen regime, when the particles do not "see" one another; it is probably of a crucial interest for cosmology, but it has also drastic consequences in handling granular media. So the natural form of the medium is dense, tending always to form a continuous network. This result tells us also (i) that the maximum density of a homogeneous gas tends to 0 as the inverse of the sample size 1/L and (ii) that concepts from continuum mechanics most likely cannot be applied to it, which means (iii) that sound waves do not propagate classically in the gas (however we show that sound propagates in dense coherent granular mediums in section 5). Then we ask what is occurring in a low-density Knudsen gas, which leads to the investigation of the problem of dynamic billiards. Indeed, this problem is considered as important in the fields of pure statistical mechanics or of quantum chaos when the balls are atoms or electrons. Here it represents simply the limiting case of a granular medium in a vibrated box, when the number of grains is quite limited. Experimental investigation shows that it behaves in a non ergodic way, due to dissipation.

It is worth noting the difficulty of studying granular gas and clustering on earth, because gravity forces the system to condense on the bottom, and to order it in layers with different densities. But a question remains: is the clustering the first stage of liquefaction; and if so, what are its dynamics?

The next sections consider dense samples on earth, but each one with some different behaviour: perfect (i.e. non viscous) liquid, streaming in plastic solid and elastic solid.

One knows for instance that dense sand filled with liquid can be liquefied during an earthquake; indeed, this can cause much damage to buildings, or can generate "craters" due to water ejection at the surface. This liquefaction has been already described in soil mechanics, using cyclic paths working at constant-volume and with the help of Darcy's law; to summarise, the medium looks viscous when the grains loose contacts while it looks plastic or solid when contacts are stressed. The second section of this chapter focuses on something quite different: it shows that a sand-liquid mixture can behaves exactly as a perfect liquid, i.e. non viscous, if inertia forces are much stronger than any other force. We will show in particular that one can generate some kind of swell at the liquid-sand interface, just by shaking the system horizontally; this relief is not ripples or dunes, because the swell flattens at once, as soon as vibration is stopped; so the granular medium is quite liquid.

The third part is concerned with the plastic flow pattern generated by cyclic quasi-static deformation. It exemplifies the plastic state of granular matter. The experiment uses periodic bulldozing. This example seems important because it recalls that vortex flow is not only generated by convection, as the recent literature on granular medium considers. The experimental conditions require that the vortex stream should be understood within classic "soil mechanics" rules only; then it makes the parallel with the problem of acoustic streaming in fluid mechanics, which does not require heating too, or any other convection. Finally it shows that the "convection" flow patterns observed in granular materials under horizontal vibration look very much as the one observed here with quasi-static cyclic forcing, so that these "convection" flows, produced by vibrations, should be governed by the same classic plastic behaviour, at least as a first approximation.

In the case of vertical vibration, the same approach should be valid if some horizontal component remains; but when the horizontal component becomes too small, or when the amplitude is too large, the modelling needs to also include fluid inside the pores to get the correct answer. That defines a scale of difficulty for computer scientists.

The last part studies the propagation of sonic and ultrasonic waves in a granular medium, since it has been argued that it can reveal the "fragile" nature of the material. In fact we will show that low-frequency sound propagates "normally" in granular media, with compression and shear waves; but it becomes scattered by the grain and diffuse then if its wavelength becomes too small, *i.e.* of the order of the grain size. Of course, the waves which we will consider are low amplitudes so that they do not disturb efficiently the network of contacts while sound propagates. This leads to the introduction of different problems such as (i) scattering, (ii) diffusion, (iii) coherence and interferences of diffused waves and (iv) wave localisation (or Anderson's localisation); and to provide some basis for understanding their rules. So, sound with a large wavelength does not 'see' the "fragile" nature of the bonds, of the contact network and of the force, because it averages the structure on a large scale, and the mean structure evolves only slightly (even when the medium is deformed) so that the medium looks quite normal; but sound with shorter wavelength is responsive to this local structure, choosing its path as a function of the heterogeneity. A recent experiment will enable us to illustrate these questions. In addition we propose a new diagnostics using ultrasonic waves to test *in situ* the evolution of the contacts network during material deforms.

I hope this will facilitate technology transfer, thus allowing either the development of new diagnostics, or the improvement of the experimental techniques.

2 Vibrating billiard, granular gas and clustering in micro-gravity

In which form can one find weightless the granular matter? At least one can give some partial answer since the recent (February 1998) ESA Minitexus 5 flight,[3,4] during which an experiment has applied a linearly polarized sinusoidal vibration (amplitude a and frequency $f = \omega/2\pi$ ranging from a = 0.3mm–2.5mm and f = 3Hz–60Hz) to 3 cubic cells of $(1cm)^3$ each, containing bronze balls in respective quantity, less than one layer of balls, 2 layers of balls and 3 layers of balls. A typical result is given in Figure 1.

As Figure 1 shows, clusters form as soon as the number of layer n is larger than 1 or 2 in weightlessness, so that a single "gas"-like phase occurs when $n \leq 1$, i.e. only when the density of its particles is sufficiently low so that the mean free path l_c of each grain is larger than or of the order of the container size L. Under

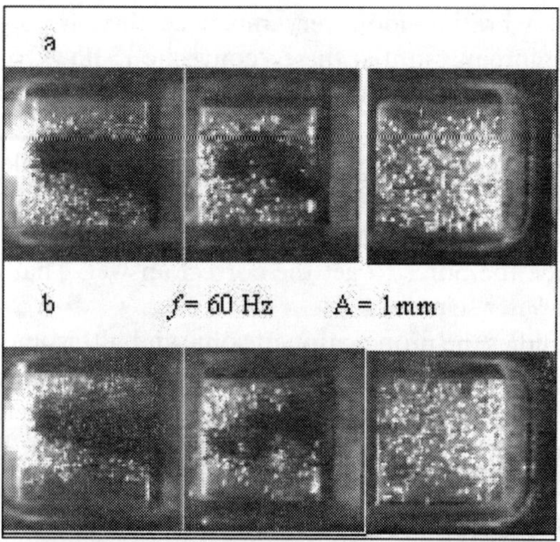

Figure 1   *Vibrated balls in weightlessness; from left to right, the cells contain: 3 layers of balls, 2 layers, 1 layer. The direction of vibration is vertical; amplitude 1mm, frequency 60Hz. Case (a): maximum "upward" position. Case (b): maximum "downward" position. One observes (i) the formation of a dense cluster surrounded by a loose gas in the case of the two densest cells, (ii) the existence of a depletion near one wall in all cases, which prove the "supersonic" nature of excitation*

these conditions the gas is extremely rarefied and the grains are almost isolated. For a usual real gas, this would correspond to the so-called Knudsen's regime ($l_c >$ L). As soon as the grain-grain collision probability increases and exceeds 1, one observes the formation of a dense cluster which remains more or less motionless in the container frame; this cluster is sandwiched between two layers of "Knudsen's gas", far from dense; a mechanisms of "evaporation" and "condensation" of grains by the cluster ensure the balance between the two phases.

Moreover, one observes that the container detaches from the gas periodically during the motion (this is visible in Figure 1); this indicates that the mean velocity of the grains in the gas phase is much lower than the maximum speed of the container, which means in turn that the conditions of excitation are supersonic[4] ($a\omega > v_{grain}$).

Finally, a pressure sensor in the direction of vibration has been used to determine the transfer of momentum due to bead collisions with the gauge, that is to say the granular pressure P, and its variations with the excitation parameters (a,ω). Using statistical analysis one finds that P distribution varies experimentally as $(a\omega)$,[1.5] which may be surprising because Boltzmann law predicts that it must be proportional to the square of the mean velocity of the balls v^2. A first explanation of this effect could be that the restitution coefficient varies with the grain speed. However one can suggest other causes; in particular, the response time τ of the sensor leads us to consider two regimes depending on the flow $N\tau$ of beads on the sensor depending on whether τ is smaller or larger than 1; indeed when $N\tau > 1$, the mean pressure P scales as $mvN\tau \approx v^2 = (a\omega)^2$, with typical fluctuations $\Delta p = mv\sqrt{(N\tau)} \approx (a\omega)^{1.5}$; on the contrary if $N\tau < 1$ the mean pressure P is alternately $mv \approx a\omega$ or 0 depending on whether a collision occurs or not within τ, and the fluctuation $\Delta p = mv \approx a\omega$, or m δv, depending on the way it is measured.

These results are still too partial to be taken as definitive, but they seem coherent. Moreover, they were partially confirmed by numerical and theoretical approaches;[5,6] in particular, the collapse of gas in a dense cluster is an intense field of theoretical research.

Nevertheless, these results show the very particular character of granular gas, and this at least for several reasons:

i) this gas can exist only in dilute regime (of Knudsen), probably with a vanishing density ρ at large cell size L, according to $l_c \approx L$, that implies $\rho = 1/(Ld^2)$, scaling as 1/L at constant grain diameter d; this is related to the dissipative character of the granular collisions. Under these conditions, it is difficult to speak about the thermodynamics of the granular gas state in the intermediate range $0<<\rho<1/L$, because it cannot scale as the volume size; so its thermodynamics functions do not have the character of an extensive quantity. On the other hand can one speak of the granular gas state, in the limit $\rho\rightarrow0$? This will be developed in point (vi) below. It would require at least that the "thermodynamics of an isolated grain" or of a "quasi-isolated particle" exists; this is quite close to the problem of billiards proposed by Sinai,[7] which is one of the foundations of chaos theory, with specific developments in statistical physics.

ii) That a granular gas exists only in the Knudsen regime also has some consequence for the description of its mechanics: it is known for a classical real gas (of atoms) that its description using concepts from continuum-media mechanics applies only for volumes L^3 larger than the mean free path, i.e. $L >> l_c$. because it requires that stress equilibrium can be written locally, with the adjacent volume elements. So, this leads to the thought that granular gas does not obey the mechanics of continuous mediums. This has a few consequences:

   – As recent theories of clustering use the continuum approach, one can wonder if they can be used.
   – As sound waves require continuum mechanics approximation, they do not propagate classically in a Knudsen gas.

iii) this experiment shows also that the excitation of a granular medium is of the "supersonic" type in general; this is likely related to the dissipative character of the grain-grain collisions, but this is not without consequence, because one knows that supersonic excitation obeys differential equations of the hyperbolic type, which admit discontinuities; under these conditions it is perhaps difficult to suppose the continuity of the state variables in any point of the medium. As this is a direct consequence of the strongly dissipative nature of collisions, this may be also observed in other situations; so it may explain the difficulty encountered in predicting the behaviour of rapid dense granular fluids when collisions dominate the dissipation, such as in the case of avalanching, of rapid surface flows, etc.

iv) Notion of granular temperature is perhaps not a panacea either. Indeed, the present vibration experiments demonstrate that the right variable is the speed of the box $a\omega$, so that the box works rather as a velostat than as a thermostat,[8,9] which means that changing the particle mass without changing the box speed and the number of layers will change the grain temperature $<mv^2/2>$ but not the mean grain velocity $<|v|>$. This is probably why the mixing of different kinds of grains will lead to problems, and generate segregation in particular.

v) On earth, the natural confinement of the medium on the bottom wall quite often hides these problems, because gravity forces the granular density to vary continuously. However, a recent experiment on "Maxwell demon" in granular gas confirms this analysis, since it demonstrates that clustering of gas occurs when the number n of layers approaches $n = 1$.[10]

vi) Vibrated billiard: As already recounted, granular gas exists only in quite dilute conditions; in the limit of fixed L and $\rho \to 0$, it corresponds to the case of a vibrating billiard, with a single ball. This is why we did simulations on a 1d box,[9] and found that the particle speed was always larger than $a\omega$, and sometimes much larger; this is in contradiction to Figure 1. Furthermore, a 3d experiment in weightlessness (CNES and ESA campaign in Airbus A300) with 1 ball in a fixed 3d box and a vibrating piston confirms this result, i.e. $v>a\omega$; better, it has shown intermittent resonance, during which the ball performs a round-trip per period synchronously with the vibration, so that $v = 2Ln>>2\pi af$, and during which the trajectory aligns along a single periodic path parallel to the vibration direction; so it corresponds exactly to 1d simulations. Intermittency of the stabilisation is also observed in 1d simulations, the motion becoming only fully synchronous at large amplitude.

This synchronisation is an important result, since it demonstrates the intermittent breaking of the ergodicity; it shows also the reduction of the dimensionality of the phase space, which jumps from 11-d (which is the number of freedom degrees for a 3d ball with time dependent excitation) to 1-d or 3-d space, corresponding to the motion of a point at constant speed, under periodic excitation. Reduction of dimensionality is classic in dissipative systems with dissipation; this explains in particular chaotic strange attractors; however, it turns out to be peculiarly efficient in the present case.

3d simulations have been performed also with a Discrete Element code from Moreau, which takes account of restitution coefficient and friction during collision; they have confirmed the experimental finding; furthermore some transient trajectories which are linear and quasi periodic and which are oriented perpendicular to the direction of vibration have also been found, for peculiar initial conditions; this reveals the existence of different cyclic trajectories with their own distinct basin of attraction, with a hierarchy of Lyapounov exponents. These transient trajectories may become steady when the tangent restitution coefficient is non zero. Experimental test of stability of the synchronous trajectory has been performed; paths remain stable around a mean position perpendicular to the faces, even when attempts are made to enforce destabilization by inclining the direction of vibration at 10° about from the cell axis. This stability explains the lost of ergodicity.

One can approach this phenomenon from the view point of the theory of chaotic quantum billiards, which is often used to test basic concepts in statistical physics. In this case classic motion corresponds to quantic motion of particle with small wavelength. One knows for instance that a cubic billiard has closed trajectories, which implies that it is not ergodic; these trajectories can be seen as "eigen modes"; so the vibration excites the ball motion on some of these trajectories; those modes which dissipate too much cannot be sustained while those which dissipate too little scatter on others (however at low enough excitation this second process does not exist); so only few eigen modes can be excited by vibration, that explains the few possible trajectories perpendicular and parallel to the faces which act as attractive basins with low dissipation. The real shape of the cavity also plays a significant role on the ergodicity/non-ergodicity of the problem. In particular, billiards having the shape of a sphere cap should not present such eigen modes. It would be interesting to demonstrate that such a cavity would result also in the destabilisation of the resonant trajectories, hence improving the ergodicity of the dynamics. In the same way, adding some hard convex fixed obstacles into the cavity should also improve the "quality of ergodicity". This is why perhaps the existence of a few balls, instead of a single one, may change the qualitative nature of the dynamics and force the problem of granular gas to be rather ergodic. However, this remains to be demonstrated.

Another conclusion is the "catastrophic" influence of dissipation during grain-grain collision, because it explains why the typical speed v of the grains is smaller than the typical boundary speed aω, or in other words, it explains the "cooling" of the granular gas below the temperature of its boundary. This dissipation imposes the process of clustering at larger density, which means that granular matter condenses spontaneously into piling; this occurs also on earth, on vibrated plates as already demonstrated by Faraday.[1] Thus condensate is the "natural state" of a granular medium, which explains, if needed, the formation of planets and asteroids, in particular. So, we turn now to the characterisation of the condensed phase.

3 Granular Medium Behaving as a "Perfect Fluid": Swell Effect at the Free Surface – The Behaviour of a Granular Medium Saturated with Liquid and Subject to Intense Horizontal Vibrations

3.1 Introduction

In this section we will show that a granular medium can behave as a perfect fluid, i.e. non viscous and frictionless. This requires the application of vibrations such that the inertial forces are significantly larger than the gravity forces and the other body forces. The proposed experiment will generate at the liquid–sand interface some kind of swell; the swell is the undulation of the sea surface due to wind; in the case of the sea, the undulation propagates because the wind always blows in the same direction; but if one forces the wind to blow back and forth periodically the undulation will remain fixed on average, the crests at some average positions and the valleys at some others, just in between. This also happens at the water–sand interface under intense horizontal vibration, as we will demonstrate.

3.2   Swell and the Kelvin–Helmholtz Instability

As reported in Figure 2, the swell is formed at the flat interface between two fluids having different speeds v_1 and v_2 and densities ρ_1, ρ_2, due to the instability of Kelvin–Helmholtz which deform the interface. These two fluids behave as if they were incompressible. Let us consider a flat interface, which is locally deformed upwards as in Figure 2, the conservation of flow imposes that the speed of fluid 1, v_1, increases above the bump and that of fluid 2, v_2, decreases below the bump. So, as Bernoulli's theorem states that $\rho v^2/2 + p$ remains constant along a flow line, it results that the local pressure p_1 decreases, while p_2 increases at the bump; this pushes the bump upward, and the bump amplifies spontaneously. The same argument applies if the bump is oriented downwards, and whatever the direction of speeds. (This can be shown by turning Figure 2 upside down and left to right). A linear stability analysis shows that the process is always unstable whatever the wavelength in the absence of capillary force (p. 154 in Reference 11), the smaller the wavelength the larger the amplification.

In fact an exact calculation can be performed in the case of two infinite layer of liquids with capillary forces. Kelvin found in this case, cf. p. 351 problem 3 of Reference 11, that there is a minimum difference U_{KH} in speed $v_1 - v_2 = U$ between the two liquids below which the plane surface remains stable; the flat interface becomes unstable above U_{KH}, with a more unstable mode characterised by its wavelength λ_{KH}. U_{KH} and λ_{KH} are found to be:

$$\|v_1 - v_2\|^2 = U^2_{KH} = 2[(\rho_1 + \rho_2)/(\rho_1\rho_2)]\,[\alpha g(\rho_2 - \rho_1)]^{1/2} \tag{1a}$$

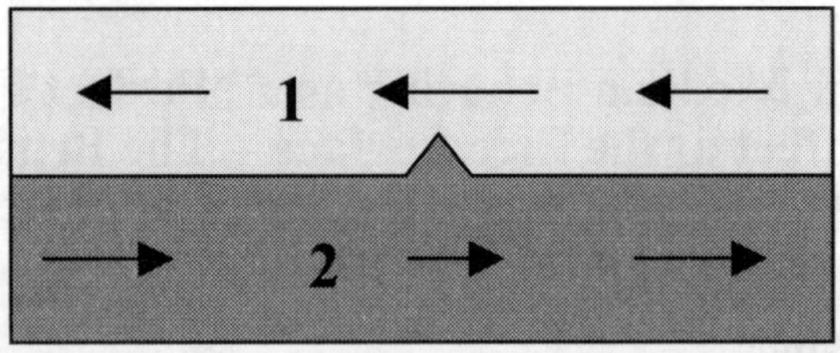

Figure 2   *The Mechanism of Kelvin–Helmholtz instability: due to the flows, the plane interface between the 2 liquids is unstable: any vertical upward (downward) disturbance is amplified because it generates a speed increase (decrease) of the flow in the upper layer which causes a pressure decrease (increase)*

$$\lambda_{KH} = 2\pi \left[\alpha / \{g(\rho_2 - \rho_1)\} \right]^{1/2} \tag{1b}$$

where α is the surface tension between the two liquids. Equation 1a confirms that instability appears as soon as the speed difference U is non zero for liquids without surface tension ($\alpha = 0$), and Equation 1b that the most unstable wavelength is for $\lambda = 0$ in this case.

When the wind always blows in the same direction the undulation which is formed propagates at the speed of capillaro-gravity waves. But when the excitation is forced by a periodic shaking, the two speeds go back and forth and the undulation does not propagate nor oscillate: rather it is fixed in the cell frame.

3.3   Vibration Excitation and Swell

Let us try and understand what occurs in the case of a vibrating cell of height $h = h_1 + h_2$ filled with two incompressible fluids of density ratio $\rho = \rho_2/\rho_1$, in the proportion $\chi = h_2/h_1$; we start from Equations 1 which include capillary forces; we then follow the approach suggested in Reference 12. We then take the limit of a null capillary force, which corresponds to the case of a cell filled with wet sand and liquid.

Let b sinωt be the horizontal periodic motion of the cell; due to inertia, this motion forces the relative motion of the two liquids, which move with a relative speed U:

$$U = b\omega \cos\omega t \, (\rho - 1)/(\rho + \chi) \tag{2}$$

As Kelvin–Helmholtz instability works independently of the direction of U, the relief alternately increases and decreases during each half period depending on the value of $U(\tau)$ compared to U_{KH}, it increases if $U(\tau) > U_{KH}$, while it decreases if $U(\tau) < U_{KH}$. So, in a first approximation one expects that the threshold will thus correspond to the conditions 1 averaged over half a period; this leads to the condition:

$$(b^2\omega^2)_{KH} = 4 \left[\alpha g/(\rho_2 - \rho_1) \right]^{1/2} \left[(\rho + 1)(\rho + \chi)^2 \right] / \left[\rho(\rho - 1)(1 + \chi)^2 \right] \tag{3}$$

Figure 3 displays such a relief.[13] It has been obtained in diphasic liquid-gas CO_2 in the vicinity of the critical point (T_c, ρ_c); it is localised at the gas–liquid interface.[13] We display this case, because it demonstrates the phenomenon is robust, since diphasic CO_2 is supposed to be hyper-compressible in the vicinity of the critical point (T_c, ρ_c); however, the timescale at which the binary system is incompressible is much longer than the period of vibration (0.1s–0.02s) in the experiment; this explains why the fluids behave as incompressible. The relief does not propagate horizontally, nor oscillate vertically; it appears frozen in the cell frame, with some rapid horizontal jittering. The relief becomes flat at once, when the vibration is stopped.

As a matter of fact, the experimental discovery of spatially periodic relief due to horizontal vibration is probably due to Wolf;[14] Lyubimov and Cherepanov[15]

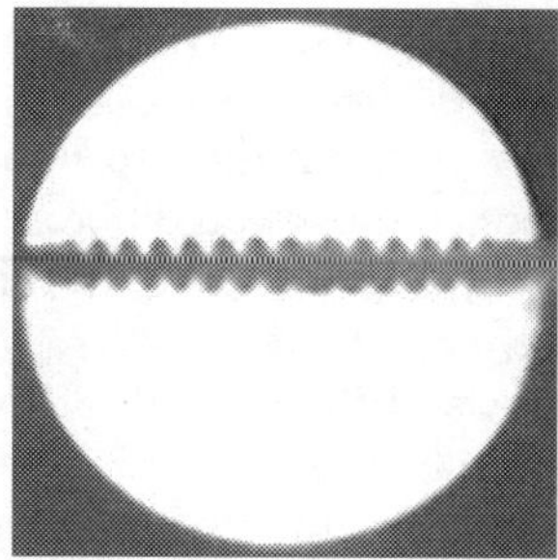

Figure 3   *Frozen relief observed in diphasic liquid-gas CO_2 in the vicinity of the critical point (T_c, ρ_c), under intense horizontal vibration*

carried out a rather complete analysis, with incompressible non viscous fluids, based on a development using multiple and hierarchical timescales ; the fast one corresponds to the frequency of excitation, the others are slower and integrate the fluid motion over several periods. The calculation in Reference 15 corresponds to the case $\chi = 1$, with $\alpha \neq 0$. These authors find:

$$(b\omega)^2 = \{(\rho_1 + \rho_2)^3/[2\rho_1\rho_2(\rho_2 - \rho_1)]\}\{\alpha k + (\rho_2 - \rho_1)g/k\}\text{th}(kh) \qquad (4)$$

with $k = 2\pi/\lambda$. This calculation also shows that the bifurcation which corresponds to passing from the flat relief to the sinusoidal relief is critical, within the meaning of the bifurcation theory, which implies that the amplitude of the relief varies close to the threshold like:

$$a_r \propto [(b^2\omega^2) - (b^2\omega^2)_{KH}]^{1/2} \propto \{2(b\omega)_{KH}[b\omega - (b\omega)_{KH}]\}^{1/2} \qquad (5a)$$

$$\text{or as } A = a_r/h \propto [(b^2\omega^2) - (b^2\omega^2)_{KH}]^{1/2}/(gh) \qquad (5b)$$

In the CO_2 gas-liquid case, we have found some discrepancy with these findings; they are likely due to working in the vicinity of the critical point.[13]

In the case of two viscous incompressible liquids ($\alpha \neq 0$), we showed in Reference 12 that the height of the relief varies linearly with $(b^2\omega^2) - (b^2\omega^2)_{threshold}$, within the accuracy of our measurements; it thus does not follow Equations 5. We suppose that it is because of not being close enough to the threshold; because relief is easily observed when a_r/λ is about 1, while Equationas 5 assume that a_r/λ is quite small. We have also observed in this case that λ remains constant close to the threshold. Then λ grows using a mechanism of period doubling.[12] Lastly, much further above the threshold, one observes that λ scales as the relief amplitude a_r. The explanation of these last three phenomena is as follows: when the system is close to the threshold, the capillary length $l_{capillar} = \{\alpha/[g(\rho_2 - \rho_1)]\}^{1/2}$ plays a significant role, because it fixes the wavelength λ of the relief according to Equation 1b): $\lambda = 2\pi\, l_{capillar}$. As the relief grows, the amplitude quickly becomes larger than $l_{capillar}$; the initial wavelength is destabilized, that provokes the period doubling:[12] indeed if $\lambda << a_r$, the flow can not penetrate more deeply into the valleys which would destabilise the relief shape; this imposes $a_r < $ (1-to-3)λ.

So, when the height of the relief becomes large in comparison to $l_{capillar}$, λ evolves. To find λ under these conditions, one can neglect α in Equation 4. If λ remains small in comparison to h, th(kh) ≈ 1, and Equation 4 leads to:

$$\lambda = 2\pi \; \{(b\omega)^2/g\}[2\rho_1\rho_2(\rho_2 - \rho_1)]/(\rho_1 + \rho_2)^3 \qquad (5c)$$

So a_r and λ grow roughly in the same way. Using h as the length unit, Equation 5c becomes:

$$\lambda/h = 2\pi\{[2\rho_1\rho_2(\rho_2 - \rho_1)]/(\rho_1 + \rho_2)^3\}\{(b\omega)^2/(gh)\} \qquad (5d)$$

This corresponds to the behaviour in Figure 3. This regime is valid until $\lambda/h<1$, then interaction of the relief with top and bottom walls occurs and the relief is stabilised with a vertical interface. Labelling W = $\{(b\omega)^2/(gh)\}$ the dimensionless vibrational parameter, Equations 5 predict that λ/h scales as W when $a_r \approx \lambda$. Or if k is the wave number, kh will scale as 1/W.

This theory applies neglecting viscosity. This is only possible when the thickness δ of the Stokes boundary layer is small, which implies $\delta = (2\nu/\omega)^{1/2} <<\lambda$, and $\delta<<h$, where $\nu = \eta/\rho$ is the kinematic viscosity of the liquid of density ρ and dynamic viscosity η. At this stage, it is probably worth recalling the definition of this viscous boundary layer: let us consider a thin plate immersed in a liquid, which oscillates in a direction parallel to its surface; this plate drags a surrounding layer of liquid whose thickness $\delta = (2\nu/\omega)^{1/2}$ depends on viscosity and frequency; this frequency dependence is due to the fact that momentum propagates due to diffusion. The effect of two liquids with different viscosities on the relief formation has been investigated recently, when δ becomes large,[12] i.e. $\delta \approx \lambda$; it makes the relief no longer symmetric.

3.4 Application to Sand–liquid Interface Under Strong Fast Horizontal Shaking

It is now time to investigate the case of liquid and sand. Let us take a rectangular closed cell (length L, height h<< L, thickness e<L) or cylindrical shape of radius R=D/2, with a horizontal axis Ox. Let us half fill it with sand, then fill it up completely with alcohol or water. The cell appears to be filled with two layers, one consisting of the wet granular layer, the other one a liquid. Let us place this cell on a vibrator which vibrates horizontally according to b sinωt. Let ν be the kinematic viscosity and $\delta = (2\nu/\omega)^{1/2}$ the boundary layer thickness.

So let us now consider the intermediate frequency range f = $\omega/(2\pi)$; it is such that δ is large compared to the grain size d, i.e. $\delta>>d$, but small compared to D or h. In this case, the grains cannot move independently but are forced to move coherently with the whole granular medium; so this medium and the liquid filling its pores appear as a single coherent deformable phase working at constant volume. Due to inertia, vibrations induce deformation of the granular layer, which in turn forces the pure liquid layer on top of it to move. So the two layers move back and forth with a dephasing equal to π. This produces a relative horizontal flow, which is able to generate an instability of the Kelvin–Helmholtz type, if the vibration is strong enough.

If this analysis is true, intense vibration should produce a space-periodic relief, immobile in the cell frame; furthermore, this relief should behave as the one which appears between two liquids without surface tension, because the liquid already wets the grains. So the relief should obey Equation 5c.

We report in Figure 4 the results of such an experiment,[16] which demonstrate that wet sand can behave as a perfect (i.e. frictionless, non viscous) liquid, if it is shaken sufficiently hard. The liquid-like nature of the sand can be demonstrated experimentally by stopping the vibrating; in this case one observes the flattening of the relief; and this flattening occurs as soon as the vibration is stopped.

So the behaviour is quite different from the one observed for dunes, for which the dune shape is stable when the wind stops blowing. It is neither similar to the case of ripple formation which occurs either below the sea during the tide, or in a river flow; because these ripples remain when the flow stops or when the tide has gone out. More information about these two mechanisms can be found in Reference 17. They appear often because of sand transport due to the generation of convection rolls which are excited by some periodic oscillation (gravity waves) in shallow water due to Schlichting mechanism.[18] When the ripples are forced, they stabilise the convection rolls and force the ripple growth. Such effect can also occur in a vibrated bed depending on the boundary conditions which are imposed.[19] The case which has been considered here correspond to the absence of periodic liquid flow when there is no periodic granular flow; but this is not always the case.

Even so, the horizontal vibration has to be intense, because the periodic relief appears only above a threshold which is large, since it correspond to $\lambda >> d$, where d is the grain diameter. At smaller amplitudes of vibration, but $b\omega^2 > g$, the grains move at the surface, but the surface is rather flat because it is sheared back and forth. One observes also convection patterns either localised near the vertical walls if the aspect ratio h/L is small or in the whole volume if $h/L \approx 1$. This kind of convection is also a general trend, which can be generated by an inertia force due to horizontal vibration, but it can also be forced by quasi-static bulldozing as it will be studied in the next section.

In the present section, the relief forms due to the combined effect of the instability of the flat horizontal interface due to the horizontal flow; it is caused by the dynamic pressure; but the cohesion of the granular medium is required to ensure that it deforms at constant volume; this is ensured by the viscosity of the liquid. For instance, if a less viscous fluid had been used, no relief would have been formed; this has been tested using a supercritical CO_2 cell with sand: CO_2 is not viscous enough to make the granular phase coherent and the relief does not appear; it is also perhaps a question of local cavitation.

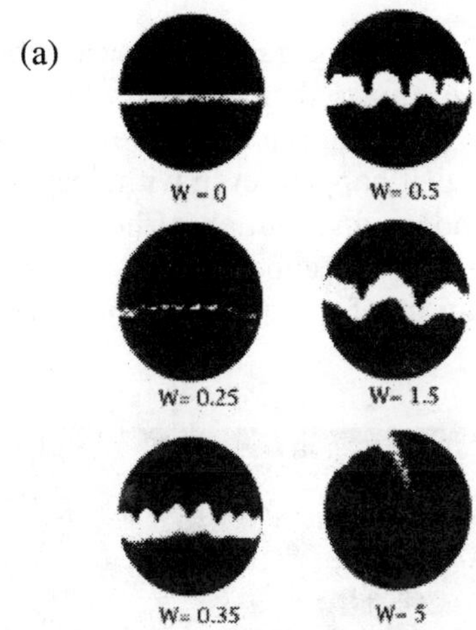

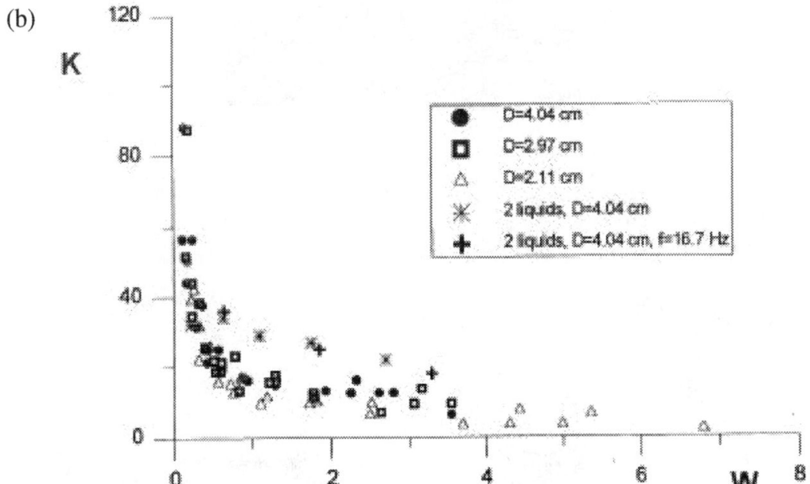

Figure 4   *Effect of intense horizontal vibration on the interface between granular material and liquid: a: typical reliefs obtained for certain values of the vibrating parameter $W = b^2\omega^2/(gD)$ in a cylinder cell filled with sand and ethyl alcohol. D = diameter of the cell. b: Dependence of the wave number of $K = 2\pi D/\lambda$ of the relief as a function of $W = b^2\omega^2/(gD)$. In fact, one finds that $K \sim 1/W$. Thus, λ varies linearly with W. Second set of data corresponds to reliefs formed with two liquids (here Fluorinert^{TM} and castor oil), when the relief amplitude a_r is large, i.e. $a_r \approx \lambda$*

3.5   Conclusion: Effect of Vertical Vibration; Other Potential Applications of Vibrations

It is also interesting to discuss the case of vertical vibration, but it exceeds the scope of the paper. One knows that vertical vibrations cause heaping in dry granular beds[20] above a given threshold. However, this is true also in the presence of water; in this case, it happens even with deep granular beds,[21] where one can observe a dense random distribution of heaps, whose typical size depends on the vibration amplitude; the cones are destabilised at larger amplitude of vibration by the formation of a liquefied sand layer with gravitation waves

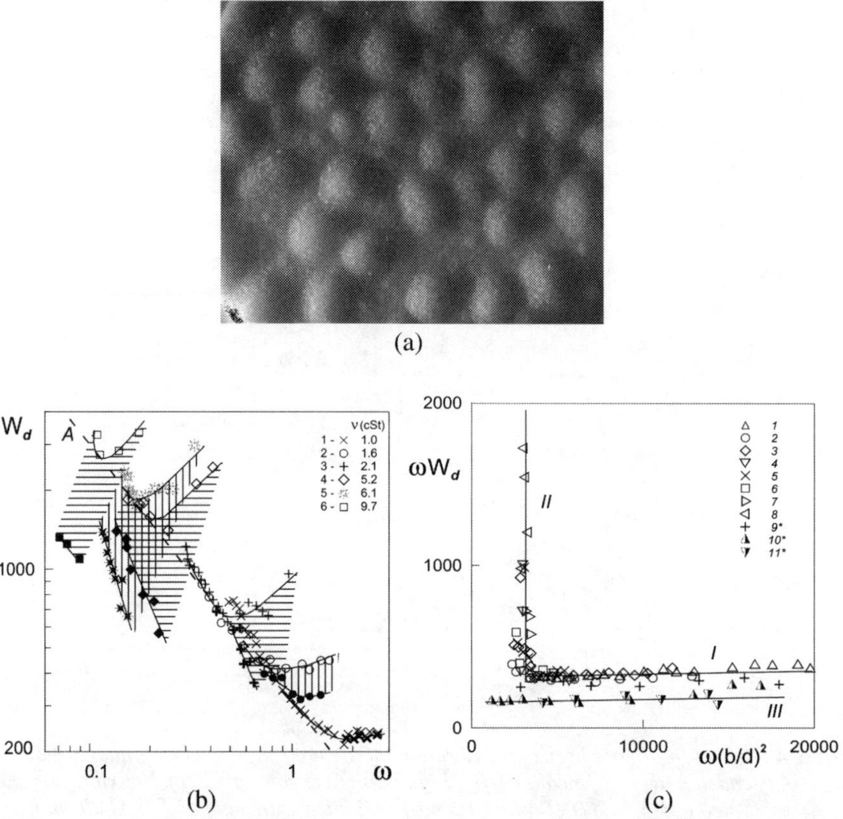

(a)

(b)   (c)

Figure 5   *Effect of a vertical vibration, $z = b\sin(\Omega t)$: **a:** Typical heaping. **b:** Limits of liquefaction (light marks) and hardening (dark) as a function of dimensionless frequency, in the $[W_d = (b^2\Omega^2)/(gd), \omega = \Omega d^2/\nu]$ plane, for a bead diameter $d = 0.11$ mm; the areas of hysteresis are shaded; line A shows the common boundary of liquefaction of granular medium. **c:** Thresholds of liquefaction in the plane $[\omega(b/d)^2 = \Omega b^2/\nu, \omega W_d = b^2 d\Omega^3/(g\nu)]$ at low frequency $(\omega < 1)$ for $d = 0.11$ mm $(1\text{–}6)$, 0.34 mm $(7, 8)$ and 0.06 mm (9). The marks 10 and 11 show the boundary of heap excitation for $d = 0.11$ and 0.06 mm*
(From References 21 and 22)

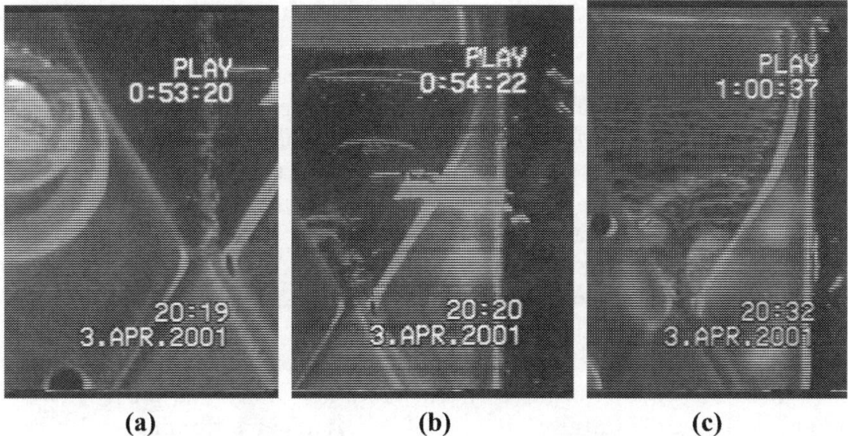

(a)   (b)   (c)

Figure 6   *(a) Normal sedimentation pattern obtained from a hole in 2d vs. (b) and (c) horizontal-layering patterns obtained during sedimentation under vertical vibration. The patterning is due to an instability of the Kelvin–Helmholtz kind,[23] which persists also near a critical phase transition.[23,24] As the beads are lighter than the liquid they are moving up during sedimentation*

excited parametrically. This happens with hysteresis[22] which demonstrates that liquid viscosity often plays an important part. Figure 5 illustrates the cone distribution (left picture), their size is 1cm diameter in the present case. The plots in Figure 5 give the liquefaction threshold in two different sets of dimensionless parameters, Figure 5c shows that the second set is the more appropriate. As this set depends on the viscosity v, this confirms the important role played by viscosity in the phenomenon.[21,22]

It is also important to state that the Bernoulli theorem may apply in many other situations; it can push the grains located near a wall towards it when the container is vibrated in a direction parallel to the wall. In the case of a drop of liquid immersed in another liquid under vibration, it can force the drop to flatten perpendicular to vibration.

This occurs also with dense clouds of sedimenting particles[23] where layering, which stops sedimentation, has been observed. Some examples in 2d are reported in Figure 6. It seems also to occur in weightlessness in liquid-gas equilibrium under vibration.[22,24]

4   Plastic Convection Flow Generated by Cyclic Quasi Static Loading – The Case of a Granular Embankment Under Cyclic Slow Bulldozing

4.1   Introduction

Granular materials deform plastically when subject to a stress increase; this is often seen as a solid-like behaviour. However, this can be viewed also as a liquid one; it depends mainly on the strain range. The purpose of this section is to show that flows generated in vibrated beds can often be interpreted within this scheme.

On the other hand, many recent studies[25–29] have been investigating the dynamics of dry granular media with a free surface and subject to vertical or horizontal vibrations. Many of them consider the flow to result from the dynamic character of the excitation, which enhance some "thermal excitation". Indeed, if the grains are "thermally" agitated in a inhomogeneous way, the diffusion coefficient of the grains varies with space, which forces flowing. Conversely, when one observes a convection, one can find in principle a temperature distribution which explains the flow, whatever the pattern.

However the real question remains: is the granular flow observed in a vibrated granular system forced by "thermal activation" really? This section aims to show that "convection" in these systems does not always require a "thermal" excitation. For this, we give an example of granular streaming under cyclic excitation with no kinetic energy involved. This may be surprising at first sight, because one tends to think that enhancement of granular flow under cyclic excitation requires the liquefaction of the material first, which then requires "heating".

But as stated in the first paragraph of this section, this is not true for all kinds of granular flow: it is for creeping avalanches, or for any other "quasi-static" deformation, for which the flow is forced by an irreversible change of the force distribution, and which is controlled by plastic behaviour.

So the real question is: can cyclic excitation force streaming in a quasi static regime? If the response is yes, and if the observed flow pattern looks similar to the one observed under "dynamic" excitation, this forces one to reconsider the problem of "dynamic" excitation and to interpret it within the scheme of plastic flow. In turn, this will necessitate a better description of plastic stress-strain behaviour.

In the first subsection a quasi static experiment is described, which generates granular flow under cyclic horizontal loading; it is found that its geometrical characteristics look similar to those of flow obtained with horizontal "dynamic" shaking. A physical explanation is proposed which uses classical soil mechanics concepts. In the second subsection, this flow generation is analysed within the scheme of fluid mechanics; a parallel is drawn with the problem of acoustic streaming in fluids. The third section extends the use of fluid mechanics and discusses the important problem of mixing. It states that a 2d continuous pattern cannot be used for mixing, as this requires at least fully developed 3d flows. This implies in particular that cylinders are never good mixers, which explains why they often generate segregation.

4.2 Experimental Results

Consider the experimental set-up of Figure 7a. It allows the applications of some cyclic bulldozing ($A \sin[2\pi ft]$) in the horizontal direction to a 2d embankment (height H, length L) of cylinders of diameter d = 5mm and length 5mm. Typically, A/H never exceeds a few percent and T = 1/f is small so that the pile deforms in quasi statics; for instance, T = [0.5s–3s] and A = [1cm,5cm] if H = 27cm and d = 5mm. Some cylinders are marked so as to be able to follow their trajectory and to measure their rotation; the very great majority of the cylinders have their axis perpendicular to the plan of the solid mass, but some others are parallel to this plane, which makes it possible to break the natural triangular order of stacking. After a few tens of cycles, the boundary conditions take a cyclic behaviour; the form is a deformable right-angled trapezoid.[30]

4.2.1   Flow Generation

The back and forth motion of the bulldozer generates a flow of the grains; we show the typical trajectory of a grain in Figurec7b, measured after each cycle. The larger the amplitude A, the larger the flow (cf following paragraph). Furthermore the grains rotate during their flow, and the rotation varies with location as shown in Figure 7d. Further details on grain rotation and motion and on their statistics can be found in[30] with some detail on diffusion and on Cosserat's mechanics. This is not the topic here.

We just note that the flow pattern of Figure 7 looks very much like the one observed under dynamic horizontal vibration.[27,29]

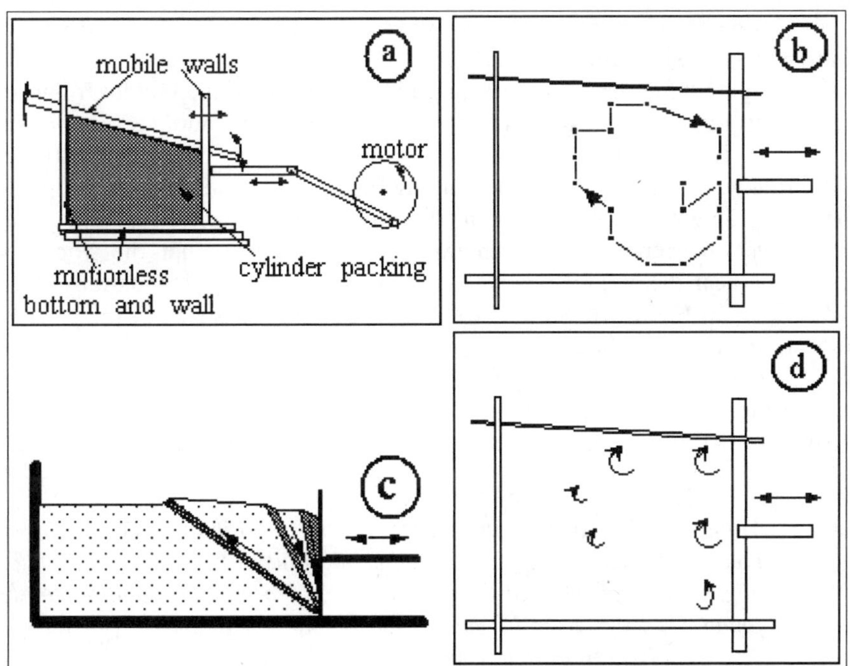

Figure 7   *(a) Experimental set-up for cyclic bulldozing; (b) typical position of a grain after each cycle; (c) orientation of passive (upward motion) and active (downward motion) failure when the free surface is horizontal; (d) typical mean amplitude of rotation of a grain after each cycle, as a function of the position in the stacking (arbitrary unit). This flow pattern looks very much like the one observed under dynamic horizontal vibration[27,29]*

4.2.2 Mechanism of Flow

Figure 7c sums up a possible explanation of this phenomenon;[29,30] it is based on the fact that passive and active Rankine states do not generate the same deformation and failure,[31] so that the back and forth motion is not symmetric and the bands of localization have different positions and orientations in these two cases. This generates some flow and forces the shape of the stacking to evolve, and the top surface inclines until it stabilises.[32]

Let α be the angle of the failure plane with the horizontal, and β be the one of the free surface (or of the lid) with the horizontal, and let φ and φ_w the internal friction angle of the medium and the friction angle of the medium with the bulldozer respectively. It turns out that the active angle α_{active} is different from the passive one $\alpha_{passive}$; but they both depend on β, φ and φ_w. One finds for instance that α_{active} decreases and $\alpha_{passive}$ increases as β increases, when φ_w is small; so there can be some inclination β_o of the free surface for which $\alpha_{passive} = \alpha_{active} = \alpha_o$. When this latter condition is satisfied, the stacking shape does not evolve anymore, which fixes the mean inclination of the lid β_o. This problem can be solved,[32] which leads to the solutions α_o and β_o of Table 1, which depend on φ and φ_w.

Indeed, concerning the failure zone, Figure 8 indicates (i) that the mean flow results from the difference of motion of the particles during the active and passive mode of deformation, (ii) that the average angles of active and passive failure are approximately identical.

Also, Figure 8 highlights the role played by the walls of the bulldozer and by the free surface: it is there that the difference between forth and back motions is the largest. It is worth noting that the grains which are in contact with the bulldozing wall remain in contact all along the cycle, but they also slide up and down, with an amplitude which is smaller upward than downward, so that they slide down on average after each cycle; one notes also that the slopes of the motion near the bulldozer correspond roughly to the two inclinations of the localizations in Figure 7c. So, due to the cyclic change of shape of the pile, the grains near the free surface or near the bulldozer seem to penetrate into the pile; but this is wrong: they move parallel to these surfaces themselves, but the surfaces themselves move.

Anyhow, the difference of grain motion during back and forth movement is

Table 1 *Values of free surface inclination β_o and of failure slope α_o corresponding to $\alpha_{active} = \alpha_{passive}$, for different values of the internal friction angle φ and of the wall friction angle φ_w. No periodic solution can be found when $\varphi = \varphi_w = 30°$ and/or for $\varphi = 30°$ for larger φ_w; nor for $\varphi = 40°$, $\varphi_w = 10°$ (from Reference 32).*

	$\varphi = 20°$ $\varphi_w = 0°$	$\varphi = 20°$ $\varphi_w = 10°$	$\varphi = 20°$ $\varphi_w = 20°$	$\varphi = 30°$ $\varphi_w = 0°$	$\varphi = 30°$ $\varphi_w = 10°$	$\varphi = 30°$ $\varphi_w = 20°$	$\varphi = 40°$ $\varphi_w = 0°$	$\varphi = 40°$ $\varphi_w = 5°$
α_o	45°	38°	35°	45°–47°	40°	36°	45°	42°
β_o	13°	15.5°	17°	25°–26°	28°	29°30′	39.5°	39.9°

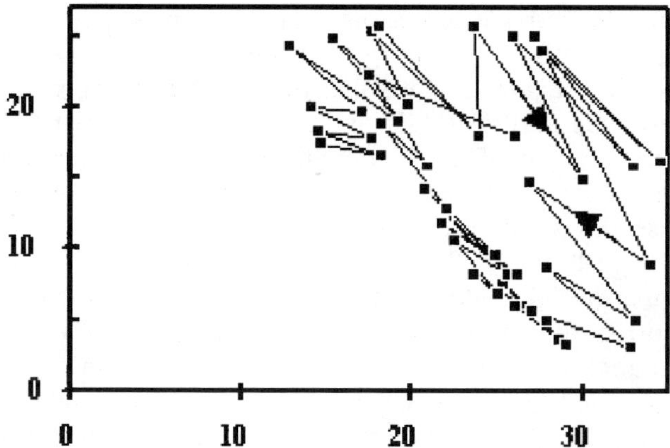

Figure 8   *Grain motion during a series of consecutive half cycles. Grain trajectory during active and passive motions are parallel mainly. Right part correspond to location near the bulldozing wall, top part to a grain near the free surface*

responsible for the streaming. This effect is also encountered in other problems of fluid mechanics. This is why it seems important to be reminded of some of them.

4.3   Use of Fluid Mechanics Concepts

4.3.1   *Parallel with the Case of Acoustic Streaming in Fluid Mechanics*

So, it is worth making the parallel between this flow generation and the one met when a fluid is contained in a container which is subjected to sonic or ultrasonic waves.[33] In this case the oscillation of the wall also forces the liquid to flow, due to viscosity. Let us now assume the wall is not perfectly flat, the flow line should adapt to the geometric constraint. This generates a gradient of speed and a space dependent dephasing. This is specially true near a corner; but it is also encountered for a sinusoid bottom. So, letting $\mathbf{u}(x,t)$ be the flow of the fluid with time t, one has $\int u(t)dt = 0$ over a period; also the speed of a particle of fluid at time t being $w(x,t)$ is, without approximation

$$\mathbf{w}(x_o,t) = \mathbf{u}(x_o + \int_{t_o}^{t} \mathbf{w}(t)dt,t) \tag{6}$$

leading to

$$\mathbf{w}(x_o,t) \approx \mathbf{u}(x_o,t) + \{[\int_{t_o}^{t} \mathbf{w}dt]\nabla\mathbf{u}(x,t)\}_{x=x_o} \tag{7}$$

which has a non zero mean, due to the variation of phase with space. In other words, because the particle moves, its mean motion is non zero, due to the variation of phase with space. This forces a coherent stream.

In the same way, Figure 8 shows that the grains perform a systematic drift after each period; this drift depends on the location, due to the boundary conditions and their variations. This generates the stream. A difference remains with fluid mechanics, because viscosity ensures non sliding conditions at the boundary in this case; this is not the case in Figures 7 and 8.

Finally, it is worth noting that the flow pattern of Figure 7 is observed similarly in granular media in a container under "dynamic" horizontal excitation, i.e. intense horizontal vibration of large frequency[29] (f > 15Hz). In this case, the mode of excitation is not quasi-static any more (for instance one observes a periodic separation of the medium from the vertical wall of the container). Anyhow, it seems that the stream results from the same combinations of mechanisms as those which provoke the quasi-static streaming of Figure 7. Some difference will probably remain between the two modes, i.e. quasi-static and dynamic, because the two stackings are submitted to different boundary conditions all along the cycles. Nevertheless, one can guess that some rather good description should be obtained using local quasi static rheology (stress-strain relation) with correct boundary conditions.

The experiment of Figure 7 also generates large fluctuations because the grain size is large compared to the sample. Does it mean that fluctuations play an important part in the streaming mechanism? Indeed, Figure 8 shows that microscopic flow rules can be defined with rather good accuracy for averaging. Anyhow, a similar effect would occur in any viscous liquid; and it is even enhanced, because molecules of liquid move randomly with a speed much faster than any flow speed, as far as this flow is much smaller than the speed of sound. This is why we believe the stream of Figure 8 is controlled completely by the macroscopic mechanics, and by the plastic behaviour of the granular material.

4.3.2   *Parallel with the Mechanics of Fluids: Chaotic Advection*

One can push further the parallel with the problem of steady flow in an unspecified fluid.[34] Let us look at the motion over several periods; in mean it is a 2d steady flow at constant volume; this requires that $\mathrm{div}(\mathbf{v}) = 0$, where $\mathbf{v}$ is the mean local velocity of a particle during a cycle; this requires that there is a stream function Ψ such that:

$$v_x = dx/dt = \partial\psi/\partial y \qquad \& \qquad v_y = dy/dt = -\partial\psi/\partial x \qquad (8)$$

where ψ is called the stream function.

Formally, the problem becomes equivalent to the problem of a particle in a field, which is governed by the system of equation:

$$dp/dt = \partial H/\partial q \qquad \& \qquad dq/dt = \partial H/\partial p \qquad (9)$$

where H is the Hamiltonian of the particle. It is known that this problem is integrable and that it cannot lead to chaos in 2d. Thus, a permanent two-dimensional flow of an incompressible fluid cannot give chaos and cannot be used for active mixing. In other words, the permanent flow lines which results from such a device can be plotted in a 2d plane; but they cannot cross, because stream lines cannot cut one another; so due to the 2d topology, these lines will look either concentric or open (if the system is open); they may separate into a few distinct zones, i.e. vortex; so the typical distance between two concentric flow line scales then as the vortex size, which is also most often the device size L itself. So, when used for mixing such a 2d device can only mix with the help of Brownian motion; this process takes place at the grain scale d and its time efficiency vanishes as $(d/L)^2$.

It should be remembered that the experimental set-up of Figure 7 cannot be used as a mixer, because its efficiency decrease strongly with the grain size. Indeed, this topology argument, on the structure of the flow line, does not remain in 3d. So, good mixing requires a fully developed 3d flow pattern, which requires at least 2 directions of excitation with 2 different frequencies.

It is worth remembering that rotating cylinders also belong to the category of 2d devices; so they can never be good mixers. So, it is quite surprising that they are used so often in plants. It is just as if one wanted to dissolve a lump of sugar in a cup of coffee just by rotating the cup, without using a spoon; this is just inefficient.

An important domain of application of convection flow for granular matter then is mixing and segregation.[35]

4.4 Conclusion

To finish this study, it is worth drawing a few conclusions. For convenience, I will term hereafter "soil mechanics" the classic stress-strain laws a granular material obeys in the quasi static regime. Of course, if "soil mechanics" exists, it should apply to a much larger domain than the one of soils.

• *1ˢᵗ Interest of the experiment: Validity of "soil mechanics" to describe flowing:* Indeed it has been demonstrated that quasi static cyclic deformation can generate flows in granular samples; we have now to prove that the deformation process obeys the classical law of plastic deformation of soil. This is done in Figure 9, which reports the experiment of Figure 7 made in the spirit of "soil mechanics": the packing is made of two kinds of duralumin cylinders (d_1 = 3mm, d_2 = 5mm, l = 50mm), and the deformation field is visualised using a superimposed square lattice. Indeed, one can observe the wedge motion, the failure zone, as in a classic text book; but one observes also the rotation of the lattice in the middle of the wedge, as it is reported in Section 4.2.

• *2ⁿᵈ Interest: Validity of the statistical approach in the mechanics of piles:* This experiment shows that cyclic excitation forces the granular medium to flow; hence it forces the grain configuration, the contact distribution and the local force network to change. Let us now consider a homogeneous sample and apply

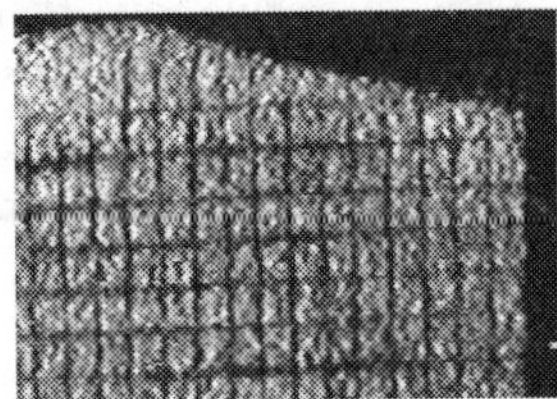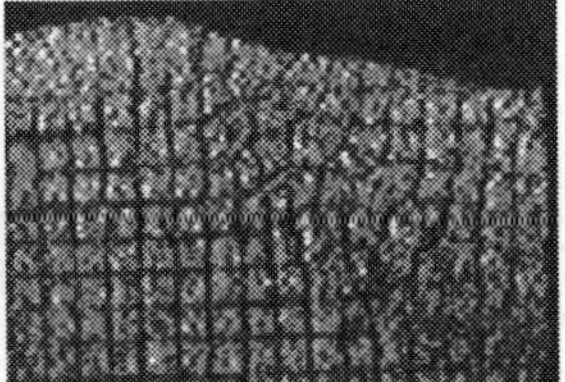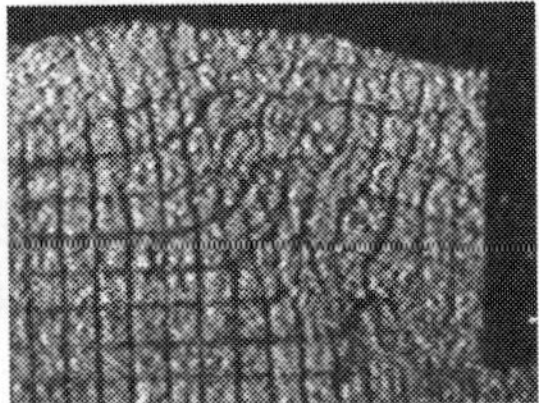

Figure 9   *Same bulldozing experiment as in Figure 7, with a larger pile and heavier grains; the initial conditions correspond to a pile to which cyclic forcing has been already applied for a while. (left) initial conditions, i.e. not deformed pile; (middle) after a cycle; (right) the pile after 1.5 cycles. One observes the localisation of the deformation at the base of the wedge; the squares in the centre of the moving wedge have rotated in the photograph on the right*

to it some cyclic deformation; as it is cyclic, the shape of the sample remains the same; but the flow which is generated permits the affiliation of a statistical approach; in particular, due to this mixing the principle of the most probable state should be valid. On the other hand, if this principle applies to the above situation, why should it not be valid at the earlier stage, or with another pile? A priori, no demon has built the pile.

• *"Fragile" behaviour and the role of force- and contact- networks.* Indeed one can still refuse to apply the principle of the most probable state and still argue that the contact and force network are quite inhomogeneous, which reveals the heterogeneous nature of the mechanics. Is this a contradiction really? One notes that the heterogeneity exists at the microscale; but at larger scale, one already speaks in terms of deformation and stress, i.e. in terms of averages which have macroscopic meaning. Furthermore if the packing was completely regular one may tell that a statistical approach would not apply to it; but its behaviour would not be regular. Of course, a disordered material would behave as a glass if the system of force was frozen, but as it deforms easily, its underlying networks of contacts and forces evolve; so if their evolutions are fast enough, they validate the description in terms of means; this justifies the existence of a stress-strain relation and the efficiency of statistical mechanics. In this case, local force should be related to the mean stress via a principle of maximum entropy; this seems to be satisfied.[36] So one has to consider the local disorder of contacts and force to be a necessity for applying the statistical mechanics. Indeed, if all the forces were equal, the packing would not be able to adapt itself to some change of stress and would break spontaneously under some specific stress excitation; this would lead to some "fragile" behaviour.[37] (Further information and discussion can be also found in the papers by K. Roberts, F. Radjai, M. Kuhn , S.J. Antony and Y. Ding in this book).

Turning now to smaller samples, one would expect that the total number of different configurations would become smaller, so that the system would become unable to adapt itself to some stress change; this predicts some noise on the stress-strain curves, and the smaller the sample, the larger the noise. This is exactly what can be observed with a triaxial test experiment[38] or in numerical simulations.

One should note on the contrary that some stress-strain curves exhibit spontaneous stick-slip. This phenomenon can be seen as revealing the "fragile" nature of the packing; it occurs in some samples, probably induced by cohesion. In the case we have studied,[39] we have shown that this fragile nature was enhanced by the macroscopic behaviour which becomes periodic for samples containing 10^9 grains or more.

• *Use of quasi static law of deformation:* So, "soil mechanics" laws should be able to predict the flow of Figure 7; but it is probably quite difficult. We see for instance in Figure 9 that the flow is primarily generated by large localised deformations; this is rather simple to compute. But these deformations provoke the flow in the bulk; in particular they provoke the slow rotation of the centre of the triangle while the triangle preserves its orientation and shape; these are drastic constrains. So it is a real "technical challenge" for soil mechanics programs. Also periodic conditions have to be found for the stress near the shear band and in the triangle. . . . The use of series expansion with two time scales may probably help; the rapid one would correspond to the cycle and would be used to describe the localisation, while the slow time would help to describe the flow, as it is commonly used in fluid mechanics; but the problem remains to identify the evolution of the boundary conditions.

Indeed, this experiment helps to demonstrate that the limit between solid-like and liquid-like behaviours is not simple; it is mainly a question of point of view. Furthermore, the difference comes from the macroscopic treatment, hence from macroscopic equations; so it has nothing to do with microscopy; this explains why discussing the Brownian motion of the particles does not help.

Finally, this experiment helps to demonstrate that the limit between static and dynamic is not discontinuous, because vibrating faster than 5–10Hz will not modify the flow efficiently.

5 Solid Behaviour Investigated with Acoustic and Ultrasonic Waves

5.1 Introduction

One knows that ultrasonic and sound waves propagate in a granular medium when they are low frequency. This explains for instance why vibration is felt at the passage of a subway train, and how Indians of America can deduce the number of riders by listening the ground. . . . From a more scientific view point, these waves reveal the elastic nature of the material response to very tiny deformations, since elastic (K) and shear (μ) moduli can be measured by measuring the speed V_p and V_s of compression and shear waves according to:

$$V_p = [(K + 4/3\mu)/\rho]^{1/2} \quad \text{and} \quad V_s = (\mu/\rho)^{1/2} \tag{10}$$

Here ρ is the density of material. This can be done in a lab on a small sample using a piezo-electric sensor; but it is also used *in situ* to determine the structure of the ground deep into the earth's crust, taking advantage of an earthquake or of an explosion and using a seismograph.

Recently,[40] Liu and Nagel have performed such lab tests on a sample containing a small number of grains confined to some extent, i.e. by gravity only: a sound pulse centred at 4kHz was then emitted by a loud speaker at some location and received a few (i.e. 10–15) grains further by a microphone of the shape of a grain (0.7cm). The received signal had an amplitude A, consisting of a fast rise followed by a trail which fluctuated quickly and then decreased slowly; it contained a high-frequency part (> 4kHz) and a low-frequency part (< 4kHz) presenting a power law distribution. This led Liu and Nagel to conclude the existence of an ambiguity in the value of sound velocity, because the shorter time of flight of the signal seemed to be incompatible with the speed of sound in the medium, defined as the group velocity $\partial A/\partial \omega$ measured by the analysis of the complete signal. Moreover the experiment showed an extreme sensitivity to tiny variations of the ball configuration.

This experiment has posed some problems of interpretation, and was considered by some scientists as the archetype of a new type of behaviour, the fragile matter one,[37] for which mean behaviour is difficult to define since it scales anomalously with distance, or can present sudden discontinuous evolution of the rhealogical law. They related it to an older experiment from Dantu[41] and others[41] which has demonstrated that the local stress network is not homogenous at the grain scale but exhibits a sub-network of strong forces propagating along chains.

In this section we want to discuss this problem and relate it to the problem of wave diffusion. It happens because the sound wavelength and the sensors used are small compared to the grain size and because the measure is too local. Of course, the bonds are "fragile" so that it changes the contact network at the local scale, the local distribution of scattering coefficient and the time response if some deformation is imposed; but what this section shows is that mean approach remains efficient and evolves much slower, because they are governed by the mean variables, so that most often the "fragile" nature is only local, not global.

So, in this section we give experimental elements[42] which reconcile the two view points at least partially; they will show that low-frequency waves propagate like traditional acoustic waves; however, when their frequency increases, and wavelength decreases their propagation can generate scattering and diffusion, leading to a speckle pattern;[43] this slows down the speed, disperses the wave in time, and attenuates the direct wave. As with any speckle, it is a variable which looks random, since its exact value is sensitive to the exact local distribution, but whose mean characteristics does not vary rapidly with the local variables This allows us to propose using acoustic speckle as a new technique to analyse the motions of the grains, and we give an example. This demonstrates that the acoustic and ultrasonic wave experiments such as the one of Liu and Nagel are in fact rather well under control.

Nevertheless, the problem of wave propagation in a heterogeneous medium can become much more complex. It can lead in particular to the mechanism of "wave localization", also called localization of Anderson.[44,45] These phenomena are significant, and complex; their field of application is vast and exceeds the mechanics of materials; it can apply equally well to seismic wave propagations, as to tidal waves.[45] It is outside the scope of this section.

5.2 Correct Experiment of Sound Propagation

A few years after Liu and Nagel,[40] an experiment of sound propagation was achieved[42] using a better controlled environment, with a larger sample confined in an oedometer at a given stress p; two ultrasonic piezoelectric sensors working in the range 500kHz were used to emit a short pulse and to receive the sound transported at a distance r much larger than the grain size d; their size D was large compared to the grain size. This has permitted the understanding of the phenomena observed by Liu and Nagel[40] and to relate them to sound scattering and sound diffusion.

Figure 10 presents typical results when the emitted signal is 2ms pulse duration centred at $\nu = 500$kHz. The signal received by the detector consists of a direct wave (marked E), of a wave (R) reflected by boundaries, then of a long train (S) which is generated by the diffusion of S. From the delay between the emission and the E signal, one can calculate the sound speed V_s which is $V_s = 750$m/s at $p = 0.75$ MPa, so that its wavelength $\lambda = V_s/\nu = 1.5$mm. The variations of V_s with p have been determined experimentally and are reported in Figure 11; they look roughly normal, reflecting a $V_s \propto p^{1/6}$ power law, in agreement with the Hertz law and spherical contacts, cf. caption of Figure 11.

Fourier analysis of the E wave for various stacking shows the existence of a high-frequency cut-off $\nu_c = \omega_c/2\pi$ which depends on the grain size and on V_s, and satisfies experimentally $1 = \nu_c d/V_s$; hence ν_c corresponds to a wavelength $\lambda_c = V_s/\nu_c$ equal to the grain diameter d.

So, frequencies ν larger than ν_c are not transmitted, but they are not diffused. Figure 10 shows that the diffused signal S gets weaker as the size of the grain gets smaller, at given wavelength; this is a normal behaviour as we will point out hereafter.

In the same way one finds that the smaller the diameter D of the receiver, the larger the amplitude S. This can be explained using a parallel with optical speckle: indeed, speckle is an interference pattern; thus it is produced by the interference of waves issued from the same source (wavenumber k_o) but after being scattered and diffused (wavenumber k_i) by different scatterers $\{\ldots, i, \ldots\}$ at locations $\{\ldots, r_i, \ldots\}$ and interfering at a particular point r; this interference produces an intensity which varies in space on a typical scale length l_s called correlation length. Indeed the local total intensity related to these interferences varies according to whether the interferences of these waves are constructive or destructive at the point considered and thus depends on the dephasing $k_i(r - r_i) + k_o r_i$ between the waves k_i from the various scatterers. The total local intensity A can be written:

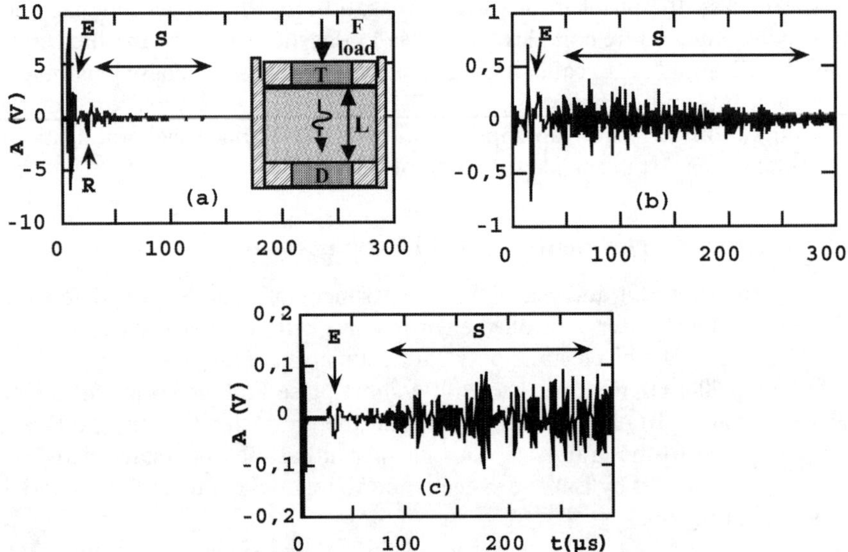

Figure 10   *Ultrasonic signal measured by a piezoelectric transducer of 12mm diameter in a pile of poly-disperse balls of glass (diameter d) under an oedometric loading with $p = 0.75$ MPa. (a) $d = 0.2$–0.3 mm and typical set-up, (b) $d = 0.4$–0.8 mm, (c) $d = 1.5 \pm 0.15$mm. (E) and (R) correspond respectively to the normal (coherent) propagation of sound emitted from the source and with its reflection on the top- and bottom- walls; (S) is a diffusion signal due to multiple scattering from scatterers. The experimental diagram is given in the insert of (A): T and D are the transmitter and receiver*[42]

$$A = \{\Sigma_i A_i \exp[-i\{\mathbf{k}_i(\mathbf{r} - \mathbf{r}_i) + \mathbf{k}_o \mathbf{r}_i\}] \exp[i\omega\tau]\} \qquad (11)$$

Equation 11 is thus a random variable of 0 mean with a typical length of correlation l_s. Its integration on the sensor surface $\pi D^2/4$ gives the signal amplitude received by the detector; this integral is also a random variable of 0 mean; but its relative fluctuations are reduced if $D \gg l_s$, because it is an average of random variables.

Let us return to the signals of Figure 10; we saw that signal S becomes significant only for the second and third experiment (Figures 10b and 10c); in Figure 10a the signal E is large, and diffusion is weak, which is in agreement with a simple estimate of the cross section of scattering σ_D using the Rayleigh wave model, which gives $\sigma_D \sim (d/\lambda)^4$.

However, the fact that σ_D becomes small does not imply that diffusion does not play an essential role in the wave propagation at long distance, i.e. in much larger samples. Indeed, diffusion is controlled by the scattering cross section σ_D which depends on the considered wavelength and the density of scatterers ρ_D; combining the two defines the mean free path l_c between two processes of diffusion, since $\rho_D \, l_c \, \sigma_D = 1$. So, whatever the material, diffusion will become the dominating phenomenon as soon as $l_c < L$, where L is the sample size. In

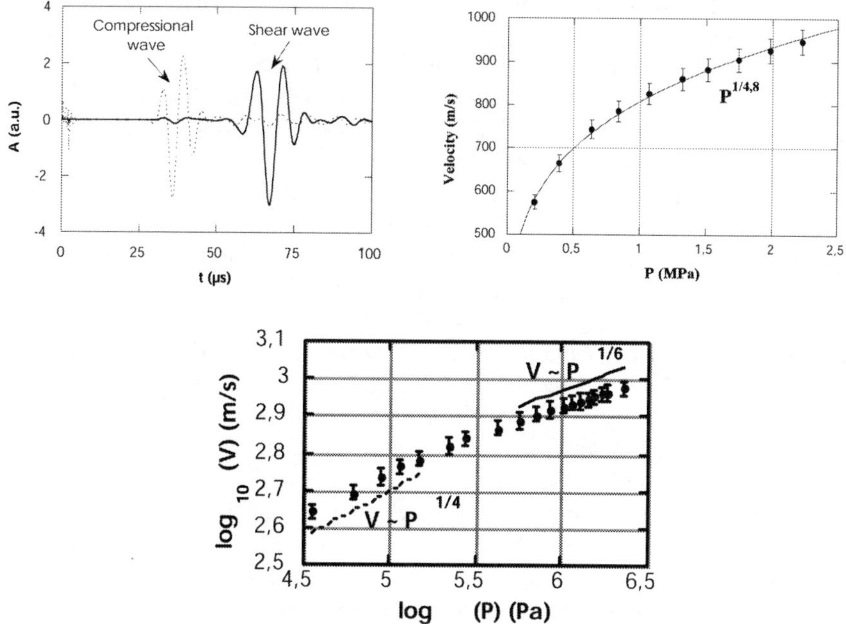

Figure 11 *From the delay δT between the direct signal E and the signal from the source of Figure 10, one can deduce the sound velocity V of the sound waves E of compression (or of shearing which are not represented here) and its variations with the stress p, here for balls with d = 0.4–0.8mm. V follows the power law predicted by Hertz contacts when p is large enough, but it deviates from it at low p. One can attribute this deviation to the increase in the real number of effective contacts when p is increased at weak p[42]*

particular, the intensity I_E of the direct wave (E) which crosses the thickness r decreases with r according to the probabilistic equation:

$$I_E = I_{Eo} \exp (-r/l_c) \tag{12}$$

This equation resembles an absorption equation; in fact, it is not; it indicates that the scatterers make the medium opaque, which is a well known phenomenon, with an analogy in optics: it explains for example why the thicker the cloud the darker it is. In the cloud case, this equation thus states that the light does not arrive on the Earth; it does not state that the light is absorbed by the cloud; in fact, the light is back-"scattered" outside the Earth.

When the transmitter is placed at the interior of the medium, and that medium diffuses strongly (l_c<<L), the whole emitted wave must leave either by the front, or by the back of the sample. But this can take a long time, even when the excitation is quite short, because it needs multiple scattering; an estimate can be obtained by treating the problem as one of simple diffusion; this is possible if l_c>>λ in this case one gets a characteristic time τ_D which scales as $\tau_D =$

$(L/l_c)^2(l_c/V)$; but the exact result is more complicated in general, depending on correlation between scatterers, . . .

When the medium absorbs the wave, one has to add a term in Equation 12: $I_E = I_{Eo} \exp(-r/l_c - \alpha r)$, where α is the absorption coefficient. A way of estimating each term is to measure the ratio of the back scattering to transmission; in fact when diffusion is important the back scattering integrated over all the directions of the half space, must be equal to the exciting intensity if $\alpha = 0$, while it will be attenuated by a factor $\exp(-2\alpha l_c)$ since the back waves will have travelled on average a length $2l_c$ in the medium. Of course, one needs to take account also of direct reflection by the front surface when this exists.

Lastly, Figure 12 shows the signal diffused from a cylindrical sample (51mm diameter, 85mm height) made of glass spheres (0.7mm diameter in mean) at different steps during a quasi-static compression test at constant volume and initial pressure $\sigma_1 = \sigma_2 = \sigma_3 = 45$kPa. at different steps. The acoustic excitation uses a 15µs pulse at 650kHz, and the diffused signal has been recorded at ($\varepsilon_{1(0)} = 0$, $\varepsilon_1 = 0.008\,\%$, $\varepsilon_{1(2)} = 0.010\%$, $\varepsilon_{1(3)} = 0.015\,\%$, $\varepsilon_{1(4)} = 0.02\,\%$) relative to the origin; time is indicated in the figure; strain rate is 10^{-5}/s, corresponding to a vertical compression speed of 0.83 µm/s. This demonstrates that the signal is quite sensitive to the contacts distribution and the forces network, and to their evolution. Furthermore, the signal at a shorter time evolves more slowly than at a longer time; this is because it corresponds to a smaller distance of diffusion, hence it corresponds to a smaller number of scattering events and is less sensitive to the change of configuration. So, one can think of using the evolution of

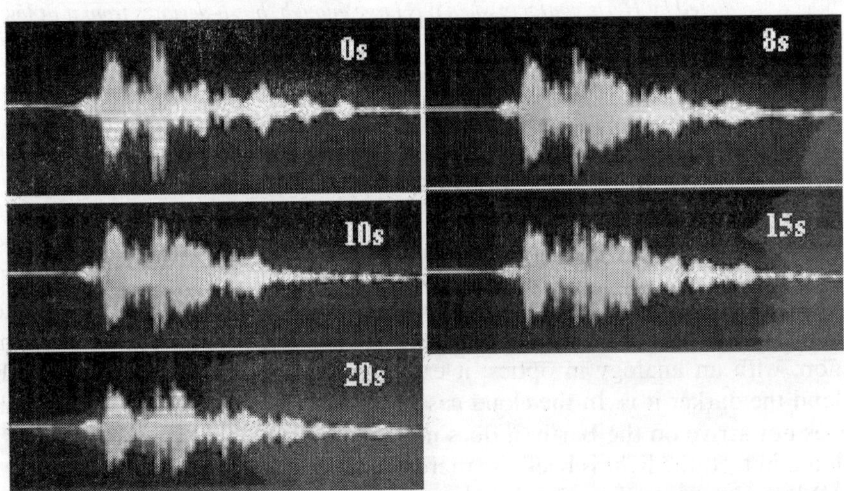

Figure 12   *Evolution of the 'speckle' with strain: 'Speckles' obtained at different times, i.e. deformation stage, (arbitrary origin) on a sample of glass spheres (d = 0.7mm,) filled with water. The confining lateral stress was 45kPa, the strain rate 10^{-5}/s = 0.001%/s; Sample: $\Phi = 51$mm, height:85mm. Sound at 650kHz with 15µs duration.[45] Sample strain is respectively $\varepsilon_0 = 0$, $\varepsilon_1 = 0.008\,\%$, $\varepsilon_2 = 0.010\%$, $\varepsilon_3 = 0.015\,\%$, $\varepsilon_4 = 0.02\%$*

'speckle',[†] to characterize the evolution of the contacts; indeed, this method has been already successfully used with the optical speckles[42] so that it can be adapted to ultrasonic waves. One can also plan to use correlation between several waves at a time to reach the distribution of more complex movements of an ensemble of grains. One can thus build a spectroscopy based on the diffusion of the acoustic waves.

Two other effects could have been studied with this experiment, but were not yet:

(*) one can observe a peak of coherent back-scattering just in the opposite direction from the excitation;[46] this is an interference effect, which one relates to the Anderson localisation problem.

(**) one can also determine the relative intensities E_S/E_P diffused into the compression (P) and shear (S) waves.

Because scattering rules allow the partial transformation of S waves into P waves and vice versa , this ratio E_S/E_P should reflect a statistical balance after a large number of processes. The constancy of this ratio is a well-known phenomenon which is observed in geo-mechanics of seism for instance. We now calculate it.

Statistical equilibrium E_S/E_P can be found easily under the three following considerations: (i) Statistical equilibrium is obtained when all the modes are actually equi-populated; (ii) scattering preserves frequency, so that if the initial spectrum is distributed in between ν and $\nu + \delta\nu$, the only eigen modes which can be populated are those ones in between ν and $\nu + \delta\nu$ at equilibrium; (iii) the density of modes can be calculated starting from any basis; so, we chose the simpler, i.e. the $\exp[i(\mathbf{k}.\mathbf{r} - 2\pi\nu\tau)]$ basis, where $\mathbf{k}$ and $\mathbf{r}$ are the wavevector and the position; one has $|\mathbf{k}| = 2\pi/\lambda = 2\pi\nu/V_j$ with V_j being the wave velocity of the $j = 1 =$ P or $j = 2 =$ S wave. It remains to count the eigen modes.

One can consider a volume $\Omega = L^3$ of material to count these modes. Any S or P eigen mode is characterised by a wave vector $\mathbf{k}$ which should satisfy boundary conditions; this requires that the scalar product k . L is equal to $2\pi(q_x + q_y + q_z)$, where the q_m are **integers** only, and each mode corresponds to a specified set (q_x, q_y, q_z). Thus within the limit k . L$>>2\pi$ and $|\delta k| = (2\pi/V_j)\delta\nu >> 2\pi/L$, the number $\delta\rho_j$ of eigen modes whose frequency lies between ν and $\nu + \delta\nu$ is given by $\delta\rho_j = G_j q^2 \delta q$, with $q = L\nu/V_j$ and $\delta q = L \delta\nu/V_j$, where G_j is the polarisation degeneracy of P and S waves; thus $G_j = 1$ for P and 2 for S. One finds[47] that the population ratio of the modes is $N_S/N_P = E_S/E_P = \delta\rho_S/\delta\rho_P = 2(V_p/V_s)^3$. One knows in addition that the V_p/V_s ratio of the sound speeds of the P and S waves are related to the Poisson's ratio η of material by $V_p/V_s = [(2 - 2\eta)/(1 - 2\eta)]^{1/2}$, cf. Reference 48.

---

† The term 'speckle' for Figure 12 is used with reference to the problem of optical speckle. Indeed, in the case of optical speckle, the interference pattern is due to the scattering of a continuous coherent excitation by a random distribution of scatters; and the spatial pattern produced looks granular due to the interference process. Here the pattern of Figure 12 is also produced by interference; it also looks granular along time, which means that the signal intensity varies slowly compared to the excitation period, but fast compared to the signal duration, i.e. the diffusion duration, so that it is made up of a series of "bursts" of intensity, which are randomly distributed in the envelope of the mean diffusion signal. (This is also why averaging over different configurations has to be used to recover normal time dependence of the diffusion intensity.)

But it is worth noting that this ratio N_S/N_P is not always accessible directly: when the experiment consists in detecting S and P waves with a sensor of surface S, the experiment measures the flow of vibration collected by the probe; this flow is proportional to $SV_{s/p}E_{s/p}$; the ratio of the intensities of vibration collected by the probe is thus $E_{probeS}/E_{probeP} = 2(V_p/V_s)^2$, cf. Reference 48.

5.3   Conclusion

In conclusion, the acoustics of granular systems is governed by four different parameters: the sample size L, the mean free path l_c and the wave length λ of the sound wave and the grain size d, (or the $\sigma_d^{1/2}$). In the preceding cases, we always considered that $\lambda << l_c$ and $\lambda << L$; in this case the diffusion formalism is well adapted, even if it is necessary to correct the speed of sound by a self-coherent term.[42] It is different when $\lambda > l_c$; in this case, interference between scattered and diffused waves is important and is difficult to compute because of the proximity of the scatterers, i.e. all $||r_i - r_j|| < \lambda$; in the general case, this multiple interference can result in forbidding propagation (exactly as happens in a laser cavity) and cause the phenomenon called wave localization or Anderson's transition.[42-44]

It seems however that this phenomenon of localisation cannot occur in the case of low-frequency acoustic waves in a compact material owing to the large wavelength; this is a consequence of the existence of a domain of quasi elastic homogeneous response at large scale. Indeed when λ and L are large, displacements are large too, even if strain is quite small; so any point of the medium should follow the movement of its neighbours at very low frequency, even if the packing is loosely tight because it is a coherent material; otherwise, the medium would present zones of infinitely low local rigidity and would break into clusters. Thus, propagation of large wavelengths will always occur in a coherent material.

6   Conclusion

These few examples show that the dynamic response of granular materials to dynamic excitation is quite diverse: the material can be a gas within very specific conditions. But it can also behave like an elastic solid, where sound propagates and diffuses before being absorbed. It can behave as a perfect (non viscous) liquid with swell and gravity waves on its top surface, or as a plastic solid, which can flow under some specific conditions. Indeed, there is probably no real difference between a plastic solid and a liquid, except the range of deformation.

All these problems can appear in very technical affiliations, because they intervene in many industrial processes. But they also raise a great number of fundamental questions, which permit (or will permit) this clarification of a certain number of effects and approaches.

Indeed, granular media are just the seat of many non linear behaviours, which can make them totally surprising and complex. But this complexity does not prove the absence of simple rules underlying their local macroscopic behaviour; the spatio-temporal chaotic behaviour is probably just the result of the non linearity, coupled to the 3d nature of any sample. It is in any case what makes credible the few examples given in this chapter, whose understanding requires using only simple concepts.

We have not developed the link with the mechanics at the microscopic scale, i.e. the scale of the grain. This factor is obviously important in understanding the macroscopic average behaviour; but once it is understood (or known), the macroscopic behaviour should be able to represent the complexity. In other words, and in the other domain of meteorology, it is true that a butterfly (or the throw of a pebble) can generate a tornado, but it does not seem important to always find which butterfly or pebble has provoked the observed tornado; the signature appears at a much larger scale. Similarly, the macroscopic trends of the mechanics of granular media come from the microscopic structure of the material, but in most cases it should not depend on the exact position of all the grains, otherwise the behaviour would remain completely unpredictable. This is just what the examples described in this chapter shows; they can all be interpreted in simple words and effects.

On the other hand, granular systems can become unpredictable in some cases, i.e. within some range of macroscopic parameters, as any other macroscopic complex system. This is "just" a question of complex evolution with complex bifurcations that occur in some domain of parameters.

Of course the examples proposed here are just a few examples of the many various facets of the dynamics of granular media. A point which would have deserved to be developed is that of the importance of the initial conditions on the development of instabilities, such as avalanches or other ruptures; this should depend on quasi static mechanics and on boundary conditions. Another point would have been segregation; but it is already the scope of a whole chapter of this book.

Acknowledgements

CNES is thanked for partial funding and ESA for supplying the rocket flight and its experimental device. Ferrari and Astrium are thanked for technical assistance and Makin Metal Powders Ltd is thanked for the gift of the bronze spheres,. This work has benefited from interesting discussions with V. Kozlov, D. Lyubimov,T. Lyubimova, S. Fauve, Y. Garrabos, D. Beysens, P.G. de Gennes and X. Jia.

References

1. M. O. Faraday, *Philos. Trans. R. Soc., London* **52**, 1831, 299.
2. *Poudres & Grains* can be found at the "Bibliothèque de France" as ISSN 1257–3957, or on the internet at: *http://prunier.mss.ecp.fr /poudres&grains/ poudres-index.htm*
3. E. Falcon, R. Wunenburger, P. Evesque, S. Fauve, C. Chabot, Y. Garrabos and D. Beysens, *Phys. Rev. Lett.*, 1999, **83**, 440–443.
4. P. Evesque, E. Falcon, R. Wunenburger, S. Fauve, C. Chabot, Y. Garrabos and D. Beysens, "Gas-cluster transition of Granular Matter under Vibration in Microgravity", in "First international Symposium on Microgravity Research & Applications in

Physical Science and Biotechnology", 10–15 Sept 2000, Sorrento, Italy; pp. 829–834; "Comparison between classical-gas behaviours and granular-gas ones in microgravity", P. Evesque, *Poudres & Grains*, Mai 2001, **12** (4), 60–82, ISSN 1257–3957, see Reference 2.

5. A. Kudrolli, M. Wolpert and J. P. Gollub, *Phys. Rev. Lett.* 1997, **78**, 1383.

6. M. A. Hopkins and M. Y. Louge, *Phys. Fluids A* 1991, **3**, 47; S. McNamara and W. R. Young, *Phys. Fluids A* 1992 **4**, 496; I. Goldhirsch and G. Zanetti, *Phys. Rev. Lett.*, 1993, **70**, 1619.

7. Y. G. Sinaï, *Chaos et Déterminisme*, 68–87, éd. Du Seuil, 1992.

8. P. Evesque, "Are Temperature and other Thermodynamics Variables efficient Concepts for describing Granular Gases and/or Flows?", *Poudres & Grains*, Avril 2002, **13** (2), 20–26, ISSN 1257–3957, see Reference 2.

9. P. Evesque, "The thermodynamics of a single bead in a vibrating container", *Poudres & Grains*, Mars 2001, **12** (2), 17–42, ISSN 1257–3957, see Reference 2.

10 P. Jean, H. Bellenger, P. Burban, L. Ponson and P. Evesque, "Phase transition or Maxwell's demon in Granular gas?", *Poudres & Grains*, Juillet–Août 2002, **13** (3), 27–39, ISSN 1257–3957, see Reference 2.

11. L. Landau and E. Lifchitz, *Mécanique des fluides*, ed. Mir, Moscow, 1989.

12. P. Evesque, A. Ivanova and V. Kozlov, in "First international Symposium on Microgravity Research & Applications in Physical Science and Biotechnology", 10–15 Sept 2000, Sorrento, Italie, 919–924.

13. R. Wunenburger, P. Evesque, C. Chabot, Y. Garrabos, S. Fauve and D. Beysens, *Phys. Rev. E*, 1999 **52**, *5440–5445*; R. Wunenburger, D. Beysens, C. Lecoutre-Chabot, Y. Garrabos, P. Evesque and S. Fauve , *Proceedings of the STAIF 2000*, January 2000, Alburque, New Mexico, USA.

14. G. H. Wolf, *Z. Physic*, B.227, N3, 1969, 291–300.

15. D. V. Lyubimov and A. A. Cherepanov, *Izv. AN SSSR, Mech. Gidk. Gaza*, 1986, **N 6**, 8–13.

16. A. Ivanova, V. Kozlov and P. Evesque, *Europhys. Lett.*, 1996, **35**, 159–64; P. Evesque, A. Ivanova, V. Kozlov, D. Lyubimov, T. Lyubimova and B. Roux, in *Powders and Grains 97*, pp. 401–4; R. P. Behringer and J. T. Jenkins ed., Balkema, 1997.

17. M. S. Longuet-Higgins, *J. Fluid Mech.*, 1981; **107**, 1–35; Tetsu Hara and Chiang C. Mei, *J. Fluid Mech.*, 1990, **211**, 183–209.

18. G. K. Batchelor, *An Introduction to Fluid Mechanics*, Cambridge Univ. Press, 1994; L. Landau and E. Lifschitz, *Mécanique des Fluides (cours de Physique tome 6)*, Mir, 1989; H. Schlichting, *Phys. Z.*, 1932, **33**, 327.

19. P. Evesque, "Influence of boundary conditions on 2-fluid systems under horizontal vibrations" , *Poudres & Grains*, Août–Septembre 2001, **12** (6), 107–114, see Reference 2.

20. P. Evesque, *Cont. Phys.* 1992, **33**, 245 & refs therein; H. M. Jaeger, S. R. Nagel, R. P. Behringer, *Rev. Mod. Phys.* 1996, **68**, 1259, & refs; see also articles in *Powders & Grains 1997*, R. P. Behringer & J. T. Jenkins eds, Balkema, Rotterdam, 1997.

21. V. G. Kozlov, A. Ivanova and P. Evesque, *Europhys. Lett.*, 1998 **42**, 413–18.

22. A. A. Ivanova, V. G. Kozlov, and P. Evesque "Fluidization of a granular medium in a viscous fluid under vertical vibration", *Fluid Dynamics*, 2000, 35(3), 406–413; P. Evesque, A. A. Ivanova and V. G. Kozlov, in *"Proceedings of the First international Symposium on Microgravity Research & Applications in Physical Science and Biotechnology"*, ESA publication 2001, 10–15 Sept 2000, Sorrento, Italy, 919–924.

23. P. Evesque, *Pour la Science*, Sept. 97, **239**, 94–96.

24. P. Evesque, D. Beysens and Y. Garrabos, "Mechanical behaviour of granular-gas and

heterogeneous-fluid systems submitted to vibrations in micro-gravity", *J. de Physique IV France*, 2001, **11**, Pr6–49 to 56.

25. H. M. Jaeger, S. R. Nagel, R. P. Behringer, *Rev. Mod. Phys.* 1996, **68**, 1259.
26. B. Thomas and A. M. Squires, *Phys. Rev. Lett.* 1998, **81**, 574, & refs therein.
27. P. Evesque, *Cont. Phys.*, 1992, **33**, 245 and refs therein; see also *Powders & Grains 1997*, 369–433, R. P. Behringer & J. T. Jenkins eds, Balkema, Rotterdam, 1997.
28. C. Laroche, S. Douady and S. Fauve, *J. Phys.* (Paris), 1989, **50**, 699; H. K. Pak and R. P. Berhinger, *Phys. Rev. Lett.* 1993, **71**, 1832; P. Evesque, E. Szmatula and J.-P. Denis, *Europhys. Lett.*, 1990, **12**, 623–627.
29. P. Evesque, P. Alfonsi, C. Stéfani and B. Barbé, *C.R. Acad. Sci. Paris*, 1990, *311* Série II, 393–98. P. Evesque, *Cont. Phys.*, 1992, **33**, 245; K. Liffman, G. Metcalfe and P. Cleary, in *Powders & Grains 1997*, pp. 405–408, R. P. Behringer and J. T. Jenkins eds, Balkema, Rotterdam, 1997; M. Medved, D. Dawson, H. M. Jaeger & S. R. Nagel, *Chaos* 1999, **9**, 691–696.
30. P. Evesque, *C. R. de l'Académie des Sciences Paris*, 1995, Série II **321**, 315–22; P. Evesque, *Physica* 1997, **D 102**, 78–92.
31. R. M. Nedderman, *Statics and kinematics of granular materials*, Cambridge Univ. Press, 1992, 47–63.
32. P. Evesque, to be published in *Poudres & Grains*, see Reference 2.
33. G. K. Batchelor, *An Introduction to Fluid Mechanics*, Cambridge Univ. Press, 1994; L. Landau and E. Lifschitz, *Mécanique des Fluides (cours de Physique tome 6)*, Mir, 1989; H. Schlichting, *Phys. Z*, 1932 **33**, 327.
34. H. Aref, *J. Fluid Mech.*, 1984, **143**, 1 (1984); A. Lichtenberg and M. A. Lieberman, *Regular and Stochastic Motion*, Springer, New York, 1983; H. C. Hilborn, *Chaos in Nonlinear Systems*, Oxford Univ. Press, 1994)
35. D. V. Khakhar, J. J. MacCarthy and J. M. Ottino, *Chaos*, 1999, **9**, 594–610; T. Shinbrot, A. Alexander, M. Moakher and F. J. Muzzio, ib. 1999, **9**, 611–620; D. V. Khakhar, J. J. McCarthy, J.-F. Gilchrist et al., *Chaos* 1999, **9** (1), 195–205.
36. P. Evesque, *Poudres & Grains*, 15 Novembre 1999, **9**, 13–19, see Reference 2, and in *Powders & Grains 2001*, Y. Kishino ed, Balkema, 2001, 153–157.
37. M. E. Cates, J. P. Wittmer, J. P. Bouchaud and P. Claudin, *Phys. Rev. Lett.*, 1998, **81**, 1841; A. J. Liu & S. R. Nagel, *Nature* 1998, **396**, 21–22.
38. F. Adjémian and P. Evesque, *Poudres & Grains*, Janvier 2002, **13** (1), 4–5, see Reference 2.
39. F. Adjémian and P. Evesque, *Eurp. Phys. J.* E, 2002, **9**, 253–259 and *Poudres & Grains*, Octobre 2001, **12** (7), 115–121, see Reference 2.
40. C. H. Liu and S. R. Nagel, *Phys. Rev. Lett.*, 1992, **68**, 2301; Phys. Rev. **B**, 1993, **48**, 15646; C. H. Liu, *Phys. Rev.* **B**, 1994, **50**, 782.
41. P. Dantu, in *Proceedings of the 4th International Conference on Soil Mechanics and Foundations Engineering*, Butterworths, London, 1957; T. Travers, M. Ammi, D. Bideau, A. Gervois, J. C. Messager and J. P. Trodec, *Europhys. Lett.*, 1987, **4**, 329.
42. X. Jia, C. Caroli and B. Velicky, *Phys. Rev. Lett.*, 1999, **82**, 1863; X. Jia, "Ultrasound propagation in disordered granular media", in *The Granular State*, S. Sen and M. Hunt (eds.), Mat. Res. Soc. Symp. Proc. Vol. 627, 351–357, MRS, San Francisco, 2000.
43. M. Monti, "Speckle photography applied to the detailled study of the mechanical behaviour of granular media", in *Powders & Grains*, J. Biarez and R. Gourvès ed., Balkema, Rotterdam, 1989, 83–89; N. Menon and D. J. Durian, *Science* 1997, **275**, 1920; N. Menon and D. J. Durian, *Phys. Rev. Lett.*, 1997, **79**, 3407.

44. P. W. Anderson, *Phys. Rev.*, 1958, **109**, 1492, P. G. de Gennes, P. Lafore, J. P. Millot, *Le Journal de Physique et le Radium*, 1959, **20**, 624.
45. P. Sheng, *Introduction to Wave Scattering, Localization, and Mesoscopic Phenomena*, Academic Press, San Diego, 1995; S. Alexander, *Phys. Rep.*, 1998, **296**, 66.
46. F. Adjémian, X. Jia, and P. Evesque, to be published.
47. C. Kittel, *Introduction à la physique de l'état solide*, Dunod, Paris, 1972.
48. R. L. Weaver, "On diffuse waves in solid media", *J. Acous. Soc. Am.*, 1982, **71** (6), 1608–1609.

CHAPTER 3

Constitutive Modelling of Flowing Granular Materials: A Continuum Approach

MEHRDAD MASSOUDI

U.S. Department of Energy, National Energy Technology Laboratory, P.O. Box 10940, Pittsburgh, PA 15236, USA
Massoudi@netl.doe.gov

Nomentclature

Symbol	Explanation
b	body force vector
D	Symmetric part of the velocity gradient
F	deformation gradient
l	identity tensor
L	gradient of the velocity vector
T	Cauchy stress tensor
u	velocity vector
x	spatial position occupied at time t
X	reference position
χ	deformation function
ν	volume fraction
ρ	bulk density
ρ_0	reference density
div	divergence operator
∇	gradient symbol
$\otimes$	outer product

1 Introduction

A recent study indicates that the commercial and large-scale solids-processing plants have an average operating reliability of 63 percent, compared to 84 percent for large-scale plants using only liquids and gases.[1] Such a poor understanding of the flow of granular materials has serious economic consequences. Because we cannot yet reliably scale up laboratory or pilot-plant designs to commercial sizes, engineers are forced to resort to costly, cut-and-try methods of design. A major challenge facing the designers of coal gasification plants is to assure reliable and efficient movement of solids into and out of high-pressure, high-temperature (fluidized-bed) processing units. Earlier studies of the flow of granular materials were mainly concerned with the engineering and structural design of bins and silos.[2,3] The inaccuracy of these theories, especially for dynamic conditions of loading or emptying, occasionally resulted in failure of the bin or silo.[4] Many situations such as discharge through bin outlets, flow through hoppers,[5] silos,[6] and chutes,[7] flow in mixers, and slurry transports require information on the flow patterns.[8] Therefore, to design equipment such as bins and silos, combustors, hoppers, chutes, hydrocyclones, etc., in an effective and economical way, a thorough understanding of the various factors governing the flow characteristics of granular materials must be obtained. Phenomena such as arching, shocks, and choking represent special challenges.[9–13] These design needs have already motivated extensive analytical and experimental investigations of the flow of granular materials. Despite wide interest and more than five decades of experimental and theoretical investigations many aspects of the behavior of flowing granular materials are not well understood. At this stage, there is still no complete understanding of the constitutive relations that govern the flow of granular materials. The general field is very much in a stage of development comparable to that of fluid mechanics before the advent of the Navier–Stokes relations. Perhaps the earliest study of granular flow is the hourglass or sand clock, which was in common use in the early 14th Century for measuring the speed of ships.[14] These devices were used, during the Middle Ages, by scholars to regulate their studies and by the clergy to time their sermons.[15] Another old example is the art of ploughing which has been used throughout centuries. The knowledge of our ancestors was more practical than theoretical; thus, with industrialization, the animals were replaced by tractors, and engineers who had designed these tractors were surprised to find that the drag of a plough is almost independent of speed.[16] That is, to everyone's surprise, it was found that ploughing at greater speeds does not require greater forces. Phenomena such as debris flow,[17] which is the agent for forming alluvial cones in the mounts of mountain canyons,[18,19] and snow or ice avalanches [20,21] are among the most threatening natural phenomena in some regions of the world. Other examples of bulk solids are coal, sand, ore, mineral concentrate, crushed oil shale, grains, cereals, animal feed, and powders. Gravitational flows of (dry) granular materials down an inclined chute has received most attention due to its simple geometry and its wide applicability in many industrial processes [see Hutter et al. (1986a,b),[22,23] Ahn et al. (1991, 1992),[24,25] Wang and Hutter (1999),[26] and Pouliquen and Chevoir (2002)[27]]. Other simple geometries of interest are vertical flows in channels [see Gudhe et al. (1994a),[28] and Spencer and Bradley (2002)[29]].

   Granular materials exhibit both the properties of a solid and a fluid as they can take the shape of the vessel containing them, thereby exhibiting fluidlike characteristics, or they can be heaped, thereby behaving like a solid. Granular materials can also sustain shear stresses in the absence of any deformation, and the critical stress at which shearing begins depends on the normal stress. The characteristics of the particles that constitute the bulk solids are probably of major importance in influencing the characteristics of the bulk solids both at rest and during flow. It is very difficult to characterize bulk solids, which are composed of a variety of materials mainly because small variations in some of the primary properties of the bulk solids, such as the size, shape, hardness, particle density, and surface roughness can result in very different behavior. Furthermore, secondary factors such as the presence or absence of moisture, the severity of prior compaction, the ambient temperature, etc., which are not directly associated with the particles, can have a significant effect on the behavior of the bulk solids. A granular material covers the combined range of granular powders and granular solids with components ranging in size from about 10 μm up to 3 mm. A powder is composed of particles up to 100 μm (diameter) with further subdivision into ultrafine (0.1 to 1.0 μm), superfine (1 to 10 μm), or granular (10 to 100 μm) particles. A granular solid consists of materials ranging from about 100 to 3,000 μm.[30] This range includes most of the materials used in laboratory experiments and whenever we use the term granular material we shall henceforth refer to this range. Brown and Richards (1970)[30] define a bulk solid as: "*An assembly of discrete solid components dispersed in a fluid such that the constituents are substantially in contact with near neighbors. This definition excludes suspensions, fluidized beds, and materials embedded in a solid mixture.*"

Also, little is known of the relationship between particle shape and flow properties in detail, although it is observed that smooth spherical particles display more favourable flow conditions than particles with a sharp angular surface, especially if they have a tendency to interlock. In addition, moisture content of the bulk solids is one of the most important factors controlling the flow properties of the granular materials. In fact, moisture content in bulk solids is mostly undesirable, because the surface moisture leads to the appearance of cohesive forces between particles of solids and of adhesive forces between particles and the walls of the container. Both retard the flow of solids and under certain conditions may stop the flow entirely. Since for the same weight of solid the total surface of solids is greater for small grains, the surface moisture content increases inversely as the particle diameter. For that reason fine particles display more cohesive and adhesive forces than the larger grain solids. Furthermore, fine particles when stored undisturbed for a certain time, have a tendency to compact. In general, the flow properties of most materials can be expected to decrease drastically as moisture content increases, particularly for finer materials.[31] There are many excellent review articles where many of the important issues relevant to granular materials are discussed. These recent articles take a general perspective and present a review of statistical theories (kinetic theory of gases, computer simulation) and continuum theories. We refer the reader to the articles by Nedderman et al. (1982),[32] Savage (1984),[15] Campbell (1990),[33] Hutter and Rajagopal (1994),[34] Jaeger et al. (1996),[35] de Gennes (1998),[36] Hermann and Luding (1998),[37] Hermann (2002),[38] and the books by Satake and

Jenkins (1988),[39] Shen et al. (1992),[40] Mehta (1994),[41] Chang et al. (1997),[42] Fleck and Cocks (1997),[43] and Duran (2000).[44]

2   Governing Equations

Although the fluid phase plays an important role in determining the dynamics of dilute suspensions, it does not have much influence on bulk solids behavior. That is, when the solid phase is dominant, the behavior of the bulk materials, in general, is governed by interparticle cohesion, friction, and collisions. In some cases the effects of the interactions between the fluid and solid constituents may be small because the interstitial fluid has relatively small density and viscosity (e.g., a gas). When the effects or the presence of the fluid phase cannot be ignored, then one has to resort to multiphase or multicomponent modeling, by considering the interaction mechanisms between the two phases. The balance laws, in the absence of chemical reactions and thermal effects, are the conservation of mass, conservation of linear momentum, and conservation of angular momentum. Conservation of mass in the Lagrangian form is:

$$\rho_o = \rho \det \mathbf{F} \tag{1}$$

where, ρ_o is the reference density of the material, ρ is the current density, and $\mathbf{F}$ is the deformation gradient which is given by:

$$\mathbf{F} = \frac{\partial \chi}{\partial \mathbf{X}} \tag{2}$$

The conservation of mass in the Eulerian form is given by:

$$\frac{\partial \rho}{\partial t} + \text{div}(\rho \mathbf{u}) = 0 \tag{3}$$

where $\frac{\partial}{\partial t}$ is the partial derivative with respect to time, and $\mathbf{u}$ is the velocity vector.
The balance of linear momentum is

$$\rho \frac{D\mathbf{u}}{Dt} = \text{div}\mathbf{T} + \rho \mathbf{b}, \tag{4}$$

where $\frac{D}{Dt}$ is the material time derivative, $\mathbf{b}$ is the body force, and $\mathbf{T}$ is the Cauchy stress tensor. The balance of angular momentum (in the absence of couple stresses) yields the result that the Cauchy stress is symmetric. The basic and fundamental question in modeling the granular materials is whether a single constitutive relation for the Cauchy stress tensor $\mathbf{T}$ is sufficient to describe the various flow regimes and geometries. In addition to this, whether one decides to use classical continuum theories or a modified version of kinetic theory of gases as applied to solid macroscopic materials, or computer simulation based on particle dynamics,[37] or experimental observations leading to phenomenological relations for the stress tensor,[45] or . . . has added to the complexity and diversity of this field of research, by producing many different forms for the stress tensor. At the present time, there is no unified theory for granular materials.

Any theory attempting to describe the behavior of flowing granular materials should embody several features, some of which are unique to granular materials. For example, a bulk solid is not exactly a solid continuum since it takes the shape of the vessel containing it; it cannot be considered a liquid for it can be piled into heaps; and it is not a gas for it will not expand to fill the vessel containing it. Perhaps the phase that the bulk solids most resemble is that of a non-Newtonian fluid. Therefore, it seems reasonable to expect a theory for flowing granular materials to exhibit characteristics unique to viscoelastic fluids such as the normal stress effects. Scientific understanding of the flow of granular materials has been hampered both by the difficulty of making measurements and by a tendency to look for immediate engineering solutions to specific problems as they arise. The flow of granular materials strongly depends upon the distribution of void space. Experiments have to be devised to quantify and describe the non-linear behavior of such materials, and theories have to be developed to explain the experimentally observed facts and predict, at least, qualitatively phenomena confirmable by further experiments.

From the perspective of continuum mechanics, there are many different approaches one can take. From the observation/experimental point of view, the pioneering work of Bagnold (1954, 1966)[46,47] has led to many formulations of non-Newtonian models.[4,48–52] For a review of this aspect of the modeling activities we refer the reader to the recent article by Elaskar and Godoy (1998).[53] The fact that granules can flow has prompted many investigators to look at the flow of particles as a fluid phenomenon, even as a compressible fluid.[54–56] At the same time, there have been many attempts to formulate or to propose rate-independent theories,[57] plasticity theories,[58] viscoelastic theories,[59] hypoplastic theories.[60] Theories with microstructure have also been proposed.[61–64] There have also been attempts to include the effect of 'fluctuation' of the particles into the stress tensor formulation.[65,66] At the same time, others have shown the similarity between the rapid flows of granular materials and the turbulent motion of a fluid.[67,68] A general continuum theory with thermodynamical restrictions was proposed by Goodman and Cowin (1971, 1972).[69,70] This work has subsequently been modified, extended, and generalized by various researchers.[71–73] In Section 3, we will look at constitutive modelling in general, and then look at particular issues which should be considered in modelling granular materials. In Section 4, we will give a review of continuum based constitutive relations used for flowing granular materials.

3 Constitutive Equations

The classical theories of continuum mechanics deal with the deformations and motions of materials that possess continuous mass densities. The general underlying assumption is the premise that any volume element, Δv, in a body can be taken to its limit, dv, without affecting the distribution of mass. According to this hypothesis, the identity of a material point in a volume element is lost, and its motion coincides with the motion of the center of mass of the body. For materials such as colloidal fluids, liquid crystals, granular, or composite materials, we need a theory that incorporates the micromotions of the particles contained in a material volume element, Δv. Materials possessing certain microstructures, e.g., with the internal couples or couple stresses, were first studied in the early twentieth century by D. Cosserat and F. Cosserat (see Truesdell and Toupin 1960[74]). Their theory was developed further and given a modern rational basis by Truesdell and Toupin,[74] Ericksen,[75,76] and Toupin[77,78] to name but a few. Eringen and his co-workers[79–82] introduced theories of microelastic solids and microfluids in which the "micromotions" of the material points contained in Δv with respect to its centroid, are taken into account in an averaged sense. In general, materials affected by such micromotions and microdeformations are called micromorphic materials.[80,81] However, theories developed for micromorphic materials are too complicated, even in the linear case, to be amenable for engineering applications. A subclass of these materials is the "micropolar" media, which exhibit microrotational effects. Eringen[82] mentions that "physically micropolar media may represent the materials that are made up of dipole atoms or dumbell molecules". In this section we will discuss some basic features for developing constitutive relations in general, and then we will mention some specific features pertaining to granular materials.

3.1 General Comments

The differences among the materials that make up different bodies are reflected in the theory by constitutive relations. In mechanics, a constitutive relation is a restriction on the forces or the motion of the body or both. This means that a body undergoes a motion when forces act on it, but the motion "caused" depends on the nature of the body. Mathematically, the purpose of the constitutive relations is to supply connections between kinematic, mechanical, and thermal fields that are compatible with the balance equations and that, in conjunction with them, provide a theory that is solvable for properly posed problems. The assumption that the body force is external is a constitutive relation. Indeed, the forces of most interest in continuum mechanics are contact forces, which are determined from the stress tensor field $\mathbf{T}$.[74] The mechanical behavior of real materials is very diverse and complex; it is impossible to formulate equations capable of describing the stress in a body under all circumstances. However, just as different figures in geometry are defined as idealizations of natural objects, continuum mechanics seeks to establish particular relations between the stress tensor and the motion of the body for "ideal materials".[83] These equations describe the most important features of the behavior of a material in a given situation. In some instances, it may be necessary to represent the same real material by different ideal materials in different circumstances. A classic example is that of the theory of incompressible viscous fluids, which gives an excellent description of the behavior of water flowing through pipes, but is useless for the study of the propagation of sound waves through water.[84] While a constitutive equation is a postulate or a definition from the mathematical standpoint, physical experience remains the first guide, perhaps reinforced by experimental data. Very rarely is it possible to formulate the basic equations of a theory from physical insight only. However, once the theorist has collected the information he wishes to use in defining the ideal materials to be used in his theory, a list of mathematical principles, some perhaps really only guidelines, become essential in formulating definite constitutive equations.[74]

Truesdell and Noll (1992, p.56)[83] postulate that (i) for a purely mechanical phenomenon, the response of the material can be described by the stress alone, and (ii) the state of the body is determined by its kinematical history. They propose the following two principles:

> "a. *Principle of determinism for the stress*: The stress in a body is determined by the history of the motion of that body.
> b. *Principle of local action*: In determining the stress at a given particle X, the motion outside an arbitrary neighborhood of X may be disregarded."

The first principle basically indicates the existence of a functional F_t such that

$$\mathbf{T}(t) = \mathbf{F}_t\,(\chi) \tag{5}$$

$$\mathbf{F}_t\,(\chi) = \mathbf{F}_t\,(\overline{\chi}) \tag{6}$$

where $\mathbf{T}(t)$ is the stress at time t. Equation 6 essentially states that for any two motions χ and $\overline{\chi}$ "that coincide in some neighborhood N (X) for all times $\tau \leq t$, the value of F_t is the same." Obviously constraints such as incompressibility, inextensibility, or other internal constraints will put further restrictions on Equations 5 and 6.

Constitutive relations are required to satisfy some general principles. First, they should hold equally in all inertial coordinate systems at any given time (often referred to as coordinate invariance requirement). This guards against proposing a relation in which a mere change of coordinate description would imply a different response in the material. Many of the so-called "power law" models used in describing non-Newtonian fluids are not coordinate-invariant. In general, this difficulty can easily be overcome by stating the equations either in tensorial form or by using direct notations not employing coordinates at all.

Another principle that is often used as a guide for selecting constitutive parameters is the principle of equipresence. Truesdell and Noll[83] state this principle as:

> *"A quantity present as an independent variable in one constitutive equation should be so present in all, unless, of course, its presence contradicts some law of physics or rule of invariance."*

Müller[85] uses this principle in proposing constitutive equations for a mixture of two fluids. However, Rivlin[86] discusses some cases where this principle seems to be contradictory. The principle of material frame-indifference (sometimes referred to as objectivity), which requires that the constitutive equations be invariant under changes of frame, is perhaps the most important of all. It is a consequence of a fundamental principle of classical physics that material properties are indifferent, that is, independent of the frame of reference of the observer. A good example of the intuitive concept of this principle is given by Truesdell and Noll:[83]

> *"A body of known weight, say one pound, when suspended by a given spring is observed to extend it by a given amount, say one inch. The spring and weight, still connected, are then laid upon a horizontal disc, to the center of which the free end of the spring is attached. The disc is then caused to spin at a steady speed such as to extend the spring again by one inch. The spectators are expected to agree that the centripetal force required to hold the weight from flying off is exactly one pound. That is, the response of the spring is unaffected by a rigid motion."*

This principle requires that constitutive relations depend only on frame-indifferent forms (or combinations thereof) of the variables pertaining to the given problem. Generally speaking there are two categories in the formulation of constitutive relations where the Principle of Material Frame-Indifference (PMFI) plays a crucial role. The first category is where one has obtained a constitutive relation based on a one-dimensional approximation of an experiment, and then one tries to generalize this correlation to a three dimensional case. An example of this category is Newton's viscosity law. This is more of an issue in coordinate transformation, rather than a change of frame. However, if the correlation involves time derivatives, then one has to be more careful. The second category is where the response of the material is time-dependent and as a result time derivatives or quantities associated with the movement and rotation of the frame (such as spin) would have to be included in the formulation.

Historically, many researchers have used the idea of "superposition" of elastic and viscous effects to describe the complex behavior of materials. Truesdell and Noll (1992, p.92)[83] give a review of the subject where ideas of 'rate-type', 'differential type', or 'integral type' materials are discussed. We can see that Equation 5 represents a class of (simple) materials depending on how the functional F depends on the motion. A *material of differential type* is one where the stress $\mathbf{T}$ depends only on a finite number of the time derivatives of $\mathbf{F}_t$. Thus,

$$\mathbf{T}(t) = \sum_{s=0}^{\infty} (\mathbf{F}^t(s)) \tag{7}$$

where $\mathbf{F}^t(s)$ is approximated for s near zero, by its Taylor expansion up to some order r such that

$$\mathbf{F}^{(t)}(0) = \mathbf{F}(t)$$
$$\dot{\mathbf{F}}^{(t)}(0) = \dot{\mathbf{F}}(t) \tag{8}$$
$$\cdots\cdots$$

A material *of rate type* is one where in addition to the general constitutive relation for simple materials

$$\mathbf{T}(t) = \sum_{s=0}^{\infty} (\mathbf{F}(t-s)) \tag{9}$$

we also have a further relationship

$$\mathbf{T}^{(p)} = \mathbf{g}(\mathbf{T}, \dot{\mathbf{T}}, \ldots, \mathbf{T}^{(p-1)}; \mathbf{F}, \dot{\mathbf{F}}, \ldots \mathbf{F}^{(r)}) \tag{10}$$

where **g** is a tensor-valued function, and (10) is subject to the initial data

$$\mathbf{T}(t_0), \dot{\mathbf{T}}(t_0), \ldots \mathbf{T}^{(p-1)}(t_0) \tag{11}$$

so that Equation 10 has a unique solution.

A *material of integral type* is one where the response functional *F* can be expressed in terms of an integral polynomial of an appropriate tensor representation of the deformation gradient **F**. For example, one can use the function

$$\mathbf{G}^*(s) = \mathbf{C}^*_{(t)}(t-s) - \mathbf{1} \tag{12}$$

to describe the deformation history, where $\mathbf{G}^*(s) \equiv \mathbf{0}$ corresponds to the rest history and $\mathbf{C}^*_{(t)}$ is related to the right Cauchy–Green tensor $\mathbf{C} = \mathbf{U}^2 = \mathbf{F}^T\mathbf{F}$.

3.2   Dilatancy

Early experimental investigations of granular materials were conducted by Hagen[87] who studied the flow of sand in tubes. Reynolds[88] observed that for a shearing motion to occur in a bed of closely packed particles, the bed must expand to increase the volume of its voids. He termed this phenomena "**dilatancy**". Reynolds[89] used the idea of 'dilatancy' to describe the capillary action in wet sand. The concept of dilatancy is generally taken to be the expansion of the voidage that occurs in a tightly packed granular arrangement when it is subjected to a deformation. Reynolds used the idea of dilatancy in describing a familiar phenomenon in sand:

"At one time the sand will be so firm and hard that you may walk with high heels without   leaving a footprint; while at others, although the sand is not

dry, one sinks in so as to make walking painful. Had you noticed, you would have found that the sand is firm as the tide falls and becomes soft again after it has been left dry for some hours. The tide leaves the sand, though apparently dry on the surface, with all its interstices perfectly full of water which is kept up to the surface of the sand by capillary attraction; at the same time the water is percolating through the sand from the sands above where the capillary action is not sufficient to hold the water. When the foot falls on this water-saturated sand, it tends to change its shape, but it cannot do this without enlarging the interstices – without drawing in more water. This is a work of time, so that the foot is gone again before the sand has yielded."

Many of the existing theories for flowing granular materials use this observation to relate the applied stress to the voidage and the velocity. One of the first and most interesting observations of the relationship between the stress in granular materials and voidage was also given by Reynolds:

"Taking a small indiarubber bottle with a glass neck full of shot and water, so that the water stands well into the neck. If instead of shot the bag were full of water or had anything of the nature of a sponge in it, when the bag was squeezed, the water would be forced up the neck. With the shot the opposite result is obtained; as I squeeze the bag, the water decidedly shrinks in the neck. . . When we squeeze a sponge between two planes, water is squeezed out; when we squeeze sand, shot, or granular material, water is drawn in."

The idea of dilatancy of granular materials can be simply explained for an idealized case: in order for a shearing motion to occur in a bed of closely packed spheres, the bed must expand by increasing its void volume. The work of Reynolds was followed by the experimental studies of Jenkin,[90] Rowe,[91] Andrade and Fox,[92] and Bolton[93] to name a few. Many attempts have been made to include the effects of dilatancy in the theory [Nixon and Chandler[94] for a plasticity theory, Mehrabadi and Nemat-Nasser[95] from a micromechanical point of view, and Goddard and Bashir[96] and Goddard[97] from a rheological perspective]. In fact, Reiner[48] was one of the first who used a non-Newtonian model to predict 'dilatancy' in wet sand. This model does not take into account how the voidage (volume fraction) affects the stress. However, using his model, Reiner showed that application of a non-zero shear stress produces a change in volume. McTigue[49] discusses the extension of the Reiner-Rivlin model to granular materials.

3.3   Cohesionless and Cohesive Materials

The mechanical properties of materials such as soils range between those of plastic clay[98] and those of clean, perfectly dry sand. Slopes of all kinds, including riverbanks and seacoast bluffs, hills, mountains, etc., remain in place because of the shearing strength possessed by the soil or rock. If we dig into a bed of dry (or completely immersed) sand, the material at the sides of the hole would slide toward the bottom. This behavior indicates a complete absence of a bond between the individual particles.[99] This sliding continues until the angle of inclination of the slopes becomes equal to a certain angle known as the "angle of repose." Brown and Richards[30] define two angles of repose as:

> *"The angle to the horizontal assumed by the free surface of a heap at rest and obtained under stated conditions:*
>
> (i) *The poured angle of repose is formed by pouring the bulk solid to form a heap below the pour point.*
> (ii) *The drained angle of repose is formed by allowing a heap to emerge as superincumbent powder is allowed to drain away past the periphery of a horizontal flat platform previously buried in the powder."*

Various techniques to measure the angle of repose are given by Weighardt.[16] Very often it is taken for granted that the angle of repose, γ is the same as the angle of internal friction, ϕ. Theory can only say that the slope of the pile of sand cannot be steeper than ϕ, or $\gamma \geq \phi$. This internal angle of friction is related to the amount of cohesion present in the material. In simple terms, the bond between the particles, cohesion, is influenced by a variety of forces including Van der Waals' forces, Coulomb forces, and capillary forces.[100] A definite angle of repose cannot be assigned to a granular material with cohesion, since the steepest angle at which such a material can stand decreases with increasing height of the slope.[99] The mechanical properties of real granular materials are so complex that a rigorous mathematical analysis of their behavior seems impossible. Therefore, many branches of applied mechanics, such as theoretical soil mechanics deal exclusively with the behavior of idealized granular materials ranging from ideal sands (cohesionless granular material) to ideal clays (ideally cohesive material, i.e., no internal friction).[101,102]

3.4   Mohr–Coulomb Criterion

Another criterion often used when devising a theory for the flow of granular materials is that the equilibrium states specified by the theory are required to coincide with the limiting equilibrium states specified by the Mohr–Coulomb criterion. The Coulomb failure criterion,[103,104] based on experiments, states that yielding will occur when

$$|S| = b_0\,T + c \tag{13}$$

Here S and T are the shear stress and normal stress, respectively, acting on a plane at a point; c is the coefficient of cohesion; and b_0 is the coefficient of static friction related to the internal angle of friction ϕ through

$$b_0 = \tan\phi \tag{14}$$

When cohesion is absent ($c = 0$), it is usual to call a granular medium an ideal one. One in which internal friction is absent ($\phi = 0$) is called an ideally cohesive medium. For dry, coarse materials, the cohesion coefficient can be neglected. Typical values for the internal angle of friction, ϕ, obtained during quasi static yielding at low stress levels are close to the angles of repose, e.g., about 24° for spherical glass beads and 38° for angular sand grains.[30] Continuum plasticity type models based on the Mohr–Coulomb criterion for failure have been proposed by Drucker & Prager,[58] Spencer,[105,106] and Jenike.[107] Some of the important issues in handling of cohesionless materials are discussed in an article by Sundaresan.[108] Other important concepts such as 'compactivity' and transmission of the stress have been discussed and introduced by Edwards [see Edwards,[109] Edwards and Oakeshott,[110] and Edwards and Grinev[111]].

4   Constitutive Modelling of Granular Materials

There are many different ways to model the behaviour of granular materials. For example, one such methodology could be using the following scheme:

- Physical and experimental models
- Numerical simulations
- Statistical mechanics approaches (e.g., extension of kinetic theory of gases)
- Standard continuum mechanics
- Ad-hoc approaches

At the same time, within the continuum mechanics approach, it is recognized that granular materials have certain 'structures' and as a result 'higher' order models or more advanced theories such as micro-mechanics, micropolar, Cossert theories, non-Newtonian models, hypoplastic or hypoelastic models, viscoelastic, turbulence models, etc., are needed. Another scheme could be the following:

- Explicit constitutive relations
- Implicit constitutive relations
- Ad-hoc relations

In this section we will give a brief review of the basic theory and the approaches used by various researchers to formulate constitutive relations. In each sub-section we start with what is usually considered the 'pioneering' works and then proceed to mention the other researchers and groups who have extended or modified that approach. We will proceed with the more popularly known and accepted schemes which can be classified as constitutive models based on (i) standard continuum mechanics, (ii) non-Newtonian fluids, (iii) micropolar and Cosserat theories, (iv) turbulence approach, (v) hypoplastic theories, and (vi) other theories. The last section includes a variety of ad-hoc approaches, which in general use the principle of superposition or some modified form of the kinetic theory of gases. The goal is not to be comprehensive, but instead to be inclusive in the sense of making an attempt to present the works of various groups, giving enough references in each approach to allow the reader the opportunity to have a sense of diverse ways which can be taken to model granular materials in such a multi-disciplinary field.

4.1 Basic Continuum Theories for Flowing Granular Materials

Goodman and Cowin[69,70] developed a continuum theory for representing the stresses that occur during the flow of granular materials. The theory is intended for situations where the stress levels are less than 10 psi. In addition, pneumatic effects are neglected; that is, the theory assumes that the material contained in the voids is a gas that does not interact with the granules. The basic idea underlying their theory is that the concept of mass distribution must be extended to admit granular materials; that is, the mass distribution must be related to the volume distribution of granules. This is achieved by introducing an independent kinematical variable called the volume distribution function. They assumed that the material properties of the ensemble are continuous functions of position. This is equivalent to assuming that the material may be divided indefinitely without losing any of its defining properties. That is, a distributed volume,

$$V_t = \int v \, dV \qquad (15)$$

and a distributed mass,

$$M = \int \rho_s \, v \, dV \qquad (16)$$

can be defined, where the function v is an independent kinematical variable called the volume distribution function and has the property

$$0 \leq v(x,t) < 1 \qquad (17)$$

The function v is represented as a continuous function of position and time; in reality, v in a granular system is either one or zero at any position and time, depending upon whether there is a granule or a void at that position. That is, the real volume distribution content has been averaged, in some sense, over the neighborhood of any given position. The classical mass density or bulk density, ρ_s, is called the distribution mass density, or simply the distributed density. The classical mass density or bulk density, ρ, is related to ρ_s and v through

$$\rho = \rho_s v \qquad (18)$$

After postulating the existence of new concepts, such as the "balance of equilibrated force" or the "balance of equilibrated inertia," Goodman and Cowin[69] proposed new balance relations in addition to the regular balance laws of continuum mechanics. Many of these ideas had already been proposed in other areas of mechanics, such as liquid crystals and micropolar materials. They also introduced a new form of the entropy inequality. They derived a constitutive equation for the Cauchy stress tensor based on the ideas of continuum mechanics, the restrictions imposed by the Clausius–Duhem inequality, the principle of frame-indifference, and incompressibility of the grains. They also assumed that the constitutive representations for the free energy, heat flux, dissipative parts of the stress, and intrinsic equilibrated body force depend linearly on temperature gradient, velocity gradients, and gradient of the volume distribution function. Thus, the equation defining a Coulomb granular material becomes:

$$\mathbf{T} = (\beta_0 - \beta v^2 + \alpha \nabla v . \nabla v + 2\alpha v \Delta v)\mathbf{1} - 2\alpha \nabla v \otimes \nabla v + \lambda(\mathrm{tr}\mathbf{D})\mathbf{1} + 2\mu\mathbf{D} \tag{19}$$

or

$$T_{ij} = (\beta_0 - \beta v^2 + \alpha v_{,k} v_{,k} + 2\alpha v v_{,kk})\delta_{ij} - 2\alpha v_{,i} v_{,j} + \lambda D_{kk}\delta_{ij} + 2\mu D_{ij} \tag{20}$$

where Δ is the Laplacian operator, $\otimes$ represents the outer (dyadic) product of two vectors. The coefficients β_0, β, and α are material constants; λ and μ are, in general, functions of ρ_s and v; and a comma denotes differentiation with respect to $\mathbf{x}$. Goodman and Cowin assumed that the stress tensor is obtained by the linear superposition of two parts: $\mathbf{T}^0$, a rate-independent (also referred to as equilibrium or non-dissipative) part, which depends on the solids fraction v and its gradients, and $\mathbf{T}^*$, a rate-dependent (viscous) part. Thus,

$$\mathbf{T} = \mathbf{T}_0 + \mathbf{T}^* \tag{21}$$

Savage[4] assumed that the stress is an isotropic function of v_0, v, grad v, and the rate of deformation, $\mathbf{D}$, where v_0 is a reference value of the volume fraction, v. That is

$$\mathbf{T} = \mathbf{T}\,(v_0, v, \nabla v, \mathbf{D}) \tag{22}$$

It should be noted that such an idea is not without precedent in continuum mechanics. An early example is that proposed by Korteweg to describe the structure of capillarity. He generalized the Navier–Stokes relation and assumed that the stress tensor was a function of the gradient of density, the second gradient of density, and the rate of deformation tensor.[83] By introducing a tensor, $\mathbf{M}$, such that

$$\mathbf{M} = \nabla v \otimes \nabla v \tag{23}$$

and observing that

$$\mathrm{tr}\mathbf{M} = |\nabla v|^2 \tag{24}$$

and

$$\mathbf{M} = |\nabla v|^2\mathbf{M} \tag{25}$$

an isotropic form for $\mathbf{T}$ can be obtained. Since $\mathbf{M}$ and $\mathbf{D}$ are symmetric tensors, according to a representation theorem,[83] the symmetric stress tensor, $\mathbf{T}$, is isotropic if

$$\mathbf{T} = a_0\mathbf{1} + a_1\mathbf{D} + a_2\mathbf{M} + a_3\mathbf{D}^2 + a_4\mathbf{M}^2 + a_5(\mathbf{DM} + \mathbf{MD}) + a_6(\mathbf{D}^2\mathbf{M} + \mathbf{MD}^2)$$
$$+ a_7(\mathbf{DM}^2 + \mathbf{M}^2\mathbf{D}) + a_8(\mathbf{D}^2\mathbf{M}^2 + \mathbf{M}^2\mathbf{D}^2) \tag{26}$$

where a_0 through a_8 are polynomials in the 10 basic invariants

$$\mathrm{tr}\mathbf{D}, \ \mathrm{tr}\mathbf{D}^2, \ \mathrm{tr}\mathbf{D}^3, \ \mathrm{tr}\mathbf{M}, \ \mathrm{tr}\mathbf{M}^2, \ \mathrm{tr}\mathbf{M}^3, \ \mathrm{tr}(\mathbf{DM}), \ \mathrm{tr}(\mathbf{DM}^2), \ \mathrm{tr}(\mathbf{MD}^2), \ \mathrm{tr}(\mathbf{D}^2\mathbf{M}^2) \tag{27}$$

This is a very general representation for $\mathbf{T}$. Equations of the type (26) are not uncommon in continuum mechanics. For example, the constitutive equation given by Truesdell and Noll[83] for a certain class of anisotropic solids capable of flow is very similar to Equation 26.

Cowin[112,113] and Savage[4] showed that by giving a special structure to the stress $\mathbf{T}_0$, one can satisfy the Coulomb failure criterion [see Equation 13]. It is basically considered that this equilibrium limit is essentially different from that when flow occurs, however slowly. Savage[4] showed that in a rough-walled channel with an angle of inclination slightly less than the angle of repose, granular material of uniform depth stays motionless in the absence of an external disturbance. However, giving a slight push the material is made to flow slowly but continuously and uniformly at constant depth. It is, therefore, evident that two different flow states exist at the same angle of inclination (but in general at different values of v). In both cases the flow is assumed to be slow enough so that the inertia forces are assumed negligible and the equilibrium condition is achieved when the shear stresses balance the components of the body force in the direction of the flow. In even the simplest experiment in which the material is pushed, the shear stresses will result from different mechanisms. In the static case, they are due to dry interparticle friction and particle interlocking; whereas, in the shear-flow case, granules override other granules and the momentum transfer associated with interparticle collisions becomes more important. There does not seem to be a smooth transition from one state to another as $\mathbf{D} \rightarrow \mathbf{0}$, and thus it does not seem possible to describe both states by a single constitutive equation. Therefore, $\mathbf{T}_0$ is regarded as an additional component of stress arising during the deformation of the granular materials due to nonuniformity of v.

Nunziato, et al.[72] re-examined the gravitational flow of granular materials with incompressible grains, in the context of a theory very similar to that of Goodman and Cowin.[69,70] They did not, however, impose a yield condition; instead, they considered smooth solutions of the governing equations of motion for the entire flow field. Based on thermodynamic arguments, conditions imposed by the Clausius–Duhem inequality, and assuming a form for the free energy which has a minimum when the volume fraction is equal to a critical volume fraction, the following constitutive equation was proposed for the stress tensor:

$$\mathbf{T} = -p\mathbf{1} - 2\alpha\nabla v \otimes \nabla v + 2\mu\mathbf{D} + [\lambda(\mathrm{tr}\mathbf{D}) + \varepsilon\dot{v}]\mathbf{1} \tag{28}$$

where p is a Lagrange multiplier, which was identified as a pressure, and the material coefficients α, μ, λ, and ε may all be nonlinear functions of the volume fraction v. They assumed the viscosity μ to be of the form

$$\mu = \hat{\mu}(v) = \mu_0\left(\frac{v_c}{v_m - v}\right)^8, \ \mu_0 \geq 0 \tag{29}$$

where v_m is the volume fraction corresponding to the densest possible packing of the material, v_c is the critical volume fraction, and $v_m \geq v_c \geq v_0 > 0$.

Cowin[113] showed that by including the gradient of the volume fraction as one of the important parameters in proposing a constitutive equation for the stress tensor, a theory could be devised for the flow of granular materials. In this theory a critical yield condition called the Mohr–Coulomb emerges naturally, as does the transition between the frictional flow regimes, characterized by the absence of deformation and the viscous flow regime, characterized by deformation. The Cauchy stress tensor $\mathbf{T}$ in a flowing granular material may depend on the manner in which the granular material is distributed, i.e., the volume fraction v and possibly also its gradient, and the symmetric part of the velocity gradient tensor $\mathbf{D}$. Based on this observation, Rajagopal and Massoudi[114] assume that:

$$\mathbf{T} = \mathbf{f}(v, \nabla v, \mathbf{D}), \tag{30}$$

Using standard arguments in mechanics, restrictions can be found on the form of the above constitutive expression based on the assumption of frame-indifference, isotropy, etc. There could be further restrictions on the form of the constitutive expression because of internal constraints, such as, incompressibility and thermodynamics restrictions due to Clausius–Duhem inequality.[115] A constitutive model that predicts the possibility of one normal stress-difference and is properly frame invariant is given by Rajagopal and Massoudi:[114]

$$\begin{aligned}\mathbf{T} = [\beta_0(v) + \beta_1(v)\,\nabla v \cdot \nabla v + \beta_2(v)\,\mathrm{tr}\,\mathbf{D}]\mathbf{1} \\ + \beta_3(v)\mathbf{D} + \beta_4(v, \nabla v)\,\nabla v \otimes \nabla v + \beta_5(v)\,\mathbf{D}^2\end{aligned} \tag{31}$$

Now if the material is 'fully' flowing, the following representations are proposed for the β's

$$\begin{aligned}\beta_0(v) &= k_1\, v \\ \beta_1(v) &= \beta_{10} + \beta_{11}\, v + \beta_{12}\, v^2 \\ \beta_2(v) &= \beta_{20} + \beta_{21}\, v + \beta_{22}\, v^2 \\ \beta_3(v) &= \beta_{30} + \beta_{31}\, v + \beta_{32}\, v^2 \\ \beta_4(v) &= \beta_{40} + \beta_{41}\, v + \beta_{42}\, v^2 \\ \beta_5(v) &= \beta_{50} + \beta_{51}\, v + \beta_{52}\, v^2\end{aligned} \tag{32}$$

where

$$\beta_{30} = \beta_{20} = \beta_{50} \tag{33}$$

due to the 'limiting principle.' This limiting principle basically indicates that as $v \rightarrow 0$ (i.e., no particles), $\mathbf{T} \rightarrow \mathbf{0}$ (i.e., no stress). In their studies, Rajagopal et al.[116] proved existence of solutions, for a selected range of parameters, when,

$$\begin{aligned} \beta_1 + \beta_4 &> 0 \\ k_1 &< 0. \end{aligned} \tag{34}$$

For other rheological parameters, i.e., β_2, β_3, and β_5, one can use the methods available in the mechanics of non-Newtonian fluids to find out more information about the signs. Obviously, since β_3 is related to the shear viscosity, it is assumed to be positive. The simulation studies of Walton and Braun[117,118] suggest a structure of β_3 similar to the one assumed by Rajagopal and Massoudi. This model has been used in a variety of simple applications such as inclined flow, flow in a vertical pipe, etc. The distinct and main feature of this model is its ability to predict the normal stress differences which are often related to the dilatancy effects. Perhaps a thermodynamic or a stability analysis would reveal further information about the signs and the relative importance of these coefficients. If the material is just about to yield, Massoudi and Mehrabadi[119] indicate that if the model is to comply with the Mohr–Coulomb criterion, the following representations are to be given to the material parameters in Equation 31:

$$\begin{aligned} \beta_0 &= c \cot \phi \\ \beta_1 &= \frac{\beta_4}{2}\left(\frac{1}{\sin \phi} - 1\right) \end{aligned} \tag{35}$$

where ϕ is the internal angle of friction and c is a coefficient measuring cohesion. The interpretation for β_2, β_3, and β_5 would stand as the previous case , if we accept that the stress tensor can be decomposed into an 'equilibrium' part and a 'dynamic' part, as discussed earlier, where β_2, β_3, and β_5 are responsible for the dynamic effects. The significance of this model as represented by Equations 31 and 32 is discussed by Massoudi.[120] Rajagopal et al.[121,122] and Baek et al.[123] discuss the details of experimental techniques using orthogonal and torsional rheometers to measure the material properties β_1 and β_4.

4.2 Non-Newtonian Models for Granular Materials

In general, individual grains behave in an elastic manner. However, the bulk solid behaves very similarly to that of a purely viscous-inelastic isotropic fluid. The motivation for such a statement originates in the early work of Bagnold[46] who performed experiments on neutrally buoyant, spherical particles suspended in Newtonian fluids undergoing shear in coaxial rotating cylinders. He was able to measure the torque and normal stress in the radial direction for various concentrations of the grains. He distinguished three different flow regimes, which he termed macro-viscous, transitional, and grain-inertia. In the so-called 'macro-viscous' region, which corresponds to low shear rates, the shear and normal stresses are linear functions of the velocity gradient. In this region, the fluid viscosity is the dominant parameter. In the region, called the 'grain-inertia region,' the fluid in the interstices does not play an important role and the dominant effects arise from particle–particle interactions. Here, the shear and the normal stresses are proportional to the square of the velocity gradient. Connecting the two limiting flow regimes was the transitional flow, in which the dependence of the stress on shear rate varied from a linear one corresponding to the macro-viscous regime to a square dependence predicted for the grain-inertia flow regime. From his experiments, Bagnold was able to define the various flow regimes in terms of dimensionless number $\overline{N}$, later referred to as the Bagnold number, given by

$$\overline{N} = \lambda_0^{1/2}\, \rho_f\, \overline{\sigma}^2\, \frac{(u_{1,2})}{\mu_f} \tag{36}$$

Here ρ_f and μ_f are the mass density and viscosity of the fluid, $\overline{\sigma}$ is the diameter of the particle, λ_0 is the linear concentration of particles, and $u_{1,2}$ is the velocity gradient. The macro-viscous regime corresponds to $\overline{N} < 40$, and the grain-inertia regime to $\overline{N} > 450$. He called the intermediate range of $\overline{N}$ the 'transitional region'. Bagnold[46] applied his cylindrical shear cell results and his analysis for the stresses to study the problems of gravity flow of particulate matter down inclines as might occur in rock falls and debris flows.[124] He observed that both the shear and the normal stresses in the grain-inertia regime are proportional to the square of the velocity gradient. The interesting phenomenon was the presence of a normal stress proportional to the shear stress, similar to that of the quasi-static behaviour of a cohesionless material obeying the Mohr–Coulomb criterion. Bagnold named this normal stress the dispersive pressure. If the mean distance between the center of the particles is defined as $b\sigma$ and the free distance between particles is e, then

$$\overline{b} = 1 + \frac{e}{\sigma} = 1 + \frac{1}{\overline{\lambda}} \tag{37}$$

where $\overline{\lambda}$ has been defined as σ/e. For isochoric motions, where variations in v are still permitted, Savage[4] assumed that the rheological coefficients depend only on the second principal invariant of $\mathbf{D}$, I_2, as well as v and v_0. It should be noted that for isochoric, two-dimensional, fully developed channel or chute flows, both I_1 and I_3 are zero. With these assumptions, Savage[4] proposed that the simplest appropriate representation for $\mathbf{T}^*$ is

$$\mathbf{T}^* = \mathbf{T} - \mathbf{T}_0 = 4\mu_0 I_2 \mathbf{1} + 4\mu_1 |I_2|^{1/2}\mathbf{D} \tag{38}$$

where the coefficients μ_0 and μ_1 are functions of v and tensile stresses are considered positive. On the basis of curve fits to Bagnold's experimental data,[46] Savage[4] assumed the following forms for μ_0 and μ_1:

$$\mu_0 = \eta_0\left(\frac{v_\infty - v_0}{v_\infty - v}\right)^8$$

$$\tag{39}$$

$$\mu_1 = \eta_1\left(\frac{v_\infty - v_0}{v_\infty - v}\right)^8$$

for $v_0 < v < v_m$, where v_∞ corresponds to the densest possible packing of particles, v_m is the maximum value of the fractional solids content at which continued shearing can occur (i.e., the v associated with the critical voids ratio), v_0 is the solids fraction at which fluidity occurs (i.e., where the residual shear resistance at zero rate disappears), and η_0 and η_1 are constants. Some values of v_∞, v_m, v_0, and the associated values of λ, taken from Bagnold,[47] are given in Table 1.

McTigue[49] proposed that $\mathbf{T}_0$ should include an isotropic thermodynamic pressure, p, and a term which would allow finite shear stresses at incipient failure. Thus, he suggested

$$\mathbf{T}_0 = -p\mathbf{1} + k(II')^{-1/2}\mathbf{D} \tag{40}$$

where p and k are functions of v and

$$II' = 1/2\, I^2 - II = 1/2\, \text{tr}\mathbf{D}^2 \tag{41}$$

In the limit as $\mathbf{D}$ vanishes, Equation 40 should give the stresses in a granular material that is at failure everywhere. This state is taken to correspond to the "critical state" of the soil mechanics literature[125] and is defined as a condition in which continuous deformation at constant volume fraction and constant shear

Table 1   *Values of critical fractional solids contents for various bulk solids (after Bagnold[47])*

Bulk Solid	v_∞ $(\lambda = \infty)$	v_m	λ_m	v_0	λ_0
Natural Angular Beach Sand (0.318–0.414 mm)	0.644	0.555	19	0.51	12.4
Spherical Lead Shot (1.6 mm Diameter)	0.74	0.63	18.5	—	—
Wax Spheres (1.32 mm Diameter)	0.74	—	—	0.60	14

stress takes place. For granular materials, the relationship between shear and normal stresses at failure is represented by the Mohr–Coulomb criterion. To obtain a specific form for k(v), McTigue[49] used the approach of invariant generalization of the Mohr–Coulomb failure criterion (due to Drucker and Prager[58]), and obtained the equilibrium part of the stress tensor $\mathbf{T}_0$ as

$$\mathbf{T}_0 = -\alpha(v^2 - v_c^2)\mathbf{1} + [c\cos\phi + \alpha\sin\phi(v^2 - v_c^2)](\mathrm{II'})^{-1/2}\mathbf{D} \tag{42}$$

where c is related to cohesion, ϕ is the angle of internal friction. To obtain a constitutive equation for the dissipative part of the stress $\mathbf{T}^*$, McTigue[49] used Bagnold's[46] idea. Assuming that the total stress, $\mathbf{T}$, can be decomposed into an equilibrium part, $\mathbf{T}_0$, and a dissipative part, $\mathbf{T}^*$, he assumed an expression for $\mathbf{T}^*$, in the form of a Reiner–Rivlin fluid.[83] Thus,

$$\mathbf{T}^* = \eta_1(v - v_m)^{-2}(\mathrm{II'})^{1/2}\mathbf{D} - \eta_2(v - v_m)^{-2}\mathbf{D}^2 \tag{43}$$

where η_1 and η_2 are material constants and v_m is the maximum volume fraction for flow. Combining Equations 42 and 43, an equation for $\mathbf{T}$ can be obtained.

Shahinpoor and Lin[126] concentrated their study on the formulation of a constitutive equation appropriate for the rapid flow regime where the effect of interparticle forces are due mainly to the exchange of momentum in particle collisions. They proposed the following constitutive equation

$$\mathbf{T}^* = -p\mathbf{1} + \mu_1 v^2 |I_2|^{1/2}\mathbf{D} + \mu_2\mathbf{D}^2 \tag{44}$$

$$I_2 = 1/2[(\mathrm{tr}\mathbf{D})^2 - \mathrm{tr}\mathbf{D}^2] \tag{45}$$

where p is a dynamic pressure, assumed to be related to the statistical average of the square of particle velocity fluctuations as well as pure hydrostatic effects due to gravitation, and μ_1 and μ_2 are constants. Equation 44 is of the Reiner–Rivlin [see Reiner[48] and Rivlin[127]] type constitutive equation except for the dependence on solid volume fraction v and the role of velocity fluctuations $\langle u_i' u_i' \rangle$ in the pressure term. Shahinpoor and Lin[126] used this model to study the problem of rapid Couette flow of cohesionless granular materials (iodized salt) between two co-axial rotating cylinders. They showed that their model is compatible with the experimental results of Bagnold[46] and that it also predicts certain normal stress effects observed in some non-Newtonian fluids. The importance of the velocity fluctuation or statistical averaging in the formulation of constitutive equations will be discussed later.

Based on his earlier studies on dissipative materials, Goddard[51] suggested that the stress tensor $\mathbf{T}$, describing the yield and flow behavior of granular materials for small deformation rates and granular kinetic energies, be of the form

$$\mathbf{T} = \mathbf{T}_s + \mathbf{T}_k \tag{46}$$

where $\mathbf{T}_s$ and $\mathbf{T}_k$ refer to the "static" and "kinetic" contributions. To obtain specific forms for $\mathbf{T}_s$ and $\mathbf{T}_k$, Goddard[51] assumed the stress $\mathbf{T}$ is given in terms of the deformation rate $\mathbf{D}$ by

$$\mathbf{T} = 2\boldsymbol{\eta}\mathbf{D}, \text{ or}$$
$$\mathbf{T}_{ij} = 2\eta_{ijkl}\mathbf{D}_{kl} \tag{47}$$

where $\boldsymbol{\eta}$ is a fourth-order viscosity tensor depending on the deformation history. Defining the modulus $|\mathbf{A}|$ of a real second rank tensor $\mathbf{A}$ by

$$|\mathbf{A}|^2 = \mathrm{tr}(\mathbf{A}^{\mathrm{T}}\mathbf{A}) \tag{48}$$

then for $|\mathbf{D}| \to \mathbf{0}$, he assumed that

$$\boldsymbol{\eta} = \boldsymbol{\mu}_0/|\mathbf{D}| + \boldsymbol{\eta}_0 + O(1) \tag{49}$$

where μ_0 and η_0 also depend on the deformation history and both are 0(1). Then, the constitutive equation for a visco-plastic material, to 0($|\mathbf{D}|$), becomes

$$\mathbf{T} = 2\boldsymbol{\mu}_0\mathbf{E} + 2\boldsymbol{\eta}_0\mathbf{D} \tag{50}$$

where $\boldsymbol{\mu}_0$ is a plastic modulus, $\boldsymbol{\eta}_0$ a viscosity, and $\mathbf{E}$ is given by

$$\mathbf{E} = \frac{\mathbf{D}}{|\mathbf{D}|} \tag{51}$$

We note that the strain-like tensor $\mathbf{E}$ is inherently rate-independent, whereas the term $\boldsymbol{\eta}_0$ allows for a rate-dependent viscous stress contribution. For an incompressible material, $\mathrm{tr}\,\mathbf{D} = 0$, and $\mathbf{T}$ is specified only up to an additive term $-\mathrm{p}_0\,\mathbf{1}$. As indicated by Goddard,[51] in the absence of history effects such as work hardening, compaction, etc., $\boldsymbol{\mu}_0$ and $\boldsymbol{\eta}_0$ depend on the isotropic invariants of $\mathbf{E}$ and $\mathbf{D}$, respectively. Thus, we have

$$2\boldsymbol{\mu}_0\mathbf{E} = 2\mu\mathbf{E} + \lambda\mathbf{E}^2$$

$$2\boldsymbol{\eta}_0\mathbf{D} = 2\eta\mathbf{D} + \kappa\mathbf{D}^2 \tag{52}$$

with additive isotropic pressure terms. The scalar coefficients (μ,λ) and (η,κ) depend, in general, on the history of deformation, and in the simplest cases, on the invariants of $\mathbf{D}$. The term associated with κ is essential to the description of rate-dependent normal-stress effects in simple shearing flows. Goddard[51] showed that by selecting appropriate isotropic forms for $\mathbf{E}$ many of the flow rules (such as Mohr–Coulomb criterion) and yield surfaces of standard plasticity would emerge. Based on these ideas, he proposed the following constitutive equations for $\mathbf{T}_s$ and $\mathbf{T}_k$:

$$\mathbf{T}_s = 2\mu_0 \mathbf{E}$$

$$\mathbf{T}_k = 2\eta_0 \mathbf{D} \tag{53}$$

where η_0 is now also a function of the grain fluctuation energy. Therefore, using (46)–(48), and (53) we can write

$$\mathbf{T} = -p\mathbf{1} + 2\mu\mathbf{E} + \lambda\mathbf{E}^2 + 2\eta\mathbf{D} + \kappa\mathbf{D}^2 \tag{54}$$

4.3   Micropolar Models

By considering the velocity and the rotational velocity of particles as two independent kinematic field variables, Kanatani[61] proposed a micropolar continuum theory for flowing granular materials. The velocity v_i and the rotation ω_{ji} of the particles are assumed to be continuous functions of position and time where ω_{ji} is a skew symmetric tensor expressing the angular velocity about an axis perpendicular to the (ji) coordinate plane. By using the variational principles and energy dissipation, Kanatani[61] obtained the following constitutive equation for the stress tensor T_{ji} for "slow flow":†

$$T_{ji} = -p\delta_{ji} + \tilde{T}_{ji} + T_{[ji]} \tag{55}$$

Where

$$\tilde{T}_{ji} = (3\sqrt{6}/10)p/\varpi \left[1/2\left(\frac{\partial v_i}{\partial x_j} + \frac{\partial v_j}{\partial x_i}\right) - 1/3\delta_{ji}\frac{\partial v_k}{\partial x_k} \right] \tag{56}$$

$$T_{[ji]} = (\sqrt{6}\mu/2)p/\varpi \left[1/2\left(\frac{\partial v_i}{\partial x_j} - \frac{\partial v_j}{\partial x_i}\right) - \omega_{ji} \right] \tag{57}$$

where

$$p = 1/3 T_{kk} \tag{58}$$

and ϖ represents the total amount of interparticle friction which includes both the shearing and the particle rotation, and m is the kinetic friction coefficient. He used both the macroscopic (or continuum) approach and the microscopic (or particulate) approach by considering that the bulk of the materials behave as a continuum whereas the individual particles are fluctuating irregularly when they are in motion. Therefore, by taking statistical averages of quantities involved in the interparticle interactions, he obtained quantitative results which are insensitive to the particle configurations.

---

† By "slow flow," it is meant a relatively slow, incipient flow where the particles are rolling over the lower layer of particles in an ordered manner. The other regime often referred to as "fast" or "rapid" flow is when the particles are following chaotic paths and interacting vigorously (e.g., collisions) with their neighbors.

Ahmadi[71] generalized the Goodman and Cowin[69,70] theory by considering the possible rotation of granules. The basic equations of balance of media with microstructure, with special considerations to granular materials with incompressible grains, were presented. This work was further generalized by the development of nonlinear constitutive equations appropriate for the rapid flow of granular materials.[128] Thermodynamic arguments and ideas from the continuum mechanics of materials with microstructure were used to derive equations of motion for compressible, as well as incompressible, granules. To obtain the constitutive equation for stress tensor in a micropolar media, it is necessary to propose a new form for the free energy. If the material is incompressible (i.e., ρ_s = constant), then the kinematic variables v and $\mathbf{u}$ cannot vary independently and, as a result, the equilibrium part of the stress for incompressible granular material is given by,

$$T^{\circ}_{ij} = \left[2\alpha v v_{,kk} + \frac{d}{dv}(\alpha v)v_{,k}v_{,k} - \frac{v^2}{2}\frac{d}{dv}\left(\frac{\gamma}{v}\right)\phi_{m,k}\phi_{m,k} \right]\delta_{ij} - 2\alpha v_{,i}v_{,j} - \gamma\phi_{m,i}\phi_{m,i} \quad (59)$$

where α and γ are scalar functions of solid volume fraction v, and ϕ_i is microrotation which is related to the gyration vector by

$$\dot{\phi}_i = v_i \quad (60)$$

To develop more general nonlinear constitutive equations appropriate to rapid flows of granular materials for an isotropic media, Ahmadi[128] proposed the following constitutive equation for the dissipative part of the stresses, $\mathbf{T}^*$:

$$T^{*}_{ij} = \lambda d_{kk}\delta_{ij} + (2\mu + \varsigma_0 |II_d|^{1/2})d_{(ij)} + \kappa d_{ji} + \varsigma_1 d_{(jm)}d_{(mj)} \quad (61)$$

where II_d is the second principal invariant of the symmetric part of the microdeformation rate tensor $d_{(ij)}$ and is given by

$$II_d = 1/2[d_{ii}d_{jj} - d_{(ij)}d_{(ji)}] \quad (62)$$

$$d_{ij} = v_{i,j} - \varepsilon_{ijk}v_k \quad (63)$$

The coefficients of viscosity λ, μ, ς_0, κ and ζ_1 are functions of volume distribution function v. The basic continuum theories do not have an 'explicit' length scale in the equations, representing for example the size of the particles, and most of these theories cannot include the effects of particle rotation. The micropolar-based theories or the Cosserat theories while introducing additional balance laws, in theory include the effects of particle rotation and some measure of length scale. Various forms of Cosserat theories have been proposed for flowing granular materials [see Tejchman and Wu,[129] Mulhaus,[130,131] Mohan et al.,[132,133] Kotera et al.,[134] and Mulhaus and Hornby[135]].

4.4   Turbulence Models

The rapid flow of granular materials is generally maintained by particle collisions causing the particles to have highly irregular paths; these irregular motions create fluctuations on all field variables (such as velocity, temperature, etc.). This suggests a similarity between the rapid flow of granular materials and the turbulent flow of a fluid. Blinowski[67] appears to have been the first to observe this similarity. He considered that any random field could be expressed as the sum of an average component and a fluctuating component. Thus,

$$\rho(\mathbf{x},t) = \overline{\rho}(\mathbf{x},t) + \rho'(\mathbf{x},t)$$
$$\mathbf{u}(\mathbf{x},t) = \overline{\mathbf{u}}(\mathbf{x},t) + \mathbf{u}'(\mathbf{x},t) \tag{64}$$
$$\mathbf{T}(\mathbf{x},t) = \overline{\mathbf{T}}(\mathbf{x},t) + \mathbf{T}'(\mathbf{x},t)$$

where, by definition

$$\overline{\rho'}(\mathbf{x},t) = \mathrm{o}$$
$$\overline{\mathbf{u}'}(\mathbf{x},t) = \mathrm{o} \tag{65}$$
$$\overline{\mathbf{T}'}(\mathbf{x},t) = 0$$

where ρ is the density, $\mathbf{u}$ is the velocity, and $\mathbf{T}$ is the stress tensor. The (ensemble) average fields are assumed to be continuous and sufficiently smooth. By introducing concepts such as "quasi-material derivative" and special assumptions on the motion of voids, Blinowski[67] obtained the differential form of equation of motion in terms of mean quantities where the effect of density fluctuation is totally omitted. He did not, however, derive an equation for the fluctuating stress tensor.

Using ensemble-averaging techniques, Ahmadi and Shahinpoor[136] derived the basic equations governing the dynamics of the mean motion in rapid flow of granular material. As a result of this averaging, a number of unknown velocity-density cross-moments arise which make the modeling efforts more difficult. However, by using mass-weighted averaging (sometimes referred to as Favre averaging), they were able to obtain simpler forms of the equations. Thus, the instantaneous velocity $\mathbf{u}$ is written as

$$\mathbf{u}(\mathbf{x},t) = \langle \mathbf{u}(\mathbf{x},t) \rangle + \mathbf{u}''(\mathbf{x},t) \tag{66}$$

where

$$\langle \mathbf{u}(\mathbf{x},t) \rangle = \frac{\overline{\rho \mathbf{u}}}{\rho} \tag{67}$$

is the mass-weighted average velocity and $\mathbf{u}''$ is the fluctuating velocity with respect to mass-weighted average velocity. The stress tensor due to mean momentum transfer caused by random motion (collision) of granules was defined by Ahmadi and Shahinpoor[136] as

$$T''_{jk} = - \overline{\rho u''_j u''_k} \qquad (68)$$

where the similarity to the Reynolds stress tensor in the classical theory of turbulence is obvious. The formulation by Ahmadi and Shahinpoor[136] assumed a basically quasi-equilibrium state of motion, i.e., the production of fluctuation energy was equal to the local energy dissipation. Subsequently, the transport and diffusion of the fluctuation energy were not considered. Ahmadi[137] considered the case where the grains are incompressible and presented a transport equation for the fluctuation stress tensor. Later Massoudi and Ahmadi[68] included the effect of the density gradient in this theory.

4.5 Hypoplastic Models

Numerous constitutive models for the mechanical behavior of granular materials have been proposed since Drucker and Prager[58] presented a theory for the dilatant behavior of granular materials. The dilatant double shearing model of Mehrabadi and Cowin[138] is based on the kinematic postulate that the deformation of granular material consists of two simple dilatant shear deformations [or, "pure stretches", Hayes[139]] along the stress characteristics. As the stress and velocity characteristics are considered to be coincident, these sets of characteristics can also be thought of as slip lines. The two constitutive equations derived on the basis of this postulate are an extension of the theory developed by Spencer[105] for incompressible granular materials. Incorporating the effect of elastic deformation and plastic work hardening, these equations become similar to a set of rate type double slip constitutive equations developed by Nemat-Nasser et al[140] for single crystals. Spencer[106] extended the planar incompressible double shearing model of Spencer[105] to three dimensions by employing a 3-dimensional generalization of Mohr–Coulomb yield condition presented by Shield[141] who proposed three yield conditions based on the relative magnitude of the principal stresses. More recently, Anand and Gu[142] have employed a similar yield condition using an elastic-plastic set of constitutive relations that are identical to those of elastic-plastic dilatant double shearing model proposed in Nemat–Nasser et al.,[140] except that they take the shear rates on the two slip systems to be equal. One consequence of this assumption is that the plastic spin vanishes.

Hypoplasticity is a generalization of hypoelasticity introduced by Truesdell.[143] Truesdell [see Truesdell and Noll,[83] 1992, p.404] defines a hypo-elastic material as one whose constitutive relation may be written in the form

$$\overset{\circ}{\mathbf{T}} = \mathbf{H}(\mathbf{T})[\mathbf{D}] \qquad (69)$$

where

$$\overset{\circ}{\mathbf{T}} = \dot{\mathbf{T}} - \mathbf{W}\mathbf{T} + \mathbf{T}\mathbf{W} \qquad (70)$$

is the co-rotational stress rate, $\mathbf{W}$ is the spin, and $\mathbf{D}$ is the symmetric part of the velocity gradient. The tensor function $\mathbf{H}(\mathbf{T})[\mathbf{D}]$ is linear in $\mathbf{D}$ and isotropic in $\mathbf{T}$

and $\mathbf{D}$. It is possible to use other (frame-invariant) time derivatives such as the convected stress rate

$$\overset{*}{\mathbf{T}} = \dot{\mathbf{T}} + \mathbf{L}^{\mathrm{T}}\mathbf{T} + \mathbf{TL} = \overset{\circ}{\mathbf{T}} + \mathbf{DT} + \mathbf{TD} \tag{71}$$

Applying the representation theorems (Spencer[144]) to Equation 69, we obtain

$$\mathbf{H(T)[D]} = [\alpha_1\mathrm{tr}\mathbf{D} + \alpha_2\mathrm{tr}(\mathbf{TD}) + \alpha_3\mathrm{tr}(\mathbf{T^2D})]\mathbf{1} + [\alpha_4\mathrm{tr}\mathbf{D} + \alpha_5\mathrm{tr}(\mathbf{TD}) + \alpha_6\mathrm{tr}(\mathbf{T^2D})]\mathbf{T}$$
$$+ [\alpha_7\mathrm{tr}\mathbf{D} + \alpha_8\mathrm{tr}(\mathbf{TD}) + \alpha_9\mathrm{tr}(\mathbf{T^2D})]\mathbf{T}^2 + \alpha_{10}\mathbf{D} + \alpha_{11}(\mathbf{DT} + \mathbf{TD}) + \alpha_{12}(\mathbf{DT}^2 + \mathbf{T^2D}) \tag{72}$$

where α_1 through α_{12} are polynomials in the principal invariants I_{T} , II_{T} , III_{T} . One can obtain special types of hypo-elastic materials by restricting the response function $\mathbf{H(T)}[\ ]$ upon $\mathbf{T}$. For example, if the response is a polynomial of degree n in the components of $\mathbf{T}$, then the material is said to be of grade n. Therefore, the constitutive relation for a hypo-elastic material of grade 1, is [Truesdell and Noll,[83] 1992, p.405]:

$$\dot{\mathbf{S}} = \mathbf{WS} - \mathbf{SW} + \left(\frac{\lambda}{2\mu} + \gamma_0 I_s\right)(\mathrm{tr}\mathbf{D})\mathbf{1} + (1 + \gamma_1 I_s)\mathbf{D} + \gamma_2(\mathrm{tr}\mathbf{D})\mathbf{S}$$
$$+ \gamma_3(\mathrm{tr}(\mathbf{SD}))\mathbf{1} + 1/2\,\gamma_4(\mathbf{SD} + \mathbf{DS}) \tag{73}$$

where $\mathbf{S} = \mathbf{T}/(2\mu)$ and λ, μ, $\gamma_0 - \gamma_4$ are material constants.

Hypoelastic constitutive equations may produce curved stress-strain curves, and in some cases these stress-strain curves reach a horizontal plateau and can thus model yielding. However, the imposed incremental linearity implies equal stiffness for loading and unloading and thus makes hypoelastic relations inappropriate to describe inelastic materials. The basic idea of hypoplasticity was developed by Kolymbas[145] by dropping the requirement that the function $\dot{\mathbf{T}} = \mathbf{h(T,D)}$ be linear in $\mathbf{D}$. Improved versions of hypoplasticity have since been proposed by Kolymbas,[145,146] Wu and Bauer,[147,148] Bauer and Wu,[149] Bauer,[150] von Wolffersdorff,[151] Wu et al.,[60] and Hill.[152] In this section, we will briefly review the basic equations of the hypoplasticity model of Wu et al.[60] The general form of the hypoplastic constitutive equation can be written as

$$\overset{\circ}{\mathbf{T}} = \mathbf{H[T,D]} + \mathbf{N(T)}\|\mathbf{D}\|, \tag{74}$$

where $\mathbf{H}$ is linear in $\mathbf{D}$, $\|\mathbf{D}\| = \sqrt{\mathrm{tr}\mathbf{D}^2}$ stands for the norm of $\mathbf{D}$, and it is obvious that the latter term is non-linear in $\mathbf{D}$. Here, we focus on a special case of this equation as given by Wu and Bauer[148] who multiplied the non-linear term $\mathbf{N(T)}\|\mathbf{D}\|$ by the factor I_e which depends on the void ratio e and becomes equal to 1 when $e = e_{\mathrm{crit}}$.

$$\overset{\circ}{\mathbf{T}} = C_1(\mathrm{tr}\mathbf{T})\mathbf{D} + C_2\frac{\mathrm{tr}(\mathbf{TD})}{\mathrm{tr}\mathbf{T}}\mathbf{T} + I_e\left(C_3\frac{\mathbf{T}^2}{\mathrm{tr}\mathbf{T}} + C_4\frac{\mathbf{T}^{*2}}{\mathrm{tr}\mathbf{T}}\right)\|\mathbf{D}\| \tag{75}$$

where

$$I_e = (1 - a) \frac{e - e_{min}}{e_{crit} - e_{min}} + a. \tag{76}$$

is assumed to fulfill the following conditions:

$$I_e|_{e=e_{crit}} = 1 \text{ and } I_e|_{e=e_{min}} = a \tag{77}$$

To account for the effects of pressure level, it is assumed that the critical void ratio and the parameter a depend on the stress level $\text{tr}\mathbf{T}$, i.e.

$$\begin{aligned} e_{crit} &= p_1 + p_2 \exp(p_3\,(\text{tr}\mathbf{T})), \\ a &= q_1 + q_2 \exp(q_3\,(\text{tr}\mathbf{T})). \end{aligned} \tag{78}$$

The parameters p_i, q_i ($I = 1, 2, 3$) are determined by fitting the experimental data. In Equation 75, C_i ($i = 1...4$) are dimensionless constants and the deviatoric stress $\mathbf{T}^*$ is defined as

$$\mathbf{T}^* = \mathbf{T} - 1/3\,(\text{tr}\mathbf{T})\mathbf{I}. \tag{79}$$

Note that the concepts of elasto-plastic theory, such as yield surface, plastic potential and decomposition of deformation into elastic and plastic parts are not used in developing the hypoplastic constitutive equation.

4.6 Other Models

Ogawa[65] and Ogawa et al.,[66] derived theories using formal procedures of continuum mechanics, while recognizing the discrete nature of granular materials. They defined two different kinds of temperature: one is the usual temperature associated with the thermal fluctuation of the molecules of each grain, and the other is related to the "random" translational and rotational fluctuations of the individual grain. These are proportional to $\langle v^2 \rangle^{1/2}$ and $\langle \omega^2 \rangle^{1/2}$ respectively, which are the root mean squares of the translational and rotational velocity fluctuations arising from interparticle collisions. They recognized that the dissipation processes for energy of flowing granular materials are different from those of a liquid or gas. In the case of gas or liquid, momentum transfer is due to the thermal motion of molecules which is depicted in the theory via the dependence of the coefficient of viscosity on the temperature. In the case of granular materials, however, the momentum transfer occurs by particle collision. Using a simple kinematical model of the microscopic behavior of granular particles, and neglecting the rotational motion of the particles and the motion of voids, they obtained the following expressions for the stress tensor T_{ij} and the energy dissipation rate γ:

$$T_{ij} = \phi[k_1\,\langle v^2 \rangle \delta_{ij} + b_c\,\langle v^2 \rangle^{1/2}\,(k_2 D_{ij} + k_3 D_{11}\delta_{ij}] \tag{80}$$

$$\gamma = -k_0\phi\langle v^2\rangle^{3/2}/b_c \tag{81}$$

where

$$b_c = \sigma(v^*/v)$$

$$\phi = \frac{\rho}{4[1-(v/v^*)^{1/3}]} \tag{82$_{a\text{-}f}$}$$

$$k_0 = (1-\alpha)[1-2(1+e)(1/3+\mu/4)+(1+e)^2(1/3+2\mu^2/15)]-1$$

$$k_1 = 2/3(1-\alpha)\{\alpha[1-(1-e)(1/3+\mu/4)+(1-\alpha)[(1-e)(1/3+\mu/4)-(1+e)^2 \\ (1/3+2\mu^2/15)]\}$$

$$k_2 = \alpha^2/2 + \alpha(1-\alpha)(1+e)(2/9+\mu/60)+(1-\alpha)^2(1+e)^2(1/9+16\mu^2/225)$$

$$k_3 = \mu/5(1-\alpha)[\alpha/4-(2\mu/45)(1-\alpha)(1+e)^2$$

where v^* corresponds to a "packed state" (for example, $v^* = 0.7405$ for a regular array of closely packed spheres), σ is the particle diameter, e is the coefficient of restitution, which is a measure of inelasticity of collisions (i.e., e is the ratio of the normal component of the relative velocity of two grains after a collision to that component immediately before the collision, $0<e<1$ where $e=1$ constitutes an elastic collision [energy is conserved]), m is the surface friction coefficient, and α is an accommodation coefficient, defined as the fraction of particles that stick to a surface during collisions. For dry, nonsticky particles, $\alpha=0$. Early works on statistical and kinetic theories approaches to study rapid flow of granular materials include works of Shahinpoor,[153] and Savage and Jeffrey.[154]

Using the ideas put forward by Ogawa,[65] i.e., the explicit incorporation of the equation governing the thermal (fluctuation) velocity, Haff[54] studied the behavior of granular materials from a continuum point of view. The fact that grain collisions are inevitably inelastic would require use of an energy equation, in addition to the conservation equations of mass and momentum. Haff[54] presented a set of complete equations, yet heuristic in the sense that they were not derived in a rigorous sense from the underlying particle interactions to model the flow of grains. He used simple "microscopic kinetic model" to derive expressions for the "coefficients" of viscosity, thermal diffusivity, and energy absorption due to collisions. He assumed that the stress tensor is given by

$$\mathbf{T} = -p\mathbf{1} + 2\mu\mathbf{D} \tag{83}$$

where

$$p = td\rho\bar{v}^2/s \tag{84}$$

and

$$\mu = qd^2\rho\bar{v}/s \tag{85}$$

where t and q are dimensionless constants, d is the grain diameter, s is the average separation distance of grain surfaces between nearest neighbors or mean free path (s << d), and $\bar{v}$ the fluctuation velocity which is the rms velocity measured in a reference frame traveling along with the flow (moving at velocity **u**). The fluctuating or thermal velocity $\bar{v}$ is determined by solving the energy equation. Later, Hui et al.[155] used these ideas to suggest boundary conditions for high-shear grain flows. Hui and Haff[156] studied the flow of granular materials in a vertical channel using the above model.

Savage[157] indicated that, by a least-square fit to the set of shear-test data that he had obtained, it was possible to represent the total measured normal stress (or shear stress) at a shear rate du_1/dx_2 as the sum of two parts.

$$\tau_{total} = \tau(v)_{Coloumb} + \mu(v)\rho\sigma^2(du_1/dx_2)^2 \tag{86}$$

The "dynamic" or "viscous" contribution is assumed to be a result of momentum transfer during collisions between particles. The rate-independent part is due to dry Coulomb friction and particle overriding during enduring contacts. The contribution from $\tau(v)$ and $\mu(v)$ are monotonic increasing functions of v such that at high concentrations and low shear-rates the frictional part is dominant and at low concentrations and high shear rates the rate-dependent term prevails. In treating the fully developed flow of granular materials down an inclined plane, Savage[157] assumed that the kinetic contribution to the stress tensor from diffusion or translation of particles from one shear layer to another is negligible. He then generalized the above equation to,

$$\mathbf{T} = \mathbf{T}_s + \mathbf{T}_c \tag{87}$$

where $\mathbf{T}_s$ is the quasi-static or rate-independent part and $\mathbf{T}_c$ is a collisional momentum flux. He assumed that

$$\mathbf{T}_s = -p_0\mathbf{1} - p_0 \sin \phi \frac{\mathbf{D}}{(1/2tr\mathbf{D}^2)^{1/2}} \tag{88}$$

where p_0 is the mean quasi-static normal stress, ϕ is the quasi-static angle of friction which in general is a function of v, and $\mathbf{D}$ is the rate of deformation tensor. For the collisional part of the stress tensor Savage[157] proposed using the micro-structural theory of Jenkins and Savage,[158] where

$$\mathbf{T}_c = [k/\sigma(\pi\theta)^{1/2} - k/5(2 + \alpha)tr\mathbf{D}]\mathbf{1} - 2k/5(2 + \alpha)\mathbf{D} \tag{89}$$

where

$$k = 2v^2(1 + e)g_0\rho_p\sigma(\theta/\pi)^{1/2} \tag{90}$$

where g_0 is the radial distribution function, s is the uniform spherical particle diameter, e is the coefficient of restitution of the particles, and

$$3\theta/2 = \langle C^2 \rangle/2 \tag{91}$$

is the fluctuation specific kinetic energy, where

$$\mathbf{C} = \mathbf{c} - \mathbf{u} \tag{92}$$

where $\mathbf{u} = \langle\mathbf{C}\rangle$ is the bulk velocity, and α is a numerical coefficient. An example of $g_0(v)$ at contact is that proposed by Carnahan and Starling[159]

$$g_0(v) = \frac{1}{(1-v)} + \frac{3v}{2(1-v)^2} + \frac{v^2}{2(1-v)^3} \tag{93}$$

Jackson[160] followed the basic procedure of Savage[157] (1983) and assumed that the stress tensor is simply the sum of frictional and collisional contributions, thus

$$\mathbf{T} = \mathbf{T}_f + \mathbf{T}_c \tag{94}$$

where

$$\mathbf{T}_c = [p - (\kappa - 2/3\mu)\mathrm{tr}\mathbf{D}]\mathbf{1} - 2\mu\mathbf{D} \tag{95}$$

$$\mathbf{T}_f = \sigma_0\mathbf{1} - \sigma_0 \sin\phi\, \frac{\mathbf{D}}{[1/2\mathrm{tr}(\mathbf{D}^2)]^{1/2}} \tag{96}$$

where $\mathbf{T}_f$ is the same as $\mathbf{T}_s$ proposed by Savage.[157]

When studying the plane shearing of granular materials, Johnson and Jackson,[161] proposed a modified version of the Lun et al.[162] stress tensor. That is, for the collisional–translational contribution to stress $\mathbf{T}_c$, they assumed

$$\mathbf{T}_c = [\rho\theta(1 + 4\eta v g_0) - \eta\mu_b(\mathrm{tr}\mathbf{D})]\mathbf{1} - (2 + \alpha)/3\left\{\frac{2\mu}{\eta(2 - \eta)g_0}(1 + 8/5\eta v g_0) + \right.$$

$$\left. [1 + 8/5\eta(3\eta - 2)v g_0] + 6/5\mu_b\eta\right\}\mathbf{S} \tag{97}$$

where $\mathbf{S}$ is the deviatoric part of the rate of deformation $\mathbf{D}$, i.e.,

$$\mathbf{S} = \mathbf{D} - 1/3(\mathrm{tr}\mathbf{D})\mathbf{1} \tag{98}$$

and

$$\eta = 1/2(1 + e), \ \mu = \frac{5m(\theta/\pi)^{1/2}}{16d^2}$$

$$\lambda = \frac{75m(\theta/\pi)^{1/2}}{8\eta(41 - 33\eta)d^2}, \ \mu_b = \frac{256\mu v^2 g_0}{5\pi}$$

$$(99)_{\text{a-d}}$$

where m is the mass of a single particle and e is the coefficient of restitution. A factor $1/3(2 + \alpha)$ appears in the deviatoric part of σ_c which is not present in Lun et al.[162] Also, instead of using g_0 based on the Carnahan–Starling form, Johnson and Jackson[161] proposed the following form

$$g_0 = \frac{1}{1 - (v/v_0)^{1/3}}$$

$$(100)$$

which ensures that $g_0 \to \infty$ when $v \to v_0$ and thus constrains v to remain smaller than v_0. For this problem, i.e., fully developed plane shear of a non-cohesive granular media, Johnson and Jackson,[161] indicate that the material is in a critical state and the shear stress is proportional to the normal stress, while the normal stress is related to the bulk density. Thus if S_f denotes the frictional contribution to the shear stress and N_f the corresponding contribution to the normal stress,

$$S_f = N_f \sin \phi$$

$$(101)$$

where ϕ is the internal angle of friction. They mention that N_f is expected to increase rapidly with bulk density and to diverge on approaching the bulk density v_0 and they propose

$$N_f = \frac{F}{(v_0 - v)^n}$$

$$(102)$$

where F and n are constants. In this paper, Johnson and Jackson[161] also discuss and motivate derivation of boundary conditions for $\mathbf{u}$ and θ (grain temperature). Variation of this approach often called the 'frictional-kinetic' models have been proposed by Prakash and Rao,[163,164] Jyotsna and Rao,[7,165] Ancey and Evesque [166] to study the flow of granular materials in hoppers and bunkers, for example.

One of the basic differences between the continuum approach and the kinetic theory approximation, when studying granular materials, is the need for additional governing equations. As a result of the averaging procedure and the fact that the fluctuations of the individual particles are considered, at least one extra equation, named the "pseudo-energy" equation is added to the list of the basic equations. Thus, even in the absence of (real) thermal effects, when solving a boundary value problem using the kinetic theory approach, we have at least four basic equations. These equations are the conservation of mass, the balance of linear and angular momentum,[167] and the pseudo-energy equation. The pseudo-energy equation in its general form is given by:

$$3/2\rho \frac{d\theta}{dt} = \mathrm{tr}(\mathbf{TD}) - \mathrm{div}\mathbf{Q}_a - \gamma_a \tag{103}$$

where $\mathbf{Q}_a$ is the flux of fluctuation energy, γ_a is the collisional rate per unit volume of energy dissipation, and θ is called the granular temperature.[168] The concept of granular temperature has its roots in turbulence modelling, where the ideas of Reynolds result in decomposing the velocity vector into an average and a fluctuating component. Thus,

$$\mathbf{u} = \bar{\mathbf{u}} + \mathbf{u}' \tag{104}$$

where $\mathbf{u}$ is the average velocity, and $\mathbf{u}'$ is the fluctuating velocity. While in classical thermodynamics, temperature has a clear meaning, making its appearance in the energy equation, there is no such role or clear meaning for the granular temperature. One can define a scalar k as a measure of the fluctuation of the flow

$$k = 1/2(\overline{\mathbf{u}'} \cdot \overline{\mathbf{u}'}) \tag{105}$$

Now the granular temperature θ is related to the kinetic energy k through

$$\theta = 2/3k \tag{106}$$

Schaffer[169] considered a different constitutive equation when studying the instability in the evolution equations of incompressible granular materials. He assumed that the material is a rigid-perfectly plastic, incompressible, cohesionless Coulomb powder with a yield surface of von Mises type, and the eigenvectors of the strain rate and stress tensor are parallel. By rigid-perfectly plastic material, it is meant that there is no viscosity, and the deformation of granular material is considered plastic in the sense that if after deformation the shearing stress is reduced, the material would not show any tendency to return to its original state. Furthermore, the requirement that the eigenvectors of stress tensor and strain rate are aligned neglects the rotation of a material element during deformation.

Another approach in postulating a yield criterion for granular materials is that of von Mises, which is basically derived from the law of sliding friction applied to the individual particles. This condition requires that

$$\sum_{i=1}^{3} (\sigma_i - \sigma)^2 \leq k^2\sigma^2 \tag{107}$$

where

$$\sigma = 1/3\,\mathrm{tr}\mathbf{T} \tag{108}$$

k is a constant characteristic of the material, and σ_i are the eigenvectors of T_{ij}. Compressive stresses are of interest so that for a material in compression, the eigenvalues σ_i of T_{ij} are positive. Similarly, the eigenvalues of the strain rate tensor D_{ij} would give the rates of compression of the material. For the material to deform, equality must hold in Equation 107, i.e.,

$$\sum_{i=1}^{3} (\sigma_i - \sigma)^2 = k^2 \sigma^2 \tag{109}$$

Schaeffer[169] showed that k is related to the angle of internal friction. Now, for a granular material to deform, the stresses in different directions must be different. Schaeffer[169] claims that "the response of the material to such unequal stresses should be to contract in the directions of greater stress and to expand in the directions of smaller stress." He then mentions that these ideas are quantitatively formulated by a flow rule that links the strain rate and stress tensor through

$$\mathbf{D} = q(\mathbf{T} - \sigma\mathbf{1}) \tag{110}$$

where q is required to be a positive scalar so that the major stress axis corresponds to an eigendirection of $\mathbf{D}$ with positive eigenvalues (i.e., contracting direction). This flow rule also contains the assumption of incompressibility, that is,

$$\mathrm{div}\mathbf{u} = -\mathrm{tr}\mathbf{D} = q\mathrm{tr}(\mathbf{T} - \alpha\mathbf{1}) = 0 \tag{111}$$

where $\mathbf{D}$ is now defined by

$$D_{ij} = -\frac{1}{2}\left(\frac{\partial u_i}{\partial x_j} + \frac{\partial u_j}{\partial x_i}\right) \tag{112}$$

To obtain a constitutive equation for $\mathbf{T}$, it is convenient to define the deviator of an nth order tensor, $\mathbf{A}$, as

$$\mathrm{dev}\mathbf{A} = \mathbf{A} - \frac{1}{n}(\mathrm{tr}\mathbf{A})\mathbf{1} \tag{113}$$

In this notation, the yield condition, Equation 109, can be rewritten as

$$|\mathrm{dev}\mathbf{T}| = k\sigma \tag{114}$$

where $|\mathbf{A}|$ indicates the magnitude of $\mathbf{A}$ and is given by

$$|\mathbf{A}| = \sqrt{\mathbf{A}.\mathbf{A}} = \sqrt{\mathrm{tr}(\mathbf{A}^{\mathrm{T}}\mathbf{A})} \tag{115}$$

Similarly, the flow rule (110) can be rewritten as

$$\mathbf{D} = q\,\mathrm{dev}\mathbf{T} \qquad (116)$$

If we use the decomposition of $\mathbf{T}$ such that

$$\mathbf{T} = \mathrm{dev}\mathbf{T} + \sigma\mathbf{1} \qquad (117)$$

and use Equations 116 and 114, we can write (117) as

$$\mathbf{T} = \sigma\left[\frac{k}{|\mathbf{D}|}\mathbf{D} + \mathbf{1}\right] \qquad (118)$$

which is the constitutive equation that Schaeffer[169] has used in his analysis. This relationship is meaningful only if $|\mathbf{D}| \neq 0$, i.e., it is implicitly assumed that the material is actually deforming. This constitutive relationship is very similar to that of viscous Newtonian fluid, with σ playing the role of the pressure. The two main differences are that the dissipation part of the stress is

- proportional to σ, rather than independent of it, as is the case in the Newtonian viscous fluid
- homogeneous of degree zero rather than degree one, in the velocity

Schaeffer[169] gives an interesting argument for justifying the use of constitutive Equation 118. In a viscous Newtonian fluid, dissipation is due mainly to momentum transfer by collisions; whereas in a granular material, dissipation is generally due to friction between sliding particles. (This is applicable at low speeds.) Thus, in this regime, dissipation would not be increased if the speed were changed. This explains the surprising phenomenon that when mechanical ploughs replaced draught animals, it was observed that ploughing at greater speeds does not require greater forces. Recently Svendsen et al.[170] have studied the effects of quasi-static frictional behaviour of granular materials.

Boyle and Massoudi[171] derived a constitutive equation by utilizing the ideas of Enskog's dense gas theory; their model can predict normal-stress differences arising due to density gradient. In recent years, many other researchers have also discussed the importance and the origin of normal stress differences in flow of granular materials.[172,173] Richman and Marciniec[168] studied the flow of granular materials down an inclined plane; they obtained a closed-form solution for the granular temperature profile, by replacing the volume fraction by its depth-averaged value in the balance equations and thereby from the constitutive relations for the normal and shear stresses they obtained the volume fraction and the velocity profiles [see also Cercignani,[174] and Jin and Slemrod[175]]. In closing this section, we would like to mention that the application of kinetic theory of gases to flows of granular materials is plagued by many assumptions, perhaps beyond what the original theory may have stood for. In the last two decades, the researchers in the field of granular materials have exploited the techniques of kinetic theory and statistical mechanics. There are certain flow regimes where the collisions of particles are rare, in the sense that the flow is so slow and the particles are so densely packed that one cannot assume the basic assumptions in the kinetic theory are valid.[176] However, certain processes such as fluidized beds present a special challenge [see Gidaspow,[177] Fan and Zhu,[178] and Jackson[179]]: before the onset of fluidization, the flow regime is perhaps more in the slow deformation range and after fluidization in the rapid flow range. It is in the rapid flow regime that the kinetic theory approach may be used. Thus, we tend to look at the kinetic-theory based models as tools which may be appropriate for some cases and irrelevant or not appropriate in other cases. Just as one can build a bridge or design a ship without ever having access to the tools in statistical theories, it is also possible to study many engineering problems involving granular materials without ever using the kinetic theory approach.

5 Conclusions

In certain industrial processes such as flows in hoppers and bins, and fluidization, the solid particles are initially in static equilibrium. Then the flow slowly starts, whether due to action of gravity as is the case for bins and chutes, or due to the upward flow of a fluid in a fluidized bed. In the case of fluidization,[180] the bed slowly expands. At this stage the yielding begins and the particles are no longer in static equilibrium. Frictional and sliding forces are the main deterrents to the flow. As the bed becomes fully fluidized, the particles begin to collide with each other and they move about rapidly. At this stage, the viscous and the interaction forces are the dominant mechanisms for flow. It is difficult to come up with a single constitutive relation which can cover the whole field of operation. In reality, in the regime where particles are colliding with each other, the fluid phase plays an important role, and thus the present model should only be used within the context of a multiphase mixture theory.[181–183]

There are still many challenging problems to study within the continuum modeling of granular materials. Flows where rotation plays a dominant role, for example rotating drums, represent special difficulties in formulating constitutive relations and in solving the complicated equations of motion [see Gray,[184] and Ding et al.[185]]. The specification of boundary conditions, especially for higher order theories remains an open problem. Free surface flows, flows with moving boundaries, or flows with slip at the surfaces, or with porous walls present special difficulties. The transition regime between slow-flow and rapid-flow with its inherent instabilities is not well understood. The dependence of material parameters, such as viscosity on surface roughness,[186] size [see Antony and Ghadiri[187]] shape and geometry of the particles, moisture content, concentration, etc., continue to be of critical concern. Furthermore, flowing granular materials represent the limiting case of two-phase flows at high solid concentration and high solid-to-fluid density ratios; this presents another set of interesting problems, namely that the motion and the material parameters of one phase could depend on the motion and the material parameters of the other phase. For such cases, of course, one needs to study the problem within the context of the theory of Interacting Continua (or Mixture Theory). Heat transfer studies, where material properties such as thermal conductivity can be variables, and where their exact structure and form is not well-known also present interesting problems. Forced convection[188] and natural convection[182] flows of granular materials have been studied for flows down an inclined plane and flow between two heated vertical plates. Vargas and McCarthy[189] discuss the heat conduction in granular materials and Hunt[190] presents a review of the important issues in heat transfer in packed beds. Important issues which we have not discussed in this essay are (i) mixing and segregation,[191] (ii) attrition,[192,193] (iii) sintering,[194] (iv) experimental techniques to measure the bulk properties and velocity or concentration profiles,[195–197] (v) micromechanical modeling,[198–201] (vi) pattern formation and sandpiles,[41,202] (vii) various numerical schemes such as the distinct elements method,[203–206] and . . .

In conclusion we would like to mention an alternative and powerful method, The Multiple Natural Configurations Theory, which has been successfully used in many applications. A general shortcoming of all higher order or higher gradient theories is the necessity of assigning boundary conditions for certain terms, which appear in the governing equations. Quite often these boundary conditions are not derived from first principles; instead they are given as ad-hoc assumptions, or they are simply specified as mathematical conveniences. Sometimes experiments have been used successfully to specify these necessary additional boundary conditions. A second shortcoming of these higher theories, whether multipolar or director theories of liquid crystals, or turbulence theory, is the need for additional balance equations. Both of these shortcomings can be overcome through the application of multiple natural configurations theory developed by Rajagopal and co-workers. This theory has been used for modeling non-Newtonian fluids,[122] anisotropic fluids,[207] polymer crystallization,[208] inelasticity and plasticity,[209,210] asphalt,[211] growth,[212] . . . The basic premise of this theory is that materials which possess certain (micro) structures, in general have numerous natural configurations associated with the deformed states of the body. As the body goes under deformation of various types, these natural configurations evolve, and it is assumed that the response of the material is elastic from each of these natural configurations, and the way these natural configurations evolve depends on the rate of dissipation. Therefore, depending on the process, i.e., the application, the same material may behave differently. For example, the viscoealstic response of a material is determined by two functions: a stored energy function characterizing the elastic response of the 'natural configuration' and a rate of dissipation function describing the rate of dissipation due to viscous effect. This theory is currently being developed for granular materials and mixtures.[213]

References

1. S. I. Plasynski, W. C. Peters, and S. L. Passman, The Department of Energy Solids Transport, Multiphase Flow Program: Proceedings of the NSF-DOE Joint workshop on Flow of Particulates and Fluids, Gaithersburg, September 16–18, 1992.
2. J. Y. Ooi, and K. M. She, 'Finite element analysis of wall pressure in imperfect silos', Int. J. Solids Structures, 1997, **34**, 2061–2072.
3. J. M. Rotter, J. M. F. Holst, J. Y. Ooi, and A. M. Sanad, 'Silo pressure predictions using discrete-element and finite-element analysis', Phil. Trans. R. Soc. Lond., 1998, A **356**, 2685–2712.
4. S. B. Savage, 'Gravity Flow of Cohesionless Granular Materials in Chutes and Channels', J. Fluid Mech., 1979, **92**, 53–96.
5. A. Drescher, 'Some aspects of flow of granular materials in hoppers', Phil. Trans. R. Soc. Lond., 1998, A **356**, 2649–2666.
6. J. Nielsen, 'Pressure from flowing granular solids in silos', Phil. Trans. R. Soc. Lond., 1998, A **356**, 2667–2684.
7. R. Jyotsna, and K. Kesava Rao, 'A frictional-kinetic model for the flow of granular materials through a wedge-shaped hopper', J. Fluid Mech., 1997, **346**, 239–270.
8. R. D. Marcus, L. S. Leung, G. E. Klinzing and F. Rizk, Pneumatic Conveying of Solids, Chapman and Hall, London, 1990.
9. S. S. Hsiau, M. H. Wu and C. H. Chen, 'Arching phenomena in a vibrated granular bed', Powder Tech., 1998, **99**, 185–193.
10. A. Samadani, L. Mahadevan and A. Kudrolli, 'Shocks in sand flowing in a silo', J. Fluid Mech., 2002, **452**, 293–301.
11. J. W. Carson and T. A. Royal, 'Modeling the flow of bulk solids', Powder Handl. Proc., 1991, **3**, 257–261.
12. A. Levy, 'Shock waves interaction with granular materials', Powder Tech., 1999, **103**, 212–219.
13. C. Wensrich, 'Dissipation, dispersion, and shocks in granular media', Powder Tech., 2002, **126**, 1–12.
14. J. F. Davidson and R. M. Nedderman, 'The hour-glass theory of hopper flow', Trans. Instn. Chem. Engrs., 1973, **51**, 29–35.
15. S. B. Savage, 'The Mechanics of Rapid Granular Flows', Adv. in Applied Mech., 1984, **24**, 289–366.
16. K. Wieghardt, 'Experiments in Granular Flow', Ann Rev. Fluid Mech., 1975, **7**, 89.
17. T. R. H. Davies, 'Large debris flows: A macro-viscous phenomenon', Acta Mech., 1986, **63**, 161.
18. T. Takahashi, 'Debris Flow', Annual Rev. Fluid Mech., 1981, **13**, 57.
19. K. Hutter, B. Svendsen and D. Rickermann, 'Debris flow modelling: A review', Continuum Mech. Thermodyn., 1996, **8**, 1.
20. J. Rajchenbach, 'Flow in powders: From discrete avalanches to continuum regime', Phys. Rev. Letters, 1990, **65**, 2221.
21. S. B. Savage and K. Hutter, 'The dynamics of avalanches of granular materials form initiation to runout', Part I; Analysis, Acta Mech., 1991, **86**, 201.
22. K. Hutter, F. Szidarovszky and S. Yakowitz, 'Plane Steady Shear Flow of a Cohesionless Granular Material Down an Inclined Plane: A Model for Flow Avalanches Part-1: Theory', Acta Mechanica, 1986, **63**, 87–112.
23. K. Hutter, F. Szidarovszky and S. Yakowitz, 'Plane Steady Shear Flow of a Cohesionless Granular Material Down an Inclined Plane: A Model for Flow Avalanches Part-II: Numerical Results', Acta Mechanica, 1986, **65**, 239–261.

24. H. Ahn, C. E. Brennen and R. H. Sabersky, 'Measurements of Velocity Fluctuation, Density, and Stresses in Chute Flows of Granular Materials', Transactions of ASME, 1991, **58**, 792–803.
25. H. Ahn, C. E. Brennen and R. H. Sabersky, 'Analysis of the fully developed Chute Flows of Granular Materials', Transactions of ASME, 1992, **59**, 109–119.
26. Y. Wang and K. Hutter, 'A constitutive theory of fluid-saturated granular materials and its application in gravitational flows', Rheol. Acta, 1999, **38**, 214–223.
27. O. Pouliquen and F. Chevoir, 'Dense flows of dry granular material', C. R. Physique, 2002, **3**, 163–175.
28. R. Gudhe, R. C. Yalamanchili and M. Massoudi, 'Flow of granular materials down a vertical pipe', Int. J. Non-Linear Mech., 1994, **29**, 1–12.
29. A. J. M. Spencer and N. J. Bradley, 'Gravity flow of granular materials in contracting cylinders and tapered tubes', Int. J. Engng. Sci., 2002, **40**, 1529–1552.
30. R. L. Brown and J. C. Richards, Principles of Powder Mechanic, Pergamon Press, London, 1970.
31. A. J. Stepanoff, Gravity Flow of Bulk Solids and Transportation of Solids in Suspension, John Wiley & Sons Inc., New York, 1969.
32. R. M. Nedderman, U. Tuzun, S. B. Savage and G. T. Houlsby, 'The flow of granular materials-I', Chem. Eng. Sci., 1982, **37**, 1597–1609.
33. C. Campbell, 'Rapid granular flows', Annual Rev. Fluid Mech., 1990, **22**, 57–92.
34. K. Hutter and K. R. Rajagopal, 'On the flows of granular materials', Continuum Mech. Thermodyn., 1994, **6**, 81–139.
35. H. M. Jaeger, S. R. Nagel and R. P. Behringer, 'Granular solids, liquids, and gases', Rev. Modern Phys., 1996, **68**, 1259.
36. P. G. de Gennes, 'Reflections on the mechanics of granular matter', Physica A, 1998, **261**, 267–293.
37. H. J. Hermann and S. Luding, 'Modelling of granular media in the computer', Continuum Mech. Thermodyn., 1998, **10**, 189.
38. H. J. Hermann, 'Granular matter', Physica A, 2002, **313**, 188–210.
39. M. Satake and J. T. Jenkins (Eds.), Micromechanics of granular materials, Elsevier Science Publishers, The Netherlands, 1988.
40. H. Shen, M. Satake, M. Mehrabadi, C. S. Chang and C. S. Campbell (Eds.), Advances in micromechanics of granular materials, Elsevier Science Publishers B. V., Amsterdam, 1992.
41. M. Mehta (Editor), Granular Matter: An Interdisciplinary Approach, Springer-Verlag, New York, 1994.
42. C. S. Chang, A. Misra, R. Y. Liang and M. Babic (Editors), Mechanics of deformation and flow of particulate materials, ASCE Publishing, New York, 1997.
43. N. A. Fleck and A. C. F. Cocks (Eds.), IUTAM symposium of Mechanics of Granular and Porous Materials, Kluwer Publishers, Dordrecht, 1997.
44. J. Duran, Sands, Powders and Grains, Springer-Verlag, New York, 2000.
45. R. P. Behringer, D. Howell, L. Kondic, S. Tennakoon and C. Veje, 'Predictability and granular materials', Physica D, 1999, **133**, 1.
46. R. A. Bagnold, 'Experiments on a Gravity Free Dispersion of Large Solid Spheres in a Newtonian Fluid Under Shear', Proc. Roy. Soc. London, 1954, **225**, 49.
47. R. A. Bagnold, 'The Shearing and Dilatation of Dry Sand and the "Singing" Mechanism', Proc. Roy. Soc. London, 1966, A **295**, 219.
48. M. Reiner, 'A Mathematical Theory of Dilatancy', American J. Math., 1945, **67**, 350–362.

49. D. F. McTigue, 'A Non-Linear Constitutive Model For Granular Materials: Applications to Gravity Flow', J. Appl. Mech., 1982, **49**, 291–296.
50. J. D. Goddard, 'Dissipative Materials as Models of Thixotropy and Plasticity', J. Non-Newtonian Fluid Mech., 1984, **14**, 141.
51. J. D. Goddard, 'Dissipative Materials as Constitutive Models for Granular Media', Acta Mech., 1986, **63**, 3.
52. T. Astarita and R. Ocone, 'Unsteady compressible flow of granular materials', Ind. Eng. Chem. Res., 1999, **38**, 1177.
53. S. A. Elaskar and L. A. Godoy, 'Constitutive relations for compressible granular materials using non-Newtonian fluid mechanics', Int. J. Mech. Sci., 1998, **40**, 1001–1018.
54. P. K. Haff, 'Grain Flow as a Fluid-Mechanical Phenomenon', J. Fluid Mech., 1983, **134**, 401.
55. G. I. Tardos, 'A fluid mechanistic approach to slow, frictional flow of powders', Powder Tech., 1997, **92**, 61.
56. T. Ivsic, A. Galovic and D. Kirin, 'Sand as a compressible fluid', Physica A, 2000, **277**, 47.
57. G. Gudehus, 'Granular materials as rate-independent simple materials: Constitutive relations', Powder Tech., 1969/70, **3**, 344.
58. D. C. Drucker and W. Prager, 'Soil Mechanics and Plastic Analysis or Limit Design', Quart. Appl. Math., 1952, **10**, 157–165.
59. F. Maddalena and M. Ferrari, 'Viscoelasticity of granular materials', Mech. Mater., 1995, **20**, 241.
60. W. Wu, E. Bauer and D. Kolymbas, 'Hypoplastic constitutive model with critical state for granular materials', Mech. Mater., 1996, **23**, 45.
61. K. I. Kanatani, 'A Micropolar Continuum Theory for the Flow of Granular Materials', Int. J. Engng. Sci., 1979, **17**, 419.
62. M. Satake, 'Consideration on the stress-dilatancy equation through the work increment tensor', in Micromechanics of Granular Materials, Edited by M. Satake and J. T. Jenkins, Elsevier, 1988.
63. J. Christoffersen, M. M. Mehrabadi and S. Nemat-Nasser, 'A micromechanical description of granular material behavior', ASME J. Appl. Mech., 1981, **48**, 339.
64. K. Bagi, 'Stress and strain in granular materials', Mech. Materials, 1996, **22**, 165.
65. S. Ogawa, 'Multitemperature Theory of Granular Materials', in Proc. U. S.-Japan Seminar on Continuum Mechanical and Statistical Approaches in the Mechanics of Granular Materials, Eds. S. C. Cowin and M. Satake, 1978, 208–217.
66. S. Ogawa, A. Umemura and N. Oshima, 'On the Equations of Fully Fluidized Granular Materials', ZAMP, 1980, **31**, 483.
67. A. Blinowski, 'On the Dynamic Flow of Granular Media', Arch. Mech., 1978, **30**, 27.
68. M. Massoudi and G. Ahmadi, 'Rapid flows of granular materials with density and fluctuation energy gradients', Int. J. Non-Linear Mech., 1994, **29**, 487.
69. M. A. Goodman and S. C. Cowin, 'Two Problems in the Gravity Flow of Granular Materials', J. Fluid Mech., 1971, **45**, 321.
70. M. A. Goodman and S. C. Cowin, 'A Continuum Theory for Granular Materials', Arch. Rat. Mech. and Anal., 1972, **44**, 249–266.
71. G. Ahmadi, 'A Generalized Continuum theory for Granular Materials', Int. J. Non-Linear Mech., 1982, **17**, 21–33.
72. J. W. Nunziato, S. L. Passman and J. P. Thomas Jr., 'Gravitational Flows of Granular Materials with Incompressible Grains', J. Rheology, 1980, **24**, 395–420.

73. K. R. Rajagopal, M. Massoudi and A. S. Wineman, 'Flow of Granular Materials between Rotating Disks', Mech. Research Comm., 1994, **21**, 629–634.

74. C. Truesdell and R. Toupin, 'The classical field theories', in Handbuck der Phsyik, III/1, Ed., Flügge, Springer-Verlag, 1960.

75. J. L. Ericksen, 'Anisotropic Fluids', Arch. Rat. Mech. Anal., 1960, **4**, 231.

76. J. L. Ericksen, 'Transversely isotropic fluids', Koll Zeits., 1960, **173**, 117–122.

77. R. A. Toupin, 'Elastic materials with couple-stresses', Arch. Rat. Mech. Anal., 1962, **11**, 385–414.

78. R. A Toupin, 'Theories of elasticity with couple-stress', Arch. Rat. Mech. Anal., 1964, **17**, 85–112.

79. A. C. Eringen and E. S. Suhubi, 'Nonlinear theory of simple micro-elastic solids – I', Int. J. Engng. Sci., 1964, **2**, 189–203.

80. A. C. Eringen, 'Mechanics of micromorphic materials', in Proc. 11th Intl. Congress Appl. Mech., Ed. H. Görther, Springer-Verlag, 1966.

81. A. C. Eringen, 'Theory of Micropolar Fluids', J. Math. and Mech., 1966, **16**, 1.

82. A. C. Eringen, 'Mechanics of micropolar continua', in Contribution to Mechanics, Ed. D. Abir, 1969.

83. C. Truesdell and W. Noll, The Non-Linear Field Theories of Mechanics, Second Edition, Springer-Verlag, New York, 1992.

84. C. Truesdell, and K. R. Rajagopal, An Introduction to the Mechanics of Fluids, Birkhauser, New York, 2000.

85. I. Müller, 'A Thermodynamic Theory of Mixtures of Fluids', Arch. Rational Mech. Anal., 1968, **28**, 1.

86. R. S. Rivlin, 'On the principles of equipresence and unification', Q. Appl. Math., 1972, **30**, 227.

87. G. Hagen, Berlin Monatsber Akad. Wiss., 1852, 35–42.

88. O. Reynolds, 'On the Dilatancy of Media Composed of Rigid Particles in Contact with Experimental Illustrations', Phil. Mag., Series 5, 1885, **20**, 469–481.

89. O. Reynolds, 'Experiments Showing Dilatancy, a Property of Granular Material, Possibly Connected with Gravitation', Proc. Roy. Inst. of Gr. Britain, 1886, **11**, 354–363.

90. C. F. Jenkin, 'The pressure exerted by granular materials: an application of the principle of dilatancy', Proc. Royal Soc. Lond., 1931, A **131**, 53.

91. P. W. Rowe, 'The stress-dilatancy relation far static equilibrium of an assembly of particles in contact', Proc. Royal Soc. Lond., 1962, A **269**, 500.

92. E. N. D. C. Andrade and J. W. Fox, 'The mechanism of dilatancy', Proc. Phys. Soc. Lond., 1949, **B62**, 483.

93. M. D. Bolton, 'The strength and dilatancy of sand', Geotech., 1986, **36**, 65.

94. S. A. Nixon and H. W. Chandler, 'On the elasticity and plasticity of dilatant granular materials', J. Mech. Phys. Solids, 1999, **47**, 1397.

95. M. M. Mehrabadi and S. Nemat-Nasser, 'Stress, dilatancy and fabric in granular materials', Mech. Mater., 1983, **2**, 155.

96. J. D. Goddard and Y. M. Bashir, 'On Reynolds dilatancy', in Recent Developments is Structured Continua, Vol. II. Edited by D. De Kee and P. N. Kaloni, Longman Scientific & Technical, New York, 1990.

97. J. D. Goddard, 'Granular dilatancy and the plasticity of glassy lubricants', Ind. Eng. Chem. Res., 1999, **38**, 820.

98. G. Mandl and R. Fernandez Luque, 'Fully Developed Plastic Shear Flow of Granular Materials', Geotech., 1970, **20**, 277.

99. K. Terzaghi, Theoretical Soil Mechanics, Wiley and Sons Inc., New York, 1943.

100. P. A. Wood, Fundamentals of Bulk Solids Flow, Report No. ICTIS/TR31, IEA Coal Research, London, 1986.

101. P. G. Nutting, 'The deformation of granular solids', J. Washington Academy of Sci., 1929, **18**, 123.

102. G. Gudehus, 'A comprehensive constitutive equation for granular materials', Soils and Foundations, 1996, **36**, 1.

103. V. V. Sokolovski, Statics of Granular Media, Pergamon, Oxford, 1965.

104. R. M. Nedderman, Statics and kinematics of granular materials, Cambridge University Press, 1992.

105. A. J. M. Spencer, 'A theory of the kinematics of ideal soils under plane strain conditions', J. Mech. Phys. Solid, 1964, **12**, 337–351.

106. A. J. M. Spencer, 'Deformation of ideal granular materials', Mechanics of Solids, Pergamon Press, Oxford and New York, pp. 607–652, 1982.

107. A. W. Jenike, 'Steady Gravity Flow of Frictional-Cohesive Solids in Converging Channels', J. Appl. Mech., 1964, **31**, 5–11.

108. S. Sundaresan, 'Some outstanding questions in handling of cohesionless particles', Powder Tech., 2001, **115**, 2–7.

109. S. Edwards, 'The rheology of powders', Rheol. Acta, 1990, **29**, 493–499.

110. S. F. Edwards and R. B. S. Oakeshott, 'Theory of powders', Physica A, 1989, **157**, 1080–1090.

111. S. F. Edwards and D. V. Grinev, 'Comapctivity and transmission of stress in granular materials', Chaos, 1999, **9**, 551–558.

112. S. C. Cowin, 'A Theory for the Flow of Granular Material', Powder Tech., 1974, **9**, 61–69.

113. S. C. Cowin, 'Constitutive Relations that Imply a Generalized Mohr--Coulomb Criterion', Acta Mechanica, 1974, **20**, 41–46.

114. K. R. Rajagopal and M. Massoudi, A Method for Measuring Material Moduli of Granular Materials: Flow in an Orthogonal Rheometer, Topical Report, DOE/PETC/TR-90/3,1990.

115. I. Müller, 'On the entropy inequality', Arch. Rat. Mech. and Anal., 1967, **26**, 118–141.

116. K. R. Rajagopal, W. C. Troy and M. Massoudi, 'Existence of Solutions to the Equations Governing the Flow of Granular Materials', European J. Mech., B/Fluids, 1992, **11**, 265–276.

117. O. R. Walton and R. L. Braun, 'Stress Calculations for Assemblies of Inelastic Spheres in Uniform Shear', Acta Mechanica, 1986, **63**, 73–86.

118. O. R. Walton and R. L. Braun, 'Viscosity, Granular-Temperature, and Stress Calculations for Shearing Assemblies of Inelastic, Frictional Disks', J. Rheology, 1986, **30**, 949.

119. M. Massoudi and M. M. Mehrabadi, 'A continuum model for granular materials: Considering dilatancy, and the Mohr–Coulomb criterion', Acta Mechanica, 2001, **152**, 121–138.

120. M. Massoudi, 'On the flow of granular materials with variable material properties', Int. J. Non-Linear Mech., 2001, **36**, 25–37.

121. K. R. Rajagopal, G. Gupta and R. C. Yalamanchili, 'A rheometer for measuring the properties of granular materials', Particulate Sci. Tech., 2000, **18**, 39–55.

122. K. R. Rajagopal and A. R. Srinivasa, 'A thermodynamic frame work for the rate type fluid models', J. Non-Newtonian Fluid Mech., 2000, **88**, 207–227.

123. S. Baek, K. R. Rajagopal and A. R. Srinivasa, 'Measurements related to the flow of granular materials in a torsional rheometer', Particulate Sci. Tech., 2001, **19**, 175–186.

124. R.A. Bagnold, 'The Flow of Cohesionless Grains in Fluids', Phil. Trans. Roy. Soc. London, Series A, 1956, **249**, 235.

125. A. N. Schofield and C. P. Wroth, Critical States Soil Mechanics, McGraw-Hill, New York, 1968.

126. M. Shahinpoor and S. P. Lin, 'Rapid Couette Flow of Cohesionless Granular Materials', Acta Mechanica, 1982, **42**, 183–196.

127. R. S. Rivlin, 'The hydrodynamics of non-Newtonian fluids, I', Proc. Royal Soc. Lond. A, 1948, **193**, 260.

128. G. Ahmadi, 'A generalized continuum theory for flow of granular mate-rials', in Advances in the mechanics and the flow of granular materials, Vol. 2, (Ed. M. Shahinpoor) II, Gulf Publishing Co., 1983.

129. J. Tejchman and W. Wu, 'Numerical study of patterning of shear bands in a Cosserat continuum', Acta Mechanica, 1993, **99**, 61–74.

130. H. B. Mulhaus, 'Shear band analysis in granular materials by Cosserat theory', Ing. Archiv., 1986, **56**, 389–399.

131. H. B. Mulhaus, 'Application of Cosserat theory in numerical solution of limit load problems', Ing. Archiv., 1989, **59**, 124–137.

132. L. S. Mohan, P. R. Nott and K. K. Rao, 'A frictional Cosserat model for the flow of granular materials through a vertical channel', Acta Mechanica, 1999, **138**, 75–96.

133. L. S. Mohan, P. R. Nott and K. K. Rao, 'A frictional Cosserat model for the slow shearing flow of granular materials', J. Fluid Mech., 2002, **457**, 377–409.

134. H. Kotera, M. Sawada and S. Shima, 'Cosserat continuum theory to simulate microscopic rotation of magnetic powders in applied magnetic field', Int. J. Mech. Sci., 2000, **42**, 129–145.

135. H. B. Muhlhaus and P. Hornby, 'Energy and averages in the mechanics of granular materials', Tectonophysics, 2001, **335**, 63–80.

136. G. Ahmadi and M. Shahinpoor, 'Towards a turbulent modeling of rapid flow of granular materials', Powder Tech., 1983, **35**, 241–248.

137. G. Ahmadi, 'A turbulence model for rapid flows of granular materials part I, Basic theory', Powder Tech., 1985, **44**, 261–268.

138. M. M. Mehrabadi and S.C. Cowin, 'Initial planar deformation of dilatant granular materials', J. Mech. Phys. Solids, 1978, **26**, 269–284.

139. M. Hayes, 'On Pure Extension', Arch. Rational Mech. Anal., 1968, **28**, 155.

140. S. Nemat-Nasser, M. M. Mehrabadi and T. Iwakuma, 'On Certain microscopic and macroscopic aspect of plastic flow of ductile materials', Three dimensional Constitutive Relations and Ductile Fracture, North Holland, Amsterdam, pp. 157–172, 1981.

141. R. T. Shield, 'On Coulomb's law of failure in soils', J. Mech. Phys. Solids, 1955, **4**, 10–16.

142. L. Anand and C. Gu, 'Granular materials: constitutive equations and strain localization', J. Mech. Phys. Solids, 2000, **48**, 1701–1733.

143. C. Truesdell, 'Hypo-elasticity', J. Rat. Mech. Anal., 1955, **4**, 83–133.

144. A. J. M. Spencer, 'Theory of invaraints', in Continuum Physics, Vol. I, Edited by A. C. Eringen, Academic Press, 1971.

145. D. Kolymbas, 'An outline of hypoplasticity', Arch. Applied Mechanics, 1991, **61**, 143–151.

146. D. Kolymbas and W. Wu, 'Introduction to hypoplasticity', in Proceedings of the Modern Approaches to Plasticity, Editor D. Kolymbas, Elsevier, Amsterdam, pp. 213–223, 1993.

147. W. Wu and E. Bauer, 'A hypoplastic model for barotropy and pyknotropy of

granular soils', in Proceedings of the Modern Approaches to Plasticity, Editor D. Kolymbas, Elsevier, Amsterdam, pp. 225–245, 1993.

148. W. Wu and E. Bauer, 'A simple hypoplastic constitutive model for sand', Int. J. Numer. Anal. Meth. Geomech., 1994, **18**, 833–862.

149. E. Bauer and W. Wu, 'A Hypoplastic constitutive model for cohesive powders', Powder Technology, 1995, **85**, 1–9.

150. E. Bauer, 'Calibration of a comprehensive hypoplastic model for granular materials', Soils Foundations, 1996, **36**, 13–26.

151. P. A. von Wolffersdorff, 'A hypoplastic relation for granular materials with a predefined limit state surface', Mech. Cohesive-frictional Materials, 1996, **1**, 251–271.

152. J. M. Hill, 'Similarity 'hot-spot' solutions for a hypoplastic granular material', Proc. R. Soc. Lond. A, 2000, **456**, 2653–2671.

153. M. Shahinpoor, 'Statistical mechanical considerations on the random packing of granular materials', Powder Tech., 1980, **25**, 163–176.

154. S. B. Savage and D. J. Jeffrey, 'The stress tensor in a granular flow at high shear rates', J. Fluid Mech., 1981, **110**, 255–272.

155. K. Hui, P. K. Haff, J. E. Ungar and R. Jackson, 'Boundary Conditions for High-Shear Grain Flows', J. Fluid Mech., 1984, **145**, 223–233.

156. K. Hui and P. K. Haff, 'Kinetic Grain Flow in a Vertical Channel', Int. J. Multiphase Flow, 1986, **12**, 289–298.

157. S. B. Savage, 'Granular Flows Down Rough Inclines – Review and Extension', in Mechanics of Granular Materials: New Models and Constitutive Relations. Eds. J. T. Jenkins and M. Stake, pp. 261–282, 1983.

158. J. T. Jenkins and S. B. Savage, 'A Theory for the Rapid Flow of Identical, Smooth, Nearly Elastic, Spherical Particles', J. Fluid Mech., 1983, **130**, 187–202.

159. N. F. Carnahan and K. E. Starling, Equations of State for Non-Attracting Rigid Spheres', J. Chem. Phys., 1969, **51**, 635.

160. R. Jackson, 'Some Features of the Flow of Granular Materials and Aerated Granular Materials', J. Rheology, 1986, **30**, 907.

161. P. C. Johnson and R. Jackson, 'Frictional-Collisional Constitutive Relations for Granular Materials with Application to Plane Shearing', J. Fluid Mech., 1987, **176**, 67–93.

162. C. K. K. Lun, S. B. Savage, D. J. Jeffrey and N. Chepurnity, 'Kinetic Theories for Granular Flow: Inelastic Particles in Couette Flow and Slightly Inelastic Particles in a General Flow Field', J. Fluid Mech., 1984, **140**, 223–256.

163. J. R. Prakash and K. K. Rao, 'Steady compressible flow of granular materials through a wedge-shaped hopper: The smooth wall radial geometry problem', Chem. Eng. Sci., 1988, **43**, 479–494.

164. J. R. Prakash and K. K. Rao, 'Steady compressible flow of cohesionless granular materials through a wedge-shaped bunker', J. Fluid Mech., 1991, **225**, 21–80.

165. R. Jyotsna and K. Kesava Rao, 'Steady incompressible flow of cohesionless granular materials through a wedge-shaped hopper: Frictional-kinetic solution to the smooth wall, radial gravity problem', Chem. Eng. Sci., 1991, **46**, 1951–1967.

166. C. Ancey and P. Evesque. 'Frictional-collisional regime for granular suspension flows down an inclined channel', Phys. Rev. E, 2000, **62**, 8349–8360.

167. J. C. Slattery, Advanced Transport Phenomena, Cambridge University Press, Cambridge, MA, 1999.

168. M. W. Richman and R. P. Marciniec, 'Gravity-Driven Granular Flows of Smooth, Inelastic Spheres Down Bumpy Inclines', Transactions of ASME, 1990, 57, 1036–1043.

169. D. G. Schaeffer, 'Instability in the evolution equations describing incompressible granular flow', J. Diff. Eq., 1987, **66**, 19–50.

170. B. Svendsen, K. Hutter and L. Laloui, 'Constitutive models for granular materials including quasi-static frictional behaviour: Toward a thermodynamic theory of plasticity', Continuum Mech. Thermodyn., 1999, **4**, 263–275.

171. E. J. Boyle and M. Massoudi, 'A theory for granular materials exhibiting normal stress effects based on Enskog's dense gas theory', Int. J. Engng. Sci., 1990, **28**, 1261–1275.

172. I. Goldhirsch and N. Sela, 'Origin of normal stress differences in rapid granular flows', Physical Rev. E, 1996, **54**, 4458.

173. N. Sela and I. Goldhirsch, 'Hydrodynamic Equations for Rapid Flows of Smooth Inelastic Spheres to Burnet Order', J. Fluid Mech., 1998, **361**, 41–74.

174. C. Cercignani, 'Shear flow of a granular material', J. Statistical Phys., 2001, **102**, 1407–1415.

175. S. Jin and M. Slemrod, 'Regularization of the Burnett equations for rapid granular flows via relaxation', Physica D, 2001, **150**, 207–218.

176. C. Truesdell and R. G. Muncaster, Fundamentals of Maxwell's Kinetic Theory of a Simple Monatomic Gas, Academic Press, New York, 1980.

177. D. Gidaspow, Multiphase flow and fluidization, Academic Press, San Diego, 1994.

178. L. S. Fan and C. Zhu, Principles of Gas-Solid Flows, Cambridge University Press, Cambridge, 1998.

179. R. Jackson, The dynamics of fluidized particles, Cambridge University Press, Cambridge, MA, 2000.

180. J. F. Davidson, R. Clift and D. Harrison (Editors), Fluidization, 2nd Ed., Academic Press, Orlando, 1985.

181. G. Johnson, M. Massoudi and K. R. Rajagopal, 'Flow of a Fluid-Solid Mixture Between Flat Plates', Chem. Engng. Sci., 1991, **46**, 1713–1723.

182. M. Massoudi and T. X. Phuoc, 'Flow and heat transfer due to natural convection in granular materials', Int. J. Non-Linear Mech., 1999, **34**, 347–359.

183. M. Massoudi, K. R. Rajagopal and T. X. Phuoc, 'On the fully developed flow of a dense particulate mixture in a pipe', Powder Tech., 1999, **104**, 258–268.

184. J. M. N. T. Gray, 'Granular flow in partially filled slowly rotating drums', J. Fluid Mech., 2001, **441**, 1–29.

185. Y. L. Ding, R. Forster, J. P. K. Seville and D. J. Parker, 'Granular motion in rotating drums: bed turnover time and slumping-rolling transition', Powder Tech., 2002, **124**, 18–27.

186. K. Craig, R. H. Buckholz and G. Domoto, 'The Effects of Shear Surface Boundaries on Stresses for the Rapid Shear of Dry Powders', ASME J. Tribology, 1987, **109**, 232.

187. S. J. Antony and M. Ghadiri, 'Size effects in a slowly sheared granular media', ASME J. Appl. Mech., 2001, **68**, 772–775.

188. R. Gudhe, K. R. Rajagopal and M. Massoudi, 'Heat transfer and flow of granular materials down an inclined plane', Acta Mechanica, 1994, **103**, 63–78.

189. W. L. Vargas and J. J. McCarthy, 'Heat conduction in granular materials', AIChE J., 2001, **47**, 1052–1059.

190. M. L. Hunt, 'Comparison of convective heat transfer in packed beds and granular flows', Annu. Rev. Heat Transfer, 1990, **3**, 163–193.

191. J. M. Ottino and D. V. Khakhar, 'Mixing and segregation of granular materials', Annu. Rev. Fluid Mech., 2000, **32**, 55–91.

192. C. R. Bembrose and J. Bridgewater, 'A review of attrition and attrition test methods', Powder Tech., 1987, **49**, 97–126.
193. M. Ghadiri, Z. Ning, S. J. Kenter and E. Puik, 'Attrition of granular solids in a shear cell', Chem. Eng. Sci., 2000, **55**, 5445–5456.
194. A. C. F. Cocks, 'Constitutive modeling of powder compaction and sintering', Prog. Materials Sci., 2001, **46**, 201–229.
195. T. Dyakowski and R. A. Williams, 'Prediction of high solids concentration regions within a hydrocyclone', Powder Tech., 1996, **87**, 43–47.
196. W. Polashenski and J. C. Chen, 'Measurement of particle phase stresses in fast fluidized beds', Ind. Eng. Chem. Res., 1999, **38**, 705–713.
197. R. A. Williams, S. P. Luke, K. L. Ostrowski and M. A. Bennett, 'Measurement of bulk particulates on belt conveyor using dielectric tomography', Chem. Eng. J., 2000, **77**, 57–63.
198. N. P. Kruyt and L. Rothenburg, 'Statistics of elastic behaviour of granular materials', Int. J. Solids Structures, 2001, **38**, 4879–4899.
199. M. R. Kuhn, 'Structured deformation in granular materials', Mech. of Mater., 1999, **31**, 407–429.
200. C. S. Chang and J. Gao, 'Kinematic and static hypotheses for constitutive modeling of granulates considering particle rotation', Acta Mechanica, 1996, **115**, 213–229.
201. C. S. Chang and L. Ma, 'A micromechanical-based microploar theory for deformation of granular solids', Int. J. Solids Structures, 1991, **28**, 67–86.
202. J. P. Bouchaud, M. E. Cates, J. R. Prakash and S. F. Edwards, 'A model for the dynamics of sandpile surfaces', J. Phys. I France, 1994, **4**, 1383–1410.
203. P. A. Cundall and O. D. L. Strack, 'A discrete model for granular assemblies', Geotech., 1979, **29**, 47–65.
204. C. Thornton and D. J. Barnes, 'Computer Simulated Deformation of Compact Granular Assemblies', Acta Mechanica, 1986, **64**, 45–61.
205. C. Thornton and S. J. Antony, 'Quasi-static deformation of particulate media', Phil. Trans. R. Soc. Lond., A, 1998, **356**, 2763–2782.
206. C. Couroyer, Z. Ning and M. Ghadiri, 'Distinct element analysis of bulk crushing: effect of particle properties and loading rate', Powder Tech., 2000, **109**, 241–254.
207. K. R. Rajagopal and A. R. Srinivasa, 'Modeling anisotropic fluids within the framework of bodies with multiple natural configurations', J. Non-Newtonian Fluid Mech., 2001, **99**, 109–124.
208. I. J. Rao and K. R. Rajagopal, 'Phenomenological modeling of polymer crystallization using the notion of multiple natural configurations', Interface Free Bound., 2000, **2**, 73–94.
209. K. R. Rajagopal, and A. R. Srinivasa, 'Mechanics of inelastic behavior of materials- Part I: Theoretical underpinnings', Int. J. Plasticity, 1998, **14**, 945–967.
210. K. R. Rajagopal and A. R. Srinivasa, 'Mechanics of inelastic behavior of materials-Part II: Inelastic response', Int. J. Plasticity, 1998, **14**, 969–995.
211. J. Murali Krishnan and K. R. Rajagopal, 'A thermodynamic framework for constitutive modeling of asphalt concrete: Theory and applications', in press ASCE J. Materials in Civil Eng., 2002.
212. D. Ambrosi and F. Mollica, 'On the mechanics of growing tumor', Int. J. Engng. Sci., 2002, **40**, 1297–1316.
213. K. R. Rajagopal, In preparation, 2002.

CHAPTER 4

High Temperature Particle Interactions

STEFAAN J. R. SIMONS and PAOLO PAGLIAI

Colloid and Surface Engineering Group, Department of Chemical Engineering, University College London, Torrington Place, London WC1E 7JE

1 Introduction

Interactions at elevated temperatures between solid particles occur in a wide range of industrial processes, for instance, in the filtering of hot gases, in the drying of pharmaceutical granules, in the curing of ceramics and in the combustion of solid fuels. Often these interactions can cause major problems in the operation of such processes. The photograph in Figure 1 is of a bed of coal-ash powder that has been solidified during gasification in a fluidised bed due to sintering between the particles at high temperature. In this example, inadequate design of the gasifier led to regions where particle motion was poor and, consequently, hot spots could develop. The lack of vigorous motion in these areas allowed time for permanent bonds between the particles to form through sintering. Obviously, the gasifier became inoperable once the powder bed had formed this solid clinker, with the only option being to shut down the process, dismantle the vessel and remove the obstruction.

It is obvious that for the reliable operation of high-temperature processes a good understanding of the fundamental mechanisms of adhesion and cohesion between particles at elevated temperatures is required. Here, adhesion is meant as the force that holds the particles together, after which they exhibit cohesive behaviour. Unfortunately, the level of understanding has been hampered by the lack of techniques available to observe and measure such interactions. However, recent developments in microscopic analysis techniques now mean that high temperature particle interactions can be studied directly, which will lead to the development of new predictive models.

This chapter will introduce the concept of the simulation of high temperature adhesion forces using liquid bridges and then will discuss the current understanding of sintering phenomena. Finally, new and novel off-line

Figure 1   *A solidified bed of coal-ash powder caused by sintering in a gasifier*

techniques will be described for the determination of actual high temperature interactions.

2   Enhanced Adhesion Forces

According to Berbner and Löffler,[1] adhesion forces between particles can be characterised by whether or not a material bridge exists between them. Those forces that develop without material bridges arise from either van der Waals or electrostatic effects or a combination of the two. However, these forces are much lower in magnitude than forces provided by material bridges that join the particles together. These bridges can form either by the addition of a liquid binder to the particle system, which usually then dries to form solid bridges, or can result from structural or chemical changes at the particle surfaces, e.g. through sintering, crystallisation or plastic deformation (Figure 2). The enhanced adhesion forces that these bridges can develop between particles mean that quite large particles, typically in the mm size range, can form strong agglomerates, which can be either advantageous, as in the pelletisation of iron ores, or can cause severe problems, such as with the defluidisation example shown in Figure 1.

High temperature adhesion forces arise from the formation of material bridges, usually through the particle surfaces changing phase due to either chemical reaction or simply melting. Thin layers of sticky material then develop which, on collisions between particles, form a bond between them. The theory of sintering will be discussed in more detail later. However, it should be noted

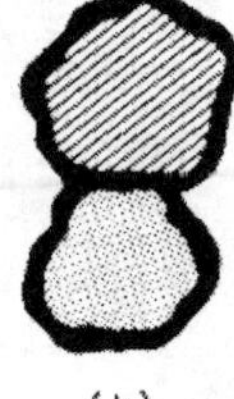
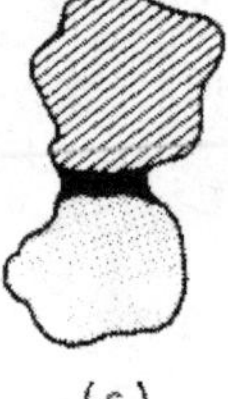

Figure 2 *Different binding mechanisms: a) solid bridge and particles are composed of the same material; b) particles and bridge have different composition; c) agglomeration of particles is characterized by the formation of binder bridges*[2]

that often researchers have attempted to simulate high temperature interactions through the addition of liquids to the particle system being studied at room temperature. Hence, a description will be given here of the adhesion forces created by liquid bridges.

2.1 Liquid Bridge Adhesion Forces

A schematic of a liquid bridge is shown in Figure 3 between two spherical particles. The pressure in the liquid bridge can be obtained from the Young–Laplace relationship in Equation 1, which relates ΔP_{br}, the pressure difference between the liquid and the atmosphere to the curvature of the liquid–vapour interface:[3,4]

$$\Delta P_{br} = \gamma_{LV} \left(\frac{1}{r_1} - \frac{1}{r_2} \right) \tag{1}$$

where γ_{LV} is the surface tension at the vapour–liquid interface. The radii of curvature r_1 and r_2 can be either positive or negative depending on the geometry of the bridge (in Figure 3, $r_1 > 0$ and $r_2 < 0$). When the Bond number $gL^2\Delta\rho/\gamma_{LV}$ (where g is the gravity acceleration, L some characteristic length of the bridge, $\Delta\rho$ the difference of densities between the two liquids) is sufficiently small, the effect of gravity is negligible and the mean curvature is nearly uniform.

In a Cartesian reference, Equation 1 can be rewritten in terms of the analytical expressions of the radii of curvature r_1 and r_2, obtaining Equation 2:[5]

$$\frac{\Delta P}{\gamma_{LV}} = \frac{1}{y(1 + \dot{y}^2)^{0.5}} - \frac{\ddot{y}}{(1 + \dot{y}^2)^{1.5}} \tag{2}$$

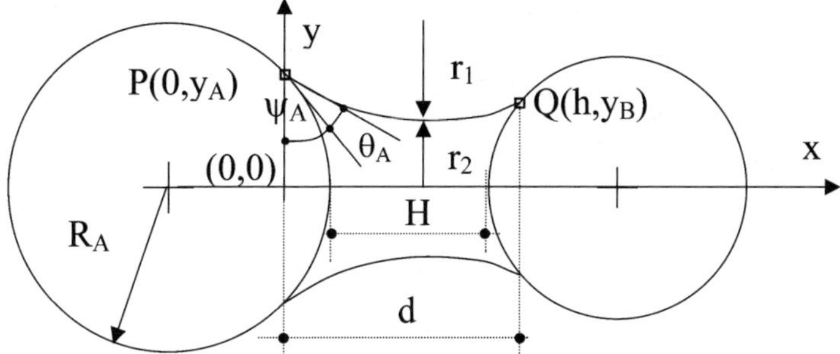

Figure 3   *Geometric parameters describing a liquid bridge*

In Equation 2, x and y are the axial and radial co-ordinates, respectively, of the liquid-vapour profile and $\dot{y}$ and $\ddot{y}$ are the first and second derivatives of y with respect to x. This equation states that the liquid bridge surface must have a constant curvature at all points. Its numerical solution is quite complex and, hence, approximations of the bridge profile are often used. The most common is the toroidal approximation. Recently, Pepin et al.[6] have developed a parabolic approximation which is much easier to use and provides predictions of bridge profiles, rupture distances and post-rupture liquid distributions with reasonable accuracy.

2.1.1   *The Toroidal Approximation*

A schematic of the toroidal approximation for a concave geometry is shown in Figure 4. Two unequally sized spheres of radius R_A and R_B, are separated by a distance H. The liquid bridge has a constant radius of curvature of r_1 in the plane of the page and has a radius of r_2 perpendicular to the page at its

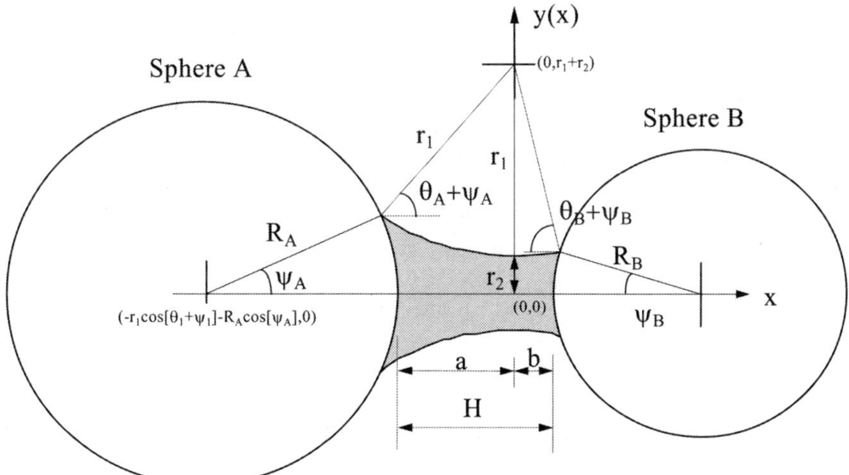

Figure 4   *Concave torroidal geometry of a pendular liquid bridge[6]*

narrowest point. The x-axis is the axis of symmetry and the origin is taken as the point where the bridge is at its narrowest. The liquid bridge contacts each sphere at a filling angle of ψ_A and ψ_B respectively, and has an apparent contact angle of θ_A and θ_B on each sphere.

Equating the vertical (y) components of the bridge geometry gives Equations 3 and 4:

$$R_A \sin \psi_A + r_1 \sin(\psi_A + \theta_A) = r_1 + r_2 \tag{3}$$

$$R_B \sin \psi_B + r_1 \sin(\psi_B + \theta_B) = r_1 + r_2 \tag{4}$$

Likewise, equating the horizontal (x) components gives Equation 5:

$$H = a + b = R_A(\cos \psi_A - 1) + r_1 \cos(\psi_A + \theta_A) + R_B(\cos \psi_B - 1) \\ + r_1 \cos(\psi_B + \theta_B) \tag{5}$$

where H is the interparticle distance. The equation for the upper toroidal bridge profile is given by (6) for a concave bridge which occurs when $(\psi_A + \psi_B + \theta_A + \theta_B) < 2\pi$, and by (7) for a convex bridge which occurs when $(\psi_A + \psi_B + \theta_A + \theta_B) > 2\pi$.†

$$y = r_2 + r_1 - \sqrt{r_1^2 - x^2} \tag{6}$$

$$y = r_2 - r_1 + \sqrt{r_1^2 - x^2} \tag{7}$$

A convex shape occurs when the two spheres are close together and/or when there is a relatively large volume of liquid. A schematic of the toroidal approximation with a convex geometry is given in Figure 5.

2.1.2   *The Parabolic Approximation*

Figure 6 is a schematic of the parabolic bridge profile approximation. The solid-liquid interface is a spherical cap, which has a maximum height of h_i. d is the length of the liquid bridge and y_{min} is the minimum liquid neck radius of the pendular bridge. The x-axis is defined as the symmetry axis of the system and $y(x)$ is a Cartesian function outlining the upper half of the liquid bridge profile. The liquid-to-solid contact points are P and Q on the two spheres, with co-ordinates of respectively $(0,y_A)$ and (d,y_B). h_i and y_i are related by Equation 8:

$$y_i = \sqrt{R_i^2 - (R_i - h_i)^2} \tag{8}$$

$y(x)$ is a second order polynomial equation in the form of Equation 9:

$$y(x) = \alpha x^2 + \beta x + \delta \tag{9}$$

Note that by varying α, β and δ, this single equation can be used to approximate

---

† It is also possible to have a conically shaped bridge (when $\psi_A + \psi_B + \theta_A + \theta_B = 2\pi$) or a cylindrical bridge (when $\psi_A + \theta_A = \psi_B + \theta_B = \pi$).

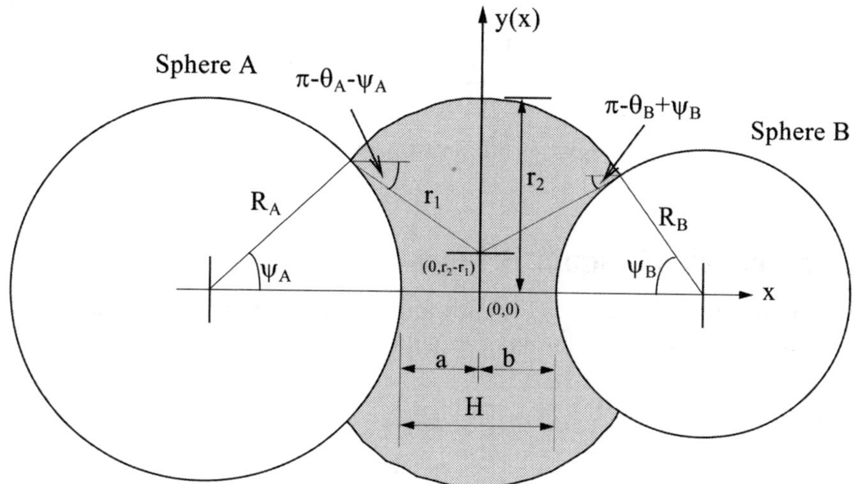

Figure 5 *Convex torroidal geometry of a pendular liquid bridge[6]*

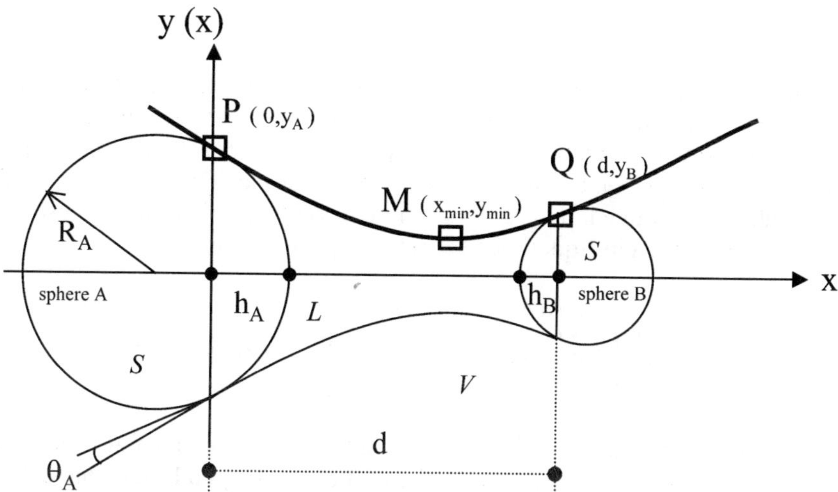

Figure 6 *Parabolic bridge profile approximation[6]*

both convex and concave bridges. In contrast the toroidal approximation involves two sets of equations, and there is also no smooth transition between the two since $r_1 \rightarrow \infty$ as the bridge approaches a conical or cylindrical shape. Hence the parabolic profile assumption is more robust and easier to use. It should also be noted that both profile approximations are three-parameter curves. The toroidal profile is defined by r_1, r_2 and H and the parabolic profile is defined by the three parameters α, β and δ.

The apparent solid–liquid contact angles can be calculated from the other parameters by Equations 10 and 11:

$$\theta_A = \frac{\pi}{2} + \tan^{-1}(y'(0)) - \sin^{-1}\left(\frac{y_A}{R_A}\right) \tag{10}$$

$$\theta_B = \frac{\pi}{2} - \tan^{-1}(y'(d)) - \sin^{-1}\left(\frac{y_B}{R_B}\right) \tag{11}$$

3   Strength of Liquid Bridges

If the bridge liquid volume V_{liq} is known, the variables α, β and δ in Equation 9 are obtained from the resolution of Equation 12:

$$\begin{cases} y(0) = y_A \\ \pi \int_{x=0}^{x=d} y^2(x)\cdot dx = V_{liq} + \sum_{i=A,B} \frac{\pi h_i}{6}(3y_i^2 + h_i^2) \\ y(d) = y_B \end{cases} \tag{12}$$

Equation 12 assumes that the pendular liquid bridges considered are axially symmetric bodies, the volumes of which can be obtained by revolution of $y(x)$ around the bridge symmetry axis. If V_{liq}, y_A and y_B are constant during separation, the evolution of the liquid bridge with interparticle distance is easily predicted. If the bridge liquid recedes from the solid particle i surface, a simple energy balance given in Equation 13 permits the calculation of y_i:[7]

$$2\pi\gamma_{LV} \times \left[R_i h_i(\cos\theta_r + 1) + \int_{x=0}^{x=d} y(x)\sqrt{1 + y'^2(x)}\cdot dx \right] = \Psi \tag{13}$$

In Equation 13, Ψ is a constant which depends on the experimental conditions and can be calculated from the knowledge of the initial liquid bridge shape. R_i is the particle i radius and θ_r is the receding contact angle of the bridge liquid on the particle i. Whether y_A and y_B are fixed or varying, apparent liquid-to-solid contact angles are calculated from Equation 10 and 11.

In Reference 6, it was shown that the rupture of liquid bridges would occur spontaneously at a certain interparticle distance. Two cases exist: one where the volume of liquid is fixed, that is when the three-phase contact line is observable on the particle surfaces during elongation, the second where there is no observable three-phase contact line and liquid covers the entire particle surfaces and the surrounding objects, creating liquid reservoirs outside the particles. Depending on the bridge elongation speed and viscosity of liquid binder, some liquid can move from the liquid reservoirs to the inter-particle gap and the liquid volume of the bridge can increase. In the former case, the shape the bridge

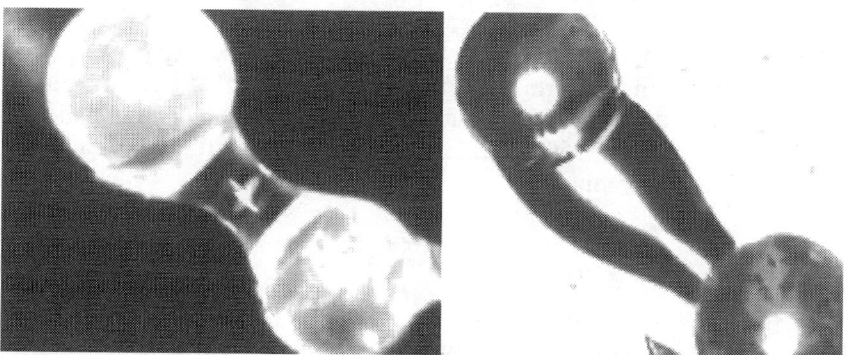

Figure 7   *Glycerol liquid bridges showing three-phase contact lines with glass spheres; (a) two non-silanised spheres (left, 126 µm; right, 111µm), (b) left, unsilanised (118 µm), right, silanised (125 µm)*

adopts depends on the wetting behaviour and the value of the three-phase contact angles.[7] In Figure 7, two examples are shown. In Figure 7(a), the liquid does not de-wet the particle surfaces, the three-phase contact line is fixed. The shape then adjusts as the contact angles change with separation (in order for the constant volume condition to be met). In Figure 7(b), the liquid de-wets the silanised particle and the contact angle on this particle remains relatively constant during separation, whereas the contact angle on the unsilanised particle, which has a fixed three-phase contact line, does change. In the case where there is no observable three-phase contact line, the liquid perfectly wets the solid and the bridge shape remains torroidal throughout separation.

The separation distance where rupture occurs can be predicted using the parabolic approximation, assuming that this occurs through the liquid's thinnest neck, y_{min} (Figure 6). For perfectly wetting liquids, liquid bridge volumes can be related to y_{min} by an empirical relationship:[8]

$$V_{liq} \approx 1.673 \times y_{min}^3 \qquad (14)$$

The capillary force F_{cap}, developed by one liquid bridge is then given by Equation 15:

$$F_{cap} = 2\pi y_{min}\, \gamma_{LV} \qquad (15)$$

For pendular bridges, a second contribution to the axial force of adhesion between particles A and B (Figure 6) must be taken into account, that arising from the reduced hydrostatic pressure in the bridge due to the curvature of the vapour–liquid interface:

$$F_{hs} = \Delta P_{br}\pi y_{min}^2 \qquad (16)$$

In a process vessel in which liquid bridges exist between particles, the liquid bridges will experience non-axial elongation at varying particle separation speeds. Hence, there will be friction and viscous contributions to the separation energy. Mazzone et al.[7] and Ennis et al.[9] described the viscous force of a pendular liquid bridge in the form of Equation 17 with v_i, the particle separation velocity, and η, the dynamic viscosity of the bridge liquid. This equation was developed for spheres of identical size. For dissimilar particles, the arithmetic average radius $\overline{R}$, is proposed:

$$F_{vis} = \frac{3\pi\eta v_i \overline{R}^2}{2H} \tag{17}$$

Hence, the total adhesion force between two spherical particles attached by a liquid bridge undergoing dynamic separation is given by:

$$F_N = 2\pi\pi_{min}\,\gamma_{LV} - \Delta P_{br}\pi y_{min}^2 + \frac{3\pi\pi\eta_i \overline{R}^2}{2H} \tag{18}$$

Pepin et al. also introduced a contact force, developed from the expression for the work of adhesion of the liquid on the solid:[8]

$$F_{fric} = \gamma_{LV}\,(\cos\theta + 1) \times \frac{A_{ab}}{H} \tag{19}$$

where A_{ab} is the interparticle contact area and θ is the apparent contact angle between the liquid and the solid surfaces (for two spheres with the same surface energies).

The force curve for the situation where the liquid wets the particles with a zero contact angle, thereby adopting a toroidal profile throughout separation, is shown in Figure 8. There is close agreement between the measured forces (measured using a micro-force balance developed at University College London and similar to the one described in Section 5.2) and those calculated using Equation 18.

3.1 Rupture Energies: Defluidisation Due to the Presence of Liquid Layers

The rupture energy, W, of a pendular liquid bridge between spherical particles in air has been modelled by Simons et al.[10] and Pitois et al.[11] In the former, the bridge geometry was modelled with a toroidal curve and, for the case of perfect wetting conditions, the integral of the quasi-static capillary force was calculated with respect to the separation distance. From the approximated solution of the non-trivial integral a simple expression was derived in terms of dimensionless parameters:

$$\tilde{W} = \frac{W}{\gamma_{LV}R^2} \approx 3.6\tilde{V}^{0.5} \tag{20}$$

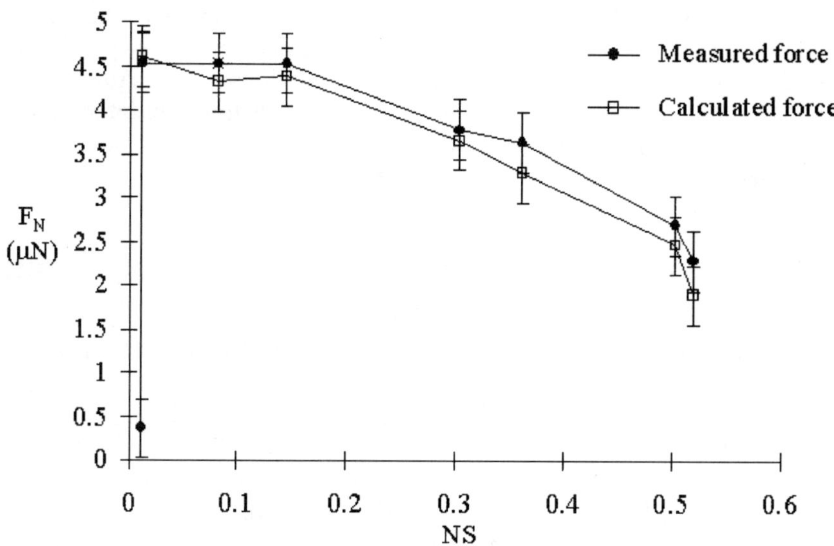

Figure 8 *Comparison of measured and calculated bridge strength for two glass spheres (69 and 56 μm diam.) versus normalised separation. Separation velocity = 29.3 μl/s[8]*

In Equation 20, R is the radius of the equi-sized spheres, V is the volume of liquid in the bridge, whilst $\tilde{V} = V/R^3$ and $\tilde{W}$ are the dimensionless volume and rupture energy of the bridge, respectively.

Pitois et al.[11] used a circular approximation for the bridge profile and obtained a simplified expression of the capillary force which, when integrated throughout separation distance, led to the expression of the dimensionless rupture energy without restrictions to the wettability exhibited by the particles:

$$\tilde{W} = 2\pi \cos\theta \left[(1 + \theta/2)(1 - A)\tilde{V}^{1/3} + \sqrt{\frac{2\tilde{V}}{\pi}} \right] \qquad (21)$$

where $A = \sqrt{\dfrac{(1 + 2\tilde{V})^{1/3}}{\pi(1 + \theta/2)^2}}$, θ is the apparent contact angle expressed in radians, whilst $\tilde{W}$ and $\tilde{V}$ are defined as above.

Simons et al.[12] then went on to compare the rupture energy calculated using Equation 20 with the kinetic energy of particles in a fluidised bed of glass ballotini doped with silicone oil. The liquid was added to simulate, at room temperatures, the increase in interparticle forces that occur at high temperatures due to sintering (see below). The kinetic energy of the powder bed was expressed as:

$$\frac{m(U^* - U_{mf})^2}{\gamma_{LV}R^2} \qquad (22)$$

where m is the particle mass, U* is the superficial gas velocity at the point of defluidisation and U_{mf} is the minimum fluidising velocity of the dry powder bed. Hence, ΔU is the excess gas velocity required to maintain fluidisation. By plotting (22) versus $\tilde{V}^{0.5}$, Simons et al.[12] found a linear relationship, indicating the importance of the liquid bridges at the particle contact points in the defluidisation process.

4   Solid Bridges: Particle Sintering at High Temperatures

Solid bridges can arise when particles come into contact at temperatures high enough to cause the surfaces to melt (approximately 60% of the absolute melting point[1]), resulting in mutual molecular diffusion at the points of contact to form sintered bridges. The initial sintering temperature is strongly dependent on particle size. However, the sintering kinetics are determined by the lowering of the free surface energy and the viscous dissipation energy. The fact that this dependence is based on chemical composition (not only of the particles, but the surrounding gases as well) and geometry makes full physical modelling of sintering processes extremely difficult. Hence, most approaches to date have been semi-empirical in nature. The important parameter of interest is the sintered neck diameter or the time taken for that neck to form.

4.1   An Example of Catastrophic Sintering – Defluidisation

As described in the introduction, poorly designed fluidised beds or insufficient fluidising velocity can lead to the presence of local non-fluidised ("dead") zones of particles that can cause hot spots and therefore particle sintering and agglomerate formation. This then leads to a rapid expansion of the dead zones and eventually the bed becomes completely defluidised.[13]

Studies by Siegell,[14] Gluckman et al.,[15] Liss[16] and Tardos et al.[17] have shown that defluidisation due to thermally induced surface cohesiveness depends on the particle physical and chemical surface properties, the heating rate of the fluidised material and the hydrodynamics of the gas and solids in the fluidised system. In particular, Compo et al.[18] and Tardos et al.[19] investigated the behaviour of different fluidised materials, relating the minimum fluidisation velocity to temperatures exceeding the initial sintering temperature T_s. They plotted operating diagrams such as Figure 9, defining regimes of fluidisation and defluidisation in the gas velocity/temperature plane, for different materials. In all cases, it was observed that the temperature at which the measured defluidisation velocity departed from the predicted behaviour in the absence of interparticle forces coincided with T_s, as measured using a dilatometer (see Section 5.2). They concluded that at such temperatures fluidisation is no longer determined by a balance of gravity, buoyancy and drag forces, but that interparticle cohesiveness and particle kinetic energy must also be taken into account.

Langston and Stephens[20] investigated the qualitative effects of temperature, fluid velocity and particle size on the fluidised bed behaviour and indicated the

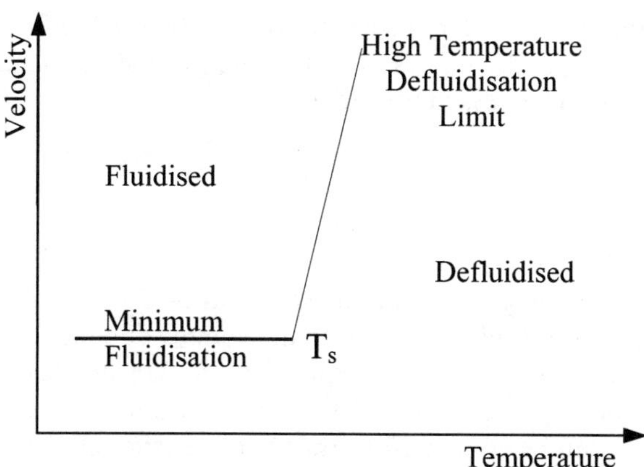

Figure 9 *High temperature regimes in a gas–solid fluidised bed*

main parameters responsible for the defluidisation phenomenon. According to the authors, the tendency to agglomerate is a function of the area of contact, particle momentum and the adhesive property of the particles, the latter being assumed to increase with excess temperature above the sintering point.

The crucial point in investigating and modelling sintering in high temperature gas fluidisation is to elucidate the mechanisms of defluidisation involved. Recent studies by Skrivfars et al.[21,22] show two main categories: in the first mechanism the stickiness of the colliding particles is caused by the flow of material between particles, forming a neck of sintered material. The type of bond is of visco-plastic nature and usually is limited by the ability of the material to flow. The strength of the neck is proportional to the temperature and the interparticle contact time in that the higher the first and the longer the second, the greater the size of the neck itself.

In the second mechanism, defluidisation is induced by the presence of a liquid phase formed by a phase change and/or chemical reaction. The viscosity of the liquid is usually low so that, in order to prevent defluidisation, it is necessary to define the relevant critical volume as a limiting condition under which the process operates.

In industrial practice it has been found that defluidisation processes can occur at very different time scales. In order to be able to intervene in the fluidisation at the right moment, it is crucial that one can detect at an early stage if the quality of the gas–solid fluidised state is changing. Modelling the sintering problem in fluidised beds has been attempted by several theoretical approaches in an effort to provide a relationship between the physical properties of the material, its sintering behaviour and the hydrodynamics of the fluidised bed.

Seville et al.[23] proposed a model based on characteristic residence times, thus avoiding the theoretical considerations on the strength and growth behaviour of sintered agglomerates, by relating the dependence of the excess fluidising velocity on temperature. During fluidisation particles are considered to remain in quiescent zones with relatively little movement until they are disturbed by the passage of bubbles. According to the 'two-phase' theory of fluidisation, Equation 23 relates the measure of the residence time in the quiescent zones to the excess fluidising velocity:

$$t_{bo} = \frac{H_{mf}}{\alpha(U - U_{mf})} \tag{23}$$

where H_{mf} is the bed height at minimum fluidisation condition, α is the ratio of the volume of solids moved by bubbles to the volume of bubbles and $(U - U_{mf})$ is the excess gas velocity (U here is equivalent to U* in (22)). The equation describes the time required to turn the contents of the bed over once, which, if sufficiently long, can be compared to the characteristic time for the sinter neck to reach a critical size such that the agglomerates cannot be disrupted by the bubble motion. By assuming that the size of a given critical neck is not dependent on the fluidising gas velocity, the following expression was derived for the time, t_c, for a sintered neck to form:

$$t_c = \left(\frac{x_c}{r}\right)^2 \frac{\eta}{k_1} \tag{24}$$

where x_c is the neck diameter, r is the particle radius, k_1 is a coefficient depending on both material properties and environmental conditions and η is the surface viscosity whose dependence on temperature is provided by the Arrhenius law:

$$\eta = \eta_0 \exp(E/RT) \tag{25}$$

By equating Equation 23 with Equations 24 and 25 the authors obtained a relationship for the temperature dependence of the minimum fluidising velocity under sintering conditions:

$$\ln \frac{(U - U_{mf})}{H_{mf}} = \ln(K_2/\alpha\eta_0) - \frac{E}{RT} \tag{26}$$

where H_{mf} is the height of the powder bed at the point of minimum fluidization. K_2 characterises the effect of the critical size of the sinter neck, as it is equal to

$$\left[\left(\frac{x_c}{r}\right)^2 1/k_1\right]^{-1}.$$

As both the surface viscosity and activation energy can be measured by means of dilatometry (see Section 5.2), Equation 26 provides predictive capability by showing an exponential trend of the minimum fluidising velocity, required to prevent de-fluidisation, with temperature.

Seville et al.[23] tested Equation 26 by fluidizing polyethylene at temperatures between 100 and 120 °C, where sintering occurs. Figure 10a is a photograph of the sintered particles, whilst Figure 10b shows the increase in minimum fluidizing velocity required as temperature increases.

Figure 11 shows the calculation of E/R from the slope of the log-log plot of Equation 5. Thus, the sintering effect can be fully predicted.

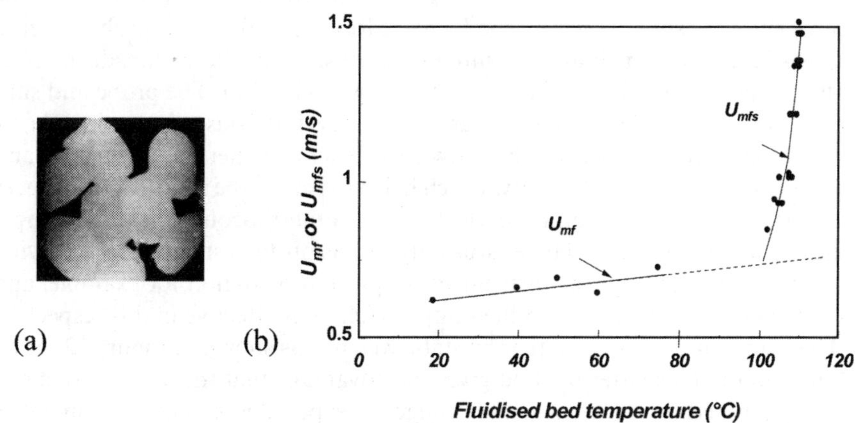

Figure 10   *(a) Photograph of sintered polyethylene particles, (b) effect of sintering on the minimum fluidisation velocity*[23]

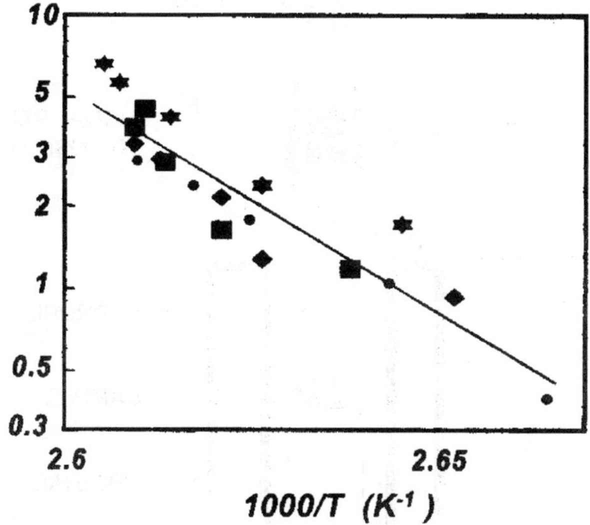

Figure 11   *Data plotted to determine the value of E/R by including the effect of bed height*[23]

5   Off-line Analysis of High Temperature Interactions

5.1   Thermo-Mechanical Analysis (TMA): Dilatometry

In this technique, the dimensional change undergone by a sample whilst being either heated, cooled, or studied at a fixed temperature, is the primary measurement. Figure 12 shows a schematic diagram of a typical TMA instrument. The sample sits on a support within the furnace. Resting upon it is a probe to sense changes in length, which are measured by a sensitive position transducer, normally a Linear Variable Displacement Transducer (LVDT). The probe and support are made from a material such as quartz glass (vitreous silica), which has a low, reproducible and accurately known coefficient of thermal expansion and also has low thermal conductivity, which helps to isolate the sensitive transducer from the changing temperatures in the furnace. A thermocouple near the sample indicates its temperature. There is usually provision for establishing a flowing gas atmosphere through the instrument, to prevent oxidation for example, and also to assist in heat transfer to the sample. Helium is effective in this respect.

The load may be applied either by static weights, as shown in Figure 12, or by a force motor. This latter method gives the advantage that the applied load can be programmed to allow a greater range of experiments. The instrument is calibrated for position measurements by heating a sample whose expansion coefficient is accurately known. When the sample carries a zero, or negligible

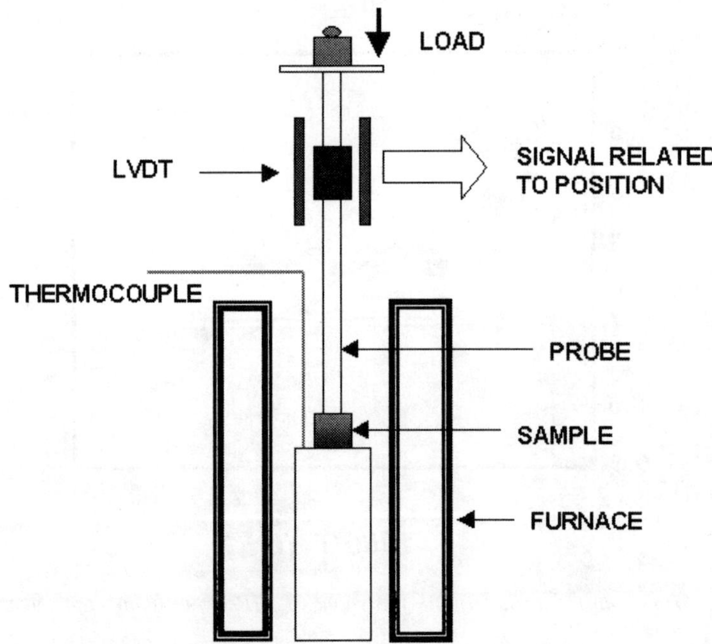

Figure 12   *Schematic of a dilatometer*
            (Adapted from Reference 24)

load, the measurement follows the free expansion or contraction of the material and accurate coefficients of thermal expansion can be routinely determined.

The materials studied are usually rigid or nearly rigid solids, as implied by most of the experimental arrangements shown in Figure 12. Liquids can be studied in a specially designed accessory, which can also be used to measure the volume changes of irregularly shaped samples or powders that are then submerged in an inert liquid such as silicone oil. Powders can also be studied as a layer with a loosely fitting lid on top, or when pressed into a pellet. Solid-state transitions can be followed in this way.

Lettieri et al.[25,26] employed the dilatometry technique in order to study the linear changes in the dimensions of a particular material, the E-Cat equilibrium catalyst, as a function of temperature. The E-cat sample analysed was heated up to 1000 °C at a rate of 10 °C per minute and was subject to a static force of 0.2 N. The thermogram obtained by dilatometry, Figure13, shows an initial expansion of the sample of E-cat up to 134 °C, after which a sharp decrease in size occurs up to about 200 °C. An even more important dimensional change occurred during the small temperature range between 414 °C and 429 °C where a relative size decrease of 11% was quantified.

Further results obtained by thermo-gravimetric analysis were used together with the dilatometry investigation in order to give an interpretation of the flow behaviour of this powder. Lettien et al. stated, as a possible explanation, that heavy hydrocarbons, which are solid at ambient conditions, might have melted with increasing temperature causing liquid to be expelled from within the pores of the catalyst and the particles to then agglomerate and defluidise. Other workers[18,19] have shown a good relation between the measurements obtained by dilatometry and the temperatures at which defluidisation occurs in fluidised beds of various particles, in particular in cases where sintering between particles takes place. However, what cannot be verified by TMA alone is the fundamental cause of the increase in interactions. This can only be done by direct observation of

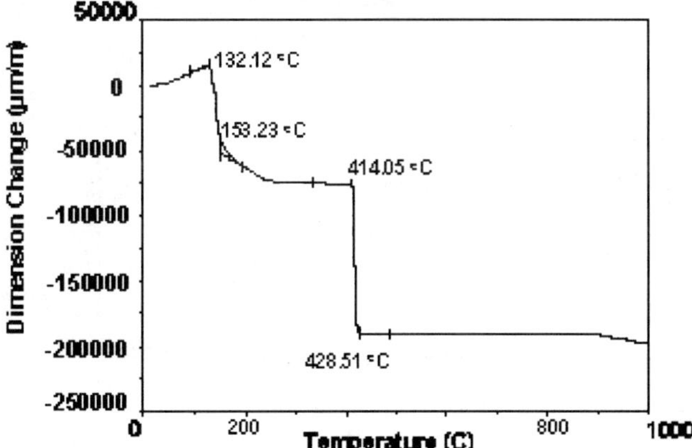

Figure 13   *Thermogram of E-Cat*[25]

the particle behaviour at high temperatures. In the following section, a new and unique device will be described that can be used not only to measure dimensional changes but to elucidate the governing mechanisms through direct observation and measurement of the particle interactions at high temperatures. For instance, the device has successfully elucidated the mechanisms governing the behaviour observed by Lettieri et al.,[25,26] which were found to be in contradiction to those postulated.

5.2   The High Temperature Micro Force Balance

A novel approach has been taken here, namely, the direct measurement and observations of particle interactions at high temperatures on a micro scale. To this end, a unique instrument has been developed, termed a High Temperature Micro-Force Balance (HTMFB), which allows for the main parameters that play a fundamental role in enhancing particulate adhesion and cohesion to be investigated.

The design of the instrument is a further improvement of the apparatus previously employed by Simons and Fairbrother.[27] The HTMFB, a schematic of which is shown in Figure 14, can be set up to measure forces of different magnitude at high temperatures (up to 1000 °C): liquid or solid bonds can be formed of different volumes and particles of different sizes and porosity can be employed.

The device works in the following manner. Particles, of diameters as low as 3 μm, are attached on the tips of two micropipettes and then aligned by the use of two micromanipulators under the focus of the microscope lens. A liquid bridge

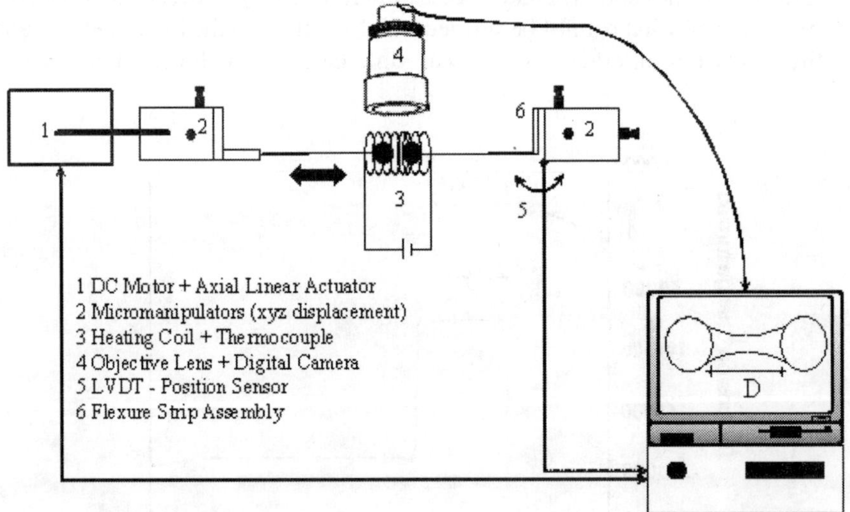

Figure 14   *Schematic of the HTMFB: 1) DC Motor + Axial Linear Actuator; 2) Micromanipulators; 3) Heating Element + Thermocouple; 4) Objective + Digital Camera; 5) LVDT displacement sensor; 6) Flexure Strip Assembly*

can be formed in two steps: a third micropipette, positioned using a third micromanipulator (not shown in the figure), is used to add liquid onto one of the particles. The particles are then positioned within the coils of a heating element and brought together using the micromanipulators until the liquid forms a bridge between them. If particle sintering is being studied, then the particles are simply positioned directly within the heating coil. The heating element is previously calibrated with a thermocouple in order to reach temperatures up to 1000 °C. Of the two micromanipulators, one is static whilst the other causes the particles to separate via a linear actuator driven by a DC motor. The linear actuator can be programmed via a software interface, either to ensure high resolution of displacement (up to 0.1 µm) or to perform cycles at different velocities. A digital camera, plugged into a personal computer (PC) and fitted in the microscope objective, grabs image sequences for each run of experiments for later analysis of, for example, the surface effects.

The separation of the particles causes a flexure strip mounted on the static micromanipulator to bend, under loads as low as 10 µN. The LVDT (Linear Variable Differential Transformer) displacement sensor provides the position of the strip with a resolution of 20 nm, and data are collected by a data logger in the PC. For each step of the actuator a defined set of electrical signals is logged and converted into displacement units so that the relevant force can be calculated by:

$$F_b = kX_b \qquad (27)$$

where k is the spring constant and X_b the displacement of the bending strip.

Pagliai et al.[28] carried out an experimental study using this device to investigate the physical changes of the particle surface with increasing temperature and to evaluate the magnitude of the force of a drying liquid bridge formed between a pair of E-Cat particles. The results were then compared to the TMA data provided by Lettieri et al.[25] and are described in the following section.

5.2.1 Direct Observation using the HTMFB

E-cat particles of diameters ranging between 60 and 500 microns were investigated using the following procedure. For each run a single particle was attached to the tip of a glass pipette and aligned inside the heating element under the lens of the microscope by means of the micromanipulator. A type K thermocouple was positioned in front of the particle's surface, its position adjusted via the second manipulator and the focus of the microscope's objective. Temperature was then increased from room values up to 200 °C, in steps of 10 °C/min. The digital camera provided images of the particle's surface at each experimental temperature, monitoring the physical changes that occurred.

Image analysis of E-cat particles with diameters up to 190 µm revealed that no expansion or shrinkage of the size took place in the range of temperatures investigated. In Figure 15 the sequence of two images of a 100 µm particle shows

 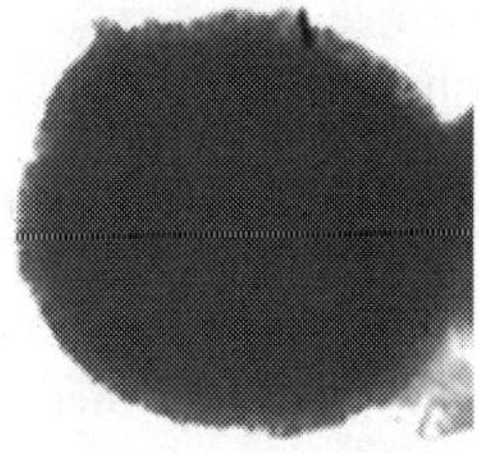

Figure 15   *100 μm E-cat particle at room temperature (left) and 200 °C (right)*

that even the particle's surface did not experience any physical macroscopic change.

Particles above 190 μm size showed behaviour more similar to that explained by Lettieri et al.;[25] at temperatures ranging between 90 °C and 130 °C, a small increase in size was observed. A typical observation is shown in Figure 16, where the sequence of images for a 250 μm E-cat particle shows different stages as temperature was varied from room temperature up to 130 °C.

It appears that low boiling compounds start to melt at 90 °C, thus forming a thin liquid layer around the particle's surface. A slight increase of the size is caused by the liquid, which flows and fills the gaps present on the irregular profile of the particle. As temperature is increased to 180 °C, the whole particle is found to be in a semi-solid state, comparable to a highly viscous bubble of liquid. Because of the evaporation of low boiling compounds, shrinkage of the drop takes place and at 200 °C some solid material is eventually found on the tip of the pipette.

In conclusion, Pagliai et al. stated that the different behaviour experienced according to particle diameter was due to the different nature of the particles themselves. According to Lettieri et al.,[25] the E-Cat provided by BP Chemicals had been employed in several cracking and regeneration cycles; moreover, no pre-treatments were performed on the material prior to the fluidisation experiments. Therefore, those particles that appeared to be of large diameter turned out to be semi-solidified agglomerates composed of fines and hydrocarbons, which underwent a phase change to become liquid droplets as the temperature reached a critical value.

 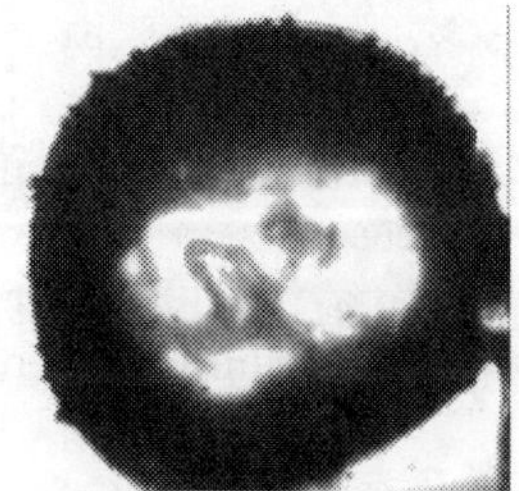 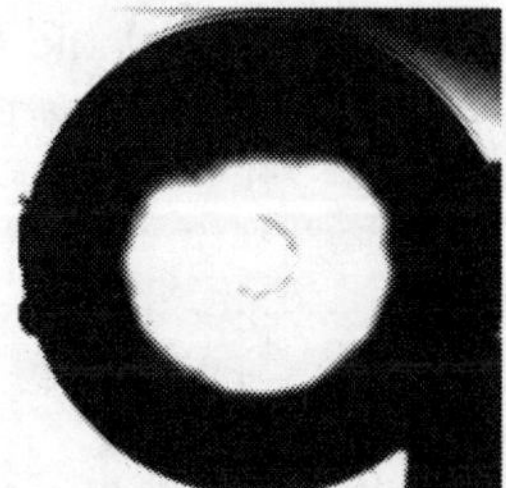

Figure 16   *250 μm E-cat particle at room temperature (left), 90 °C (centre) and 130 °C (right)*

5.2.2   Force Measurement

In the experiments described in the previous section, a liquid bridge was formed between a 250 μm E-cat particle and a 100 μm diameter particle of the same material (see Figure17). Temperature was then increased up to 100, 150 and 200 °C, respectively.

During drying the pulling force exerted by the bridge, F_b, was evaluated using Equation 27 where the measured displacement was caused by the shrinking of the bridge as it dried. Physical changes, such as the initial and final separation distance, were monitored by means of image analysis.

Results obtained from the force measurements of the drying liquid bridge are shown in Figure 18 where the force has been made dimensionless by dividing it by the maximum force that would be obtained on complete shrinkage. During drying, two main forces enhance the strength of the bond according to the operative temperature: the surface tension and the resultant force caused by the formation of a solid bridge. The solidification process is considered terminated when a steady value was reached within 90–100% of the maximum force.

The surface tension, which is the driving force at 100 °C, can be considered negligible as compared to the force arising during the solidification process. The

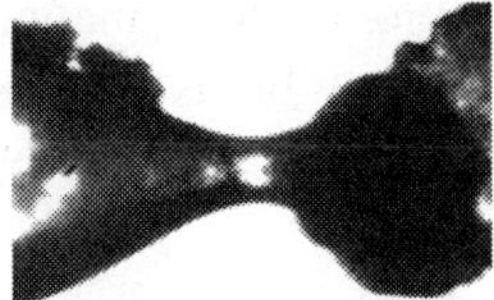

Figure 17   *Typical experimental set up during measurement of the pulling force exerted by a drying liquid bridge formed between a 250 μm (left) and 100 μm E-Cat particle*

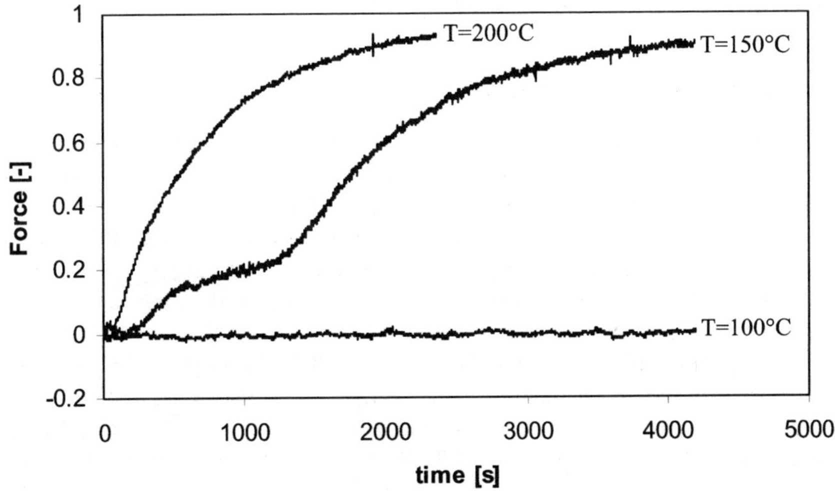

Figure 18   *Summary plot of the experimental drying curves*

analysis at 100 °C revealed that the solidification process is strongly dependant on both the volume of liquid and the temperature. The evaporation rate of the liquid at such a temperature is fairly low, say very close to the equilibrium, and so it can be taken as the rate of consolidation of the solid bridge.

Temperature is the driving mechanism in the solidification process, since by increasing this parameter the evaporation rate and therefore the solidification rate rise up to the maximum limit in a shorter time. Surface tension is quickly overcome at 200 °C, thus permitting the solidifying force to take over almost instantaneously.

In conclusion, new diagrams have been plotted where the strength of the drying liquid bridge and the formation of a solid bond can be read as a function of temperature and time. Some considerations on the relevant behaviour of the fluidised bed studied by Lettieri et al.[25,26] have been drawn as follow:

- At 100 °C the fluidised bed is a mixture of solids and liquid: the time taken to form the solid bonds is extremely long and temperature and liquid volume sensitive. This explains the sluggish flow behaviour as well as the formation of channels and rat holes within the bed.
- By increasing the temperature from 100 to 200 °C, the rate of solidification is extremely high. Liquid bridges may thus be able to generate solid bonds before rupture, causing agglomerate formation and subsequent defluidisation. In this work, every attempt to break the solid bond at the end of each experiment failed. This result explains the inability of Lettieri et al.[25,26] to re-fluidise the catalyst after the bed had been allowed to cool down.

Current work is focussed on the development of a laser-based device to enhance the resolution of the displacement measurement down to the nm length scale, thereby increasing the force resolution of the instrument to that comparable with Atomic Force Microscopy (AFM). An environmental chamber is also being designed to allow the effects of the surrounding gas medium to be studied.

5.3 Atomic Force Microscopy

Atomic force microscopy is a technique used to study colloidal properties. Yalamanchili et al.,[29] highlighted the use of AFM for a variety of applications including: study of almost any solid surface, examination of surfaces in air and liquid and measurement of interparticle forces such as bubble/particle interactions,[30] film stability, wetting and dispersion. Depending on the various applications, different customised designs are available in the literature.

The atomic force microscope operates on a principle similar to that of the flexure strip assembly described above for the HTMFB. It probes a sample surface with a sharp tip fastened to a cantilever spring with typical lateral dimensions of 100–200 μm and a thickness of 1 μm. Cantilevers usually give spring constants ranging from 0.1 to 3 N/m so that an AFM is able to measure forces in the range of 10^{-6}–10^{-9} N. As the cantilever bends due to the interaction with the sample surface, its position is detected by the optical deflection of a

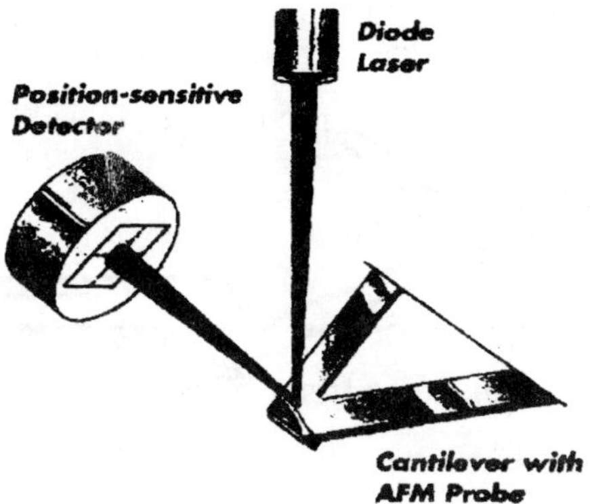

Figure 19   *Schematic of the optical-deflection technique for detecting cantilever deflection*[29]

laser beam using a Position Sensitive Detector (PSD), Figure 19. The high resolution of such a device permits evaluation of displacements in the range of 10 Å. This sensitivity is sufficient for measuring adhesion and very short range forces between molecular-sized tips and surfaces, but not longer range forces.

AFM studies can be broadly split in two categories: topography and probe-to-surface interactions. Topographic measurements allow for the determination of friction coefficients, elastic moduli, roughness, normal and shear stresses of sampled surfaces at both room and high temperature. In this mode of operation, a cantilever tip scratches the solid surface to be investigated and follows the asperities. Movements of the cantilever tips are interpreted by dedicated software and used to evaluate the surface properties. AFM accessories are available on the market (Digital Instrument) to heat the sample surface.

In the probe technique, a particle (probe) is glued to the tip of the cantilever and is approached to a either a solid, liquid or fluid surface, allowing the determination of either short range (e.g. van der Waals) or adhesion forces, depending on whether the probe and the sample are put into contact or simply narrowed to very small separation distances.

The force applied after the tip makes contact with the surface can provide a measurement of the stiffness or compliance of the sample, whilst the force required to pull the sample up from the surface is a measure of the adhesion between the tip and the sample. This is illustrated in Figure 20.

The crucial factor for the probe technique is the determination of the cantiliver spring constant that cannot be assumed to be the same as that declared by the manufacturer. The probe that is glued on the AFM cantilever in fact changes the resonant frequency of the system (mass + cantilever) and therefore further measurements are required in order to calculate the spring constant of the cantilever. The spring constant can be calculated, for example, through the investigation of the shifting of the resonant frequency of the cantilever before and after the probe is glued, or by means of the thermal noise method.[31] The need for a re-calculation of the cantilever spring constant has lead to the increasing use of

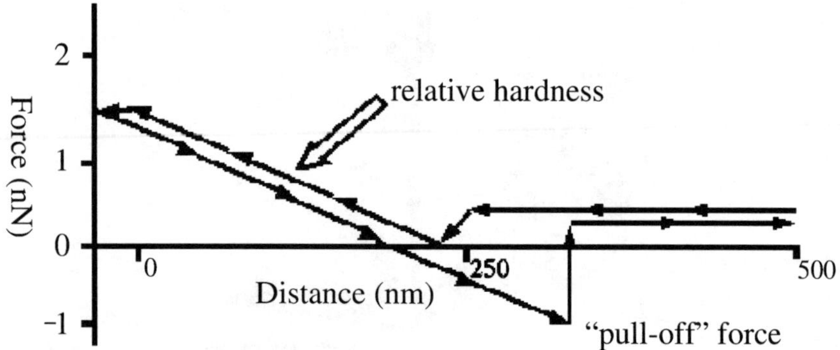

Figure 20   *Typical Force-Distance chart obtained by AFM*

commercial AFM instruments (Nanoscope IIIa, Digital Instrument) that have this feature built in.

Commercial atomic force microscopes cannot always meet all the requirements for a project and thus need to be customized for specific tasks, particularly when used for high temperature investigations using the probe technique. This involves the design of a proper environment in which the particle and the sample interact withouth limiting the movement of the cantilever. When the temperature is increased to higher values, further problems can be experienced because of physical-chemical modifications to the cantilever, which can greatly influence the determination of the spring constant and ultimately affect the force evaluation.

Toikka et al.[32] highlighted the limitations of AFM in the measurement of adhesion forces between a particle and a sample in a heated environment. It is very difficult with a conventional atomic force microscope to place a particle, attached to the cantilever tip, into contact with a sample surface and then to maintain that position over an extended period of time in a controlled environment. Atomic force microscopes usually make several contacts over short time periods which, in itself, can lead to changes in the interaction geometry and can introduce cross-contamination of the sample. Toikka et al.[32] therefore designed an apparatus based on AFM but with the capability of not only placing but maintaining a micron-sized particle in contact with a heated sample and then measuring the pull-off force.

A schematic of the rig is shown in Figure 21. As in AFM, the particle is attached to the tip of the cantilever, which is placed in a holder located at the end of the displacement arm. Vertical displacement between the particle and a fixed sample, located directly below the particle, is made by using an Inchworm (Burleigh). This design allows considerable movement with submicron resolution and the holding of position, when required. Movement is controlled, via a PC, by expanding and contracting a Piezo-Electric Crystal (PEC) in the vertical z-direction whilst the Inchworm is sequentially reclamped at each end (the action is described by the authors as being similar to the crawl of a caterpillar). Conventional atomic force microscopes suffer from PEC hysteresis and this is

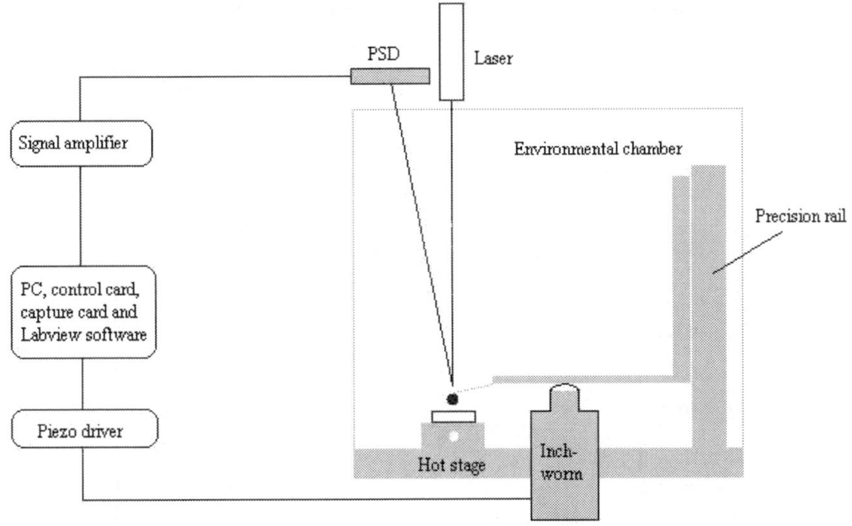

Figure 21 *Schematic of the force rig designed to measure fine particle adhesion via their pull-off force*
(Redrawn from Reference 32)

why they cannot hold position over extended time periods. However, the Inchworm technique also has its limitations. The clamping causes a vertical glitch of around ± 1 μm, leading to occasional shear forces being detected in the measured data.

The heating device takes the form of a hot stage, which consists of a resistive wire inside a ceramic tube (rated to 800 °C). The temperature of the sample surface is measured directly using a thermocouple and voltmeter, similar to the HTMFB described earlier. An optical encoder ensures the accuracy of all measured displacements, whilst the cantilever deflection is detected using a laser diode and a PSD. The environmental chamber is constructed so that the laser can be directed through a glass top, whilst plastic walls made for ease of construction and include inlets for the heating device and thermocouple. During the experiments the chamber can be maintained at low pressures.

Toikka et al.[32] used their device to study the effects of temperature on the fracture adhesion energy of an inorganic particle (zirconia sphere, 10.6 μm diameter) in contact with a polymer surface. Figure 22 shows the pull-off force data measured after 2 minutes in contact at decadal temperature intervals between 20 and 70 °C. The force can be seen to increase moderately up to 40 °C, then by almost two orders of magnitude at 50 °C, reaching a maximum before decreasing again at higher temperatures. Separate Dynamic Mechanical Analysis (DMA) shows a glass transition temperature of the polymer of approximately 40 °C (Figure 23), giving a good correlation with the AFM device. The maximum adhesion energy was due to the visco-elastic and plastice deformation of the polymer surface during contact and removal of the particle, whilst the subsequent decrease was due to the decrease in viscosity of the surface at higher

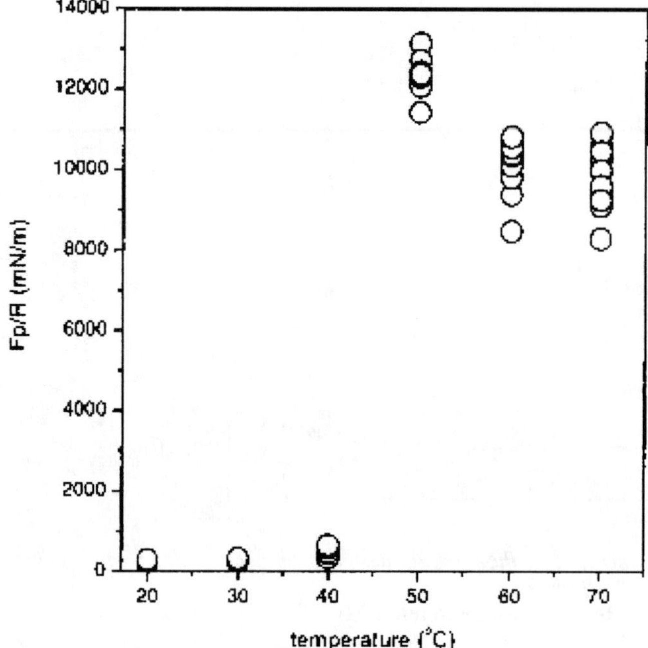

Figure 22   *Pull-off force between a zirconia particle and a polyester surface measured as a function of temperature*[32]

temperatures. Toikka et al.[32] did find, however, that the measured data fell below theoretical predictions and they attributed this to the effect of surface asperities reducing the actual contact area.

In conclusion, AFM can return good results when measuring forces at the particle-sample contact at room temperature, but the customisation of the instrument and the control of the environment are still constraints for high temperature experiments. In contrast, the design of the HTMFB developed at UCL allows the flexure strip to bend at room temperature remotely from where the particle-particle interactions take place at high temperatures, thereby over-coming the problems associated with force measurement and also allowing both surfaces at the contact point to be heated.

6   Conclusion

The fundamental mechanisms of particle adhesion (and cohesion) at elevated temperatures are still poorly understood. Significant advances have been made in recent years by combining micro-scale data with macro-scale observations, par-ticularly in the area of defluidisation of fluidised beds due to particle sintering. The increase in sophistication of AFM and the development of novel micro-mechanical devices is now leading to new possibilities in the determination of the fundamental mechanisms through direct observation and measurement of the interactions as they occur.

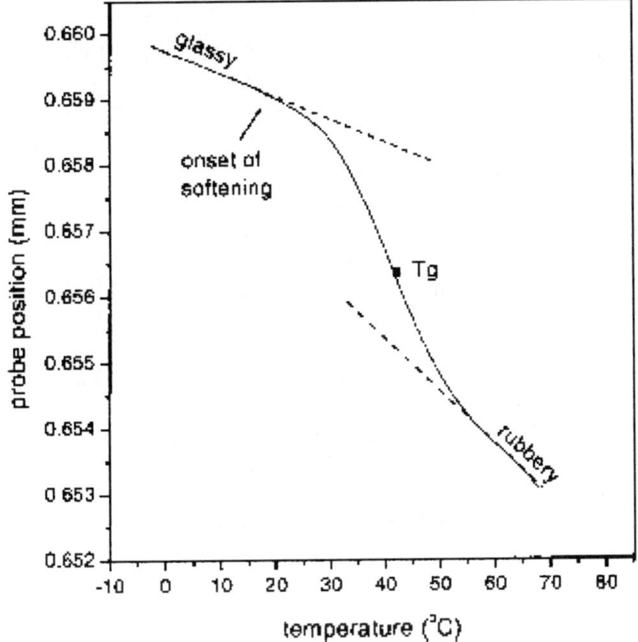

Figure 23 *Dynamic thermal analysis data measured on the polyester film showing the glass transition temperature of 43 °C at the point of inflection*[32]

References

1. S. Berbner and F. Löffler, *Powder Tech.*, 1994, **78**, 273.
2. W. Pietsch, *Size Enlargement by Agglomeration*, Wiley, New York, 1991.
3. G. Lian, C. Thornton and M. J. Adams, *J. Coll. Int. Sci.*, 1993, **161**, 138.
4. K. S. Hwang, R.M. German and F. V. Lenel, *Metall. Trans.* A, 1987, **18A**, 11.
5. F. R. E. De Bisschop, W. J. L. Rigole, *J. Coll. Int. Sci.*, 1982, **88**, 117.
6. X. Pepin, D. Rossetti, S. M. Iveson and S. J. R. Simons, *J. Coll. Int. Sci.*, 2000, **232**, 289.
7. D. N. Mazzone, G. I. Tardos and R. Pfeffer, *Powder Tech.*, 1987, **51**, 71.
8. X. Pepin, S. J. R. Simons, S. Blanchon, D. Rossetti and G. Couarraze, *Powder Tech.*, 2001, **117**, 123.
9. B. J. Ennis, G. Tardos and R. Pfeffer, *Powder Tech.*, 1991, **65**, 257.
10. S. J. R. Simons, J. P. K. Seville and M. J. Adams, *Chem. Eng. Sci.*, 1994, **49**, 14, 2331.
11. O. Pitois, P. Moucheront, and X. Chateau, *J. Colloid Interface Sci.* 2000, **231**(1), 26.
12. S. J. R. Simons, J. P. K. Seville and M. J. Adams, Proceedings of the Sixth International Symposium on Agglomeration, Nagoya, Japan, 1993, 117.
13. J. C. Song, L. T. Fan and N. Yutani, *Chem. Eng. Comm.*, 1984, **25**, 105.
14. J. H. Siegell, J. H., *Powder Tech.*, 1984, **38**, 13.
15. M. J. Gluckman, J. Yerushalmi and A. M. Squires, Defluidization characteristics of sticky or agglomerating beds, in "Fluidization Techology", ed. D. L. Keairns, Hemisphere Publishing Corp., New York, 1976, **2**, 395.
16. B. Liss, dissertation presented at the 87th AIChE meeting, Boston, 1979.

17. G. I. Tardos, D. Mazzone and R. Pfeffer, *Can. J. Chem. Eng.*, 1985, **63**, 384.

18. P. Compo, R. Pfeffer and G. I. Tardos, *Powder Tech.*, 1987, **51**, 85.

19. G. I. Tardos, D. Mazzone and R. Pfeffer, *Can. J. Chem. Eng.*, 1984, **62**, 884.

20. B. G. Langston and F. M. Stephens, *J. Met.*, 1960, **12**, 312.

21. B. J. Skrifvars, M. Hupa and M. Hiltunen, *Ind. Eng. Chem. Res.*, 1992, **31**, 1026.

22. B. J. Skrifvars, M. Hupa, R. Backman and M. Hiltunen, *Fuel*, 1994, **73**, 171.

23. J. P. K. Seville, H. Silomon-Pflug and P. C. Knight, *Powder Tech.*, 1998, **97**, 160.

24. http://www.anasys.co.uk/library/tma1.htm

25. P. Lettieri, J. G. Yates and D. Newton, *Powder Tech.*, 2000, **110**, 117.

26. P. Lettieri, J. G. Yates and D. Newton, *Powder Tech.*, 2001, **120**, 34.

27. S. J. R. Simons and R. J. Fairbrother, *Powder Tech.*, 2000, **110**, 44.

28. P. Paglia and S. J. R. Simons, Proceedings of the World Congress on Particle Technology 4, Sydney, Australia, 2002.

29. M. R. Yalamanchili, S. Veeramasuneni, M. A. D. Azevedo and J. D. Miller, *Colloids and Surfaces A*, 1998, **133**, 77.

30. M. Preuss and H. J. Butt, *Int. J. Min. Process.*, 1999, **56**, 99.

31. D. A. Walters, J. P. Cleveland, N. H. Thomson and P. K. Hansma, *Rev. Sci. Instr.*, 1996, **67** (10), 3583.

32. G. Toikka, G. M. Spinks and H. R. Brown, *Langmuir*, 2001, **17**, 6207.

CHAPTER 5

Critical State Behaviour of Granular Materials Using Three Dimensional Discrete Element Modelling

T. G. SITHARAM, S. V. DINESH, and B. R. SRINIVASA MURTHY

Department of Civil Engineering, Indian Institute of Science, Bangalore – 560 012, INDIA.

1   Introduction

The critical state concept from the Cambridge group is a major step for modeling the constitutive behavior of soils by combining density and plasticity criteria. The term critical state is used to describe a condition at which plastic shearing of soil would continue indefinitely without changes in volume or effective stresses Wood.[1] The critical state concept has been found to be of relevance for clays but its application for sands has been less successful and is still debated because of the difficulty in obtaining a unique Normal Consolidation Line (NCL). However many attempts have been made to explain the behaviour of granular materials within the critical state framework (Been et al.,[2] Vesic and Clough,[3] and Lee and Seed[4]). Many investigators have addressed whether the critical state and steady state lines for sands are identical (Casagrande,[5] Poulos,[6] and Sladen et al.[7]). Castro et al.[8] have obtained a unique steady state line for sand irrespective of the method of sample preparation, initial stress state, and stress path. However they did not investigate the steady state under extension loading. Bishop,[9] Miura and Toki,[10] Hanzawa,[11] and Vaid et al.[12] have concluded that the steady state depends on the mode of deformation. Also Alarcon-Guzmann[13] concludes that the particle matrix structure and test type affect whether the sand will reach the critical state or steady state line. This implies non-uniqueness of critical state. The state parameter defined by Been and Jefferies[14] to describe the behavior of sands is based on Critical State Line (CSL) as the reference line. The critical state line was assumed as a straight line but the studies have shown it to be a curved line (Lee and Seed,[4] and Vesic and Clough[3]). Been et al.[2] suggest a unique bilinear critical state line for Erksak sand. It needs to be established whether there is a unique CSL for a particular granular material irrespective of its initial state. The influence of fabric on the critical state behavior of sand has not been addressed because of the difficulty in quantifying it.

In this work, we will highlight the above aspects using numerical simulations carried out by three-dimensional Discrete Element Method (DEM). DEM offers a better opportunity to understand the micromechanical behavior of granular materials considering particulate assembly using micro-parameters such as average coordination number, induced anisotropy coefficients and fabric tensors. Further, the critical state behaviour for both loose and dense assembly is analysed from the numerical results under different stress paths. An attempt has been made to understand and explain the critical state behavior micromechanically by tracing the evolution of microstructure and contact forces in the assembly.

2 Discrete Element Method and Micromechanics

2.1 Discrete Element Method

To simulate 3-Dimensional assembly of particles the version TRUBAL developed by Cundall and Strack[15] and Strack and Cundall[16] with cubic periodic boundaries, further updated and modified by Chantawarungal[17] has been used. Each particle is identified by properties of density and size. Contact type is identified by contact properties such as normal stiffness, tangential stiffness, coefficient of friction and adhesion between types of particles. The subroutines to extract micro-macro parameters has been programmed in TRUBAL to automate the extraction of the data.

2.2 Micromechanics

Micromechanics is a science, which deals with the relationship between external stresses and strains, average internal forces and displacements. This approach requires complete information of all characteristics such as contact forces, average microscopic geometry, contacts distribution and coordination number for all elements in the assembly. Micromechanics describes the macro behaviour through statistical integration at micro level. The micro-macro relationship consists of fabric (spatial arrangement) as an essential term. Methods to quantify fabric however are not well established. The presence of fabric terms in the relationship between force and stresses and similarly between displacements and strains can be easily visualized. These micro-macro relationships require a complete knowledge of the number of contacts in a representative volume and also the distribution of contact orientations for the purpose of relating average forces to stresses and displacements to strains. In this direction experimental work with assemblies of optically sensitive materials has provided a better understanding of micro-macro relationship (Dantu,[18] De Josselin de Jong and Virruijt,[19] Oda,[20-22] Dresher and De Josselin de Jong,[23] and Oda and Konishi[24,25]).

A representative volume is defined as an assemblage which contains a large number of particles to be representative of the granular materials at the continuum level. The micro-parameters such as average coordination number, contact density, contact normals and vectors and their spatial distribution, which are associated with characterization of microstructure, are described below. Average coordination number (N) indicates the number of contacts per particle. It is defined as the ratio of total number of contacts (twice the physical contacts) in the assembly to total number of particles. Contact density (m_v), indicating the number of contacts per unit volume of the assembly is defined as the ratio of total number of contacts (twice the physical contacts) to total volume of the assembly. A contact normal (n^c) is a vector directed normal to the tangent plane at the point of contact between two particles. The contact vector (l^c) is a vector directed from the mass center of a particle to the point of contact with a neighboring particle. The contact force vector (f^c) describes interparticle contact force when two particles interact through a contact point. The orientation of contacts for 3-D assemblies is characterized by a distribution function E(Ω) defining the number of contacts ΔM (Ω^g) falling with in an elemental solid angle $\Delta\Omega$. Polar histograms can be used to characterize the contact orientations.

Detailed microscopic information in a numerically simulated assembly of spheres can be used to trace the evolution of the microstructure and the contact forces during shear deformations. Numerical simulations using DEM on plane assemblies of discs by several researchers have been used to study such relationships (Strack and Cundall,[26] Rothenberg,[27] Bathurst,[28] Cundall,[29] Thornton and Barnes,[30] Rothenberg and Bathurst,[31] and Sitharam[32]). Later with the development of 3D simulation code many researchers have reported the numerical simulation of quasi static shear deformation (Chantawarungul,[17] Thornton and Sun,[33,34] Thornton,[35] Thornton et al.,[36] Thornton and Antony,[37] Robertson,[38] Thornton and Antony,[39] Sitharam et al.,[40] McDowell and Harireche,[41,42] Dinesh,[43] and Antony[44]). The DEM offers a unique opportunity to obtain complete quantitative information on all microscopic features of the assembly of particles.

In particulate media each particle will be in contact with several of its neighbors, forming a group where each particle interacts with the neighbors at the contact point and through contact forces. Several such groups form an assemblage. The average stress carried by the assembly within volume V can be calculated from the applied forces on the boundary particles. One such relationship to calculate the macroscopic applied average stress tensor for the entire assemblage is given by Landau and Lifshitz:[45]

$$\sigma_{ij} = \frac{1}{V}\sum_{\beta\varepsilon s} f_i^\beta\, l_j^\beta \ldots\ldots i, j = 1,2 \tag{1}$$

In the above expression, f^β_i represents the equivalent assembly boundary force components, and l^β_j represents the co-ordinates of the intersection points of these forces on the boundary. The volume term V in the equation corresponds to the entire assembly and is approximated by the area contained by the convex polygon of line segments connecting adjoining boundary sphere centroids. The expression given above is simply a statement of the force balance between boundary loads in terms of the stress tensor and the boundary forces.

In a granular assembly, boundary loads are distributed among the intergranular contacts. The balance between boundary loads and internal forces leads to the expression for the stress tensor in the form of (Rothenburg[27]):

$$\sigma_{ij} = \frac{1}{V} \sum_{c\varepsilon v} f_i\, l_j \ldots \ldots \ldots \ldots \ldots i, j = 1, 2 \tag{2}$$

where the sum is with respect to all contact forces within a volume V, with Cartesian components of f_i multiplied by components of contact vector, l_j. The above equation is the same as the one presented by Weber,[46] Christoffersen et al.[47] and Bathurst.[28]

In order to understand load transfer in granular materials in terms of a few simple averages of forces acting on different contacts, for an assemblage of spheres the procedure is simple and has been described by Rothenburg[27] and Chantawarangul[17] in depth. They have shown that the macroscopic stress (Expression 2) can be related to the microscopic force and fabric parameters by the expression:

$$\sigma_{ij} = m_v\, \bar{l}_0 \int_\Omega \bar{f}^c_i\, (\Omega) n_j E(\Omega) d\Omega \tag{3}$$

where,
σ_{ij} = Macroscopic stress,
$\bar{f}^c_i$ = Average contact force acting at contacts with orientations Ω,
$\bar{l}_0$ = Assembly average contact vector length,
m_v = Contact density, and
$E(\Omega)$ = Distribution of contact normals.

The distribution of contact normals in a granular system can be described by a second order fabric tensor. For an infinite system, the distribution traces three dimensional surface with certain axes of symmetry, with the general form given by

$$E(\Omega) = \frac{1}{4\pi} \{1 + a^r_{ij} n^c_i n^c_j\} \tag{4}$$

where a^r_{ij} = symmetric second order deviatoric tensor representing coefficient of contact normal anisotropy.

The average contact forces at contacts can be decomposed into average normal contact forces $\overline{f}^n(\Omega)$ and average tangential contact forces $\overline{f}^t(\Omega)$.

The distribution of average normal contact force is given by

$$\overline{f}^n(\Omega) = \overline{f}^n_0 \left[1 + a^n_{ij} n_i n_j\right] \tag{5}$$

where,

$\overline{f}^n(\Omega)$ = Distribution of average normal contact force in the segment,

$\overline{f}^n_0$ = Average contact normal force over all contacts in the assembly,

a^n_{ij} = Symmetric second order deviatoric tensor representing co-efficient of normal contact force anisotropy coefficients and defines the directional variation of the normal contact force.

The average tangential contact force distribution is given by

$$\overline{f}^t_i(\Omega) = \overline{f}^n_0 \left[a^t_{ij} n_j - (a^t_{kl} n_k n_l)\, n_i\right]$$
$$\overline{f}^t(\Omega) = \sqrt{\left[\overline{f}^t_i \overline{f}^t_i\right]} \tag{6}$$

where,

a^t_{ij} = symmetric second order deviatoric tensor representing tangential contact force anisotropy coefficient,

$\overline{f}^t(\Omega)$ = Distribution of average tangential contact force in the segment,

$\overline{f}^n_0$ = Average normal contact force in the assembly.

Tensor a_{ij} is a deviatoric invariant of the symmetric second order tensor describing the distribution of contact normals, normal contact force and tangential contact force. Upon substituting Expression 4 for the contact normal distribution and Expressions 5 and 6 for average contact forces into the general expression for the stress tensor (3), we have a general expression relating stress, force and fabric tensors under static equilibrium of granular assemblies as below (Chantawarungal[17]):

$$\sigma_{ij} = \frac{m_v \overline{f}^n_0 \overline{l}_0}{3} \left[\delta_{ij} + \frac{2}{5}\left(a^r_{ij} + a^n_{ij} + \frac{3}{2} a^t_{ij}\right) + \frac{2}{35}\left\{ \left(a^n_{kl} - a^t_{kl}\right) a^r_{kl}\, \delta_{ij} + \left(4a^n_{il} + 3a^t_{il}\right) a^r_{lj} \right\} \right] \tag{7}$$

where,

σ_{ij} = Macroscopic stress in terms of microstructural parameters, δ_{ij} = Kronecker delta,

$\overline{f}^n_0$ = Measure of average normal contact force in the assembly, and

$\overline{l}_0$ = Assembly average contact vector length.

The above equation represents the stress tensor in terms of the microstructural parameters of granular materials. With this expression one can clearly see that the stress tensor can be split up in to four parts, spherical stress component, deviatoric tensor of fabric, deviatoric tensor of forces, and product of force and fabric terms. It can be seen from the above expressions that the hydrostatic stress carrying capacity of a granular medium is mainly related to the contact density m_v and the average normal contact force $\overline{f_0^n}$ while the deviatoric stress carrying capacity is related to the ability to develop anisotropy in contact orientations and contact forces. For the microscopic anisotropy coefficient tensor, a_{ij}^r, a_{ij}^n, and a_{ij}^t scalar parameters of a_r^d, a_n^d, and a_t^d are generated from their invariants as given by the following equation.

$$a_r^d = \sqrt{\frac{3}{2}\, a_{ij}^r a_{ij}^r}\,; \quad a_n^d = \sqrt{\frac{3}{2}\, a_{ij}^n a_{ij}^n}\,; \quad a_t^d = \sqrt{\frac{3}{2}\, a_{ij}^t a_{ij}^t} \tag{8}$$

In the present work the critical state behaviour of granular materials will be examined in terms of these micro parameters.

3 Numerical Testing Programme

The three dimensional assembly consisting of 1000 sphere particles is generated in a random manner in accordance with pre-set particle size, gradation and packing criteria with no initial contacts to represent the granular material. Twenty-one different sphere diameters in the range of 20 to 100 mm corresponding to a log normal distribution are used for the simulations. Each sphere and contact have prescribed properties including a radius, density, normal and tangential contact stiffness and coefficient of inter-particle friction. The input parameters used in the numerical simulations are shown in Table 1. The contact is modeled with linear springs in normal and tangential directions. The objective of this work is to study the critical state behavior of loose and dense polydisperse assemblages of granular materials under undrained and drained stress path conditions.

The number of particles considered in this study are 1000 and since the numerical scheme employs a periodic space the assembly is free from boundary effects, also the opposite faces of the periodic boundary are numerically connected. Any assembly with a finite number of particles constitutes an infinite system. However simulations on 300, 1000, and 3000 sphere particle assembly have shown very similar results, though there is considerable noise in the 300

Table 1 *Input parameters selected for the numerical simulations*

Properties	Symbols	Numerical Values Used
Normal contact stiffness	K_n	1.0×10^5 N/m
Tangential contact stiffness	K_t	1.0×10^5 N/m
Particle density	γ	2000 kg/m^3
Cohesion at particle contact	c	0.0
Contact friction	μ	0.0 and 0.5
Servo gain	G	*1e–8*

sphere particle assembly. On the basis of this and also keeping the computational run time in mind a 1000 sphere particle assembly was chosen for the numerical simulations.

Loose and dense assemblies are generated to study the effect of density by assigning a coefficient of contact friction of 0.5 and 0.0 respectively to all particle contacts. Simulations of isotropic compression are performed to compress the initially generated loose and dense assemblies by distorting the periodic cell and changing its volume using a servo control as $\dot{\varepsilon}_{kk} = \dot{\varepsilon}_{kk} + \dfrac{G}{3}(\sigma_{kk}^{specified} - \sigma_{kk}^{measured})$ up to 200 kPa under isotropic loading conditions of $\sigma_{11} = \sigma_{22} = \sigma_{33}$. Where G is the specified servo gain. This parameter represents the ratio of change in grid strain rate for an error in the controlled stress specified on the CYCLE input command. The error in the stress value is the difference between specified value to that measured for the assembly. To ensure a stable system at the specified stress state the calculation cycles were continued, till the void ratio, average coordination number and stress values are constant. The isotropic stress is increased incrementally to 5, 10, 25, 50, 100 and 200 kPa. Figure 1a shows the 3-D view of the initially generated loose assembly. Figure 1b shows the 3D view of the loose assembly compacted to 200 kPa.

Starting from the initial isotropically compressed stable assembly at different confining pressures [5kPa, 25 kPa, 50 kPa, 100 kPa and 200 kPa], a series of numerical triaxial compression tests ($\sigma_1 > \sigma_2 = \sigma_3$) are carried out under drained and undrained stress paths on both loose and dense assemblies. To carry drained tests the servo control as described above was used, and for undrained tests a constant volume condition was maintained by a strain controlled mode, in which the increment in grid velocity is proportional to the measured and specified values. The dense assemblages which are isotropically compacted with friction coefficient of zero are equilibrated with friction coefficient of 0.5 for subsequent shear tests. All shear tests have been carried out with a contact friction coefficient of 0.5. While carrying out drained shear tests, in one series the ratio of $\Delta q/\Delta p$ was maintained at a ratio of 3 to 1. All simulations were continued until a critical state of deformation of the specimen was reached at about 25 to 30% axial strain. Table 2 shows the program of numerical experiments carried out.

4   Results and Discussions

The effective stress parameters used are mean normal stress, (p) and deviator stress (q), which are defined as follows:

$$p = \frac{\sigma'_{11} + \sigma'_{22} + \sigma'_{33}}{3}; \quad q = \sqrt{\frac{3}{2}\sigma'_{ij}\sigma'_{ij}} \tag{9}$$

Where σ'_{ij} is the deviatoric stress tensor. The typical macroscopic and micromechanical results in an undrained triaxial compression test at a confining pressure

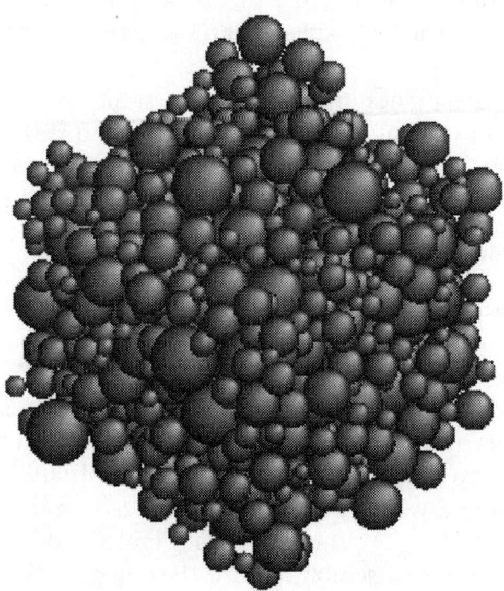

1a. View of the initially generated assembly.

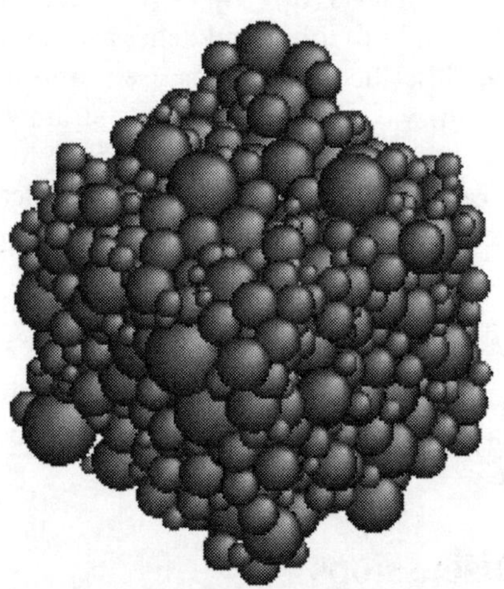

1b. View of the compacted assembly at 200 kPa.

Figure 1   *3-D View of polydisperse assembly in loose state*

of 25 kPa are shown in Figures 2 to 7. Figure 2 shows the plot of normalized deviator stress (q/p) and excess pore-water pressure increment versus axial strain. The excess pore-water pressure was computed by taking the difference in drained (total) and undrained stress paths and pore water was not modelled in

Table 2 *Numerical simulations of Monotonic Triaxial Shear Tests carried out on loose and dense polydisperse sample at 25 kPa*

Test No.	Initial sample conditions	Test path during shear test	Initial state of the sample	Confining pressure in kPa.
		Undrained	0.6	5
			0.55	25
			0.6	5
1	Loose		0.55	25
		Drained	0.5	50
			0.42	100
			0.32	200
		Undrained	0.47	5
			0.43	25
			0.47	5
2	Dense		0.43	25
		Drained	0.39	50
			0.32	100
			0.24	200

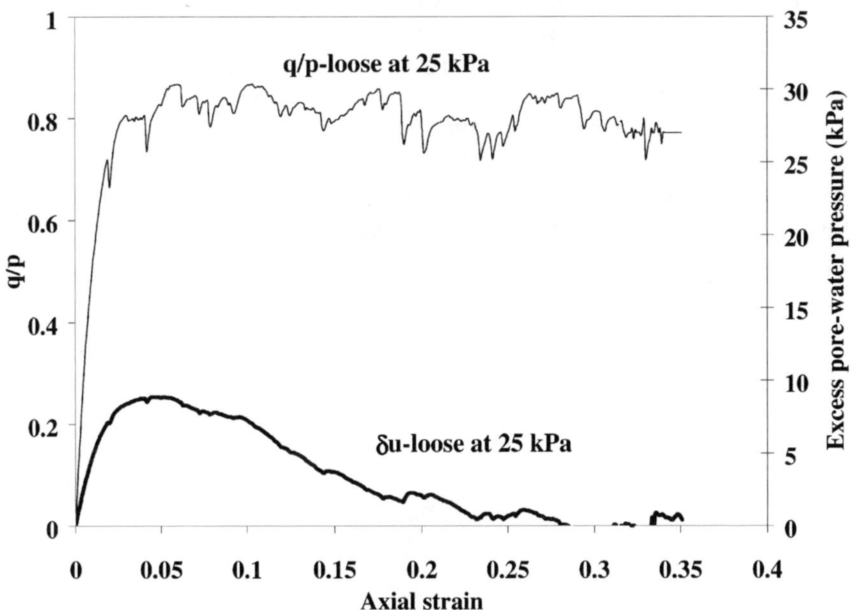

Figure 2 *Plot of q/p and pore-water pressure increment Vs axial strain in undrained test on loose assembly at a confining pressure of 25 kPa*

the analysis. It is observed that both q/p and excess pore-water pressure increase rapidly at low axial strain levels. The q/p ratio value stabilizes after reaching the peak and remains constant until large strain levels (30–35%). The excess pore-water pressure shows an initial increase to suppress the volumetric compression of the assembly and later decreases to prevent the dilation. But beyond 20% strain level it stabilises indicating critical state and remains constant until very large strains of up to 35%. This large strain behaviour is indicative of plastic shearing under constant volume and effective stresses. Figure 3 shows the plot of deviator stress (q) versus mean stress p. The assembly shows an initial compressive behaviour with decrease in mean p, but it later shows phase transformation and then exhibits dilation with an increase in mean stress. Though the real particle interaction in a three-dimensional assembly is truly non-linear very similar to Hertz's mechanism, a linear contact model was used. Dinesh[43] has shown that in a drained shear test on assemblies at constant confining pressure the value of q/p in the initial stages and also the peak stress ratio (q/p) for the Hertzian model is slightly lower than that of the linear contact model. However the behavior is identical in both cases and the linear contact model captures the behavior similarly to that of the non-linear Hertz model. However the e-log p relationships are significantly different for linear and Hertz contact models in isotropic compression tests. Figure 4 shows the plot of average coordination number versus axial strain. It is observed that the average coordination number decreases marginally in the initial stages and remains fairly constant thereafter. The decrease indicates an initial rearrangement of fabric reflected by a loss of contacts. Figure 5 shows the plot of stress ratio (q/p) from applied boundary stress and from the stress-force-fabric relationship [from the microparameters in Equation 7, which were measured during the simulation] versus axial strain. It is evident that the stress ratio from the stress-force-fabric relationship is very similar to that of the stress ratio from the applied boundary stress. Equation 7 shows how the microparameters contribute to the stress ratio. It is clear that the deviatoric stress-carrying capacity of a granular medium at any loading stage consists of components from: (1) contact normal distribution accounting for

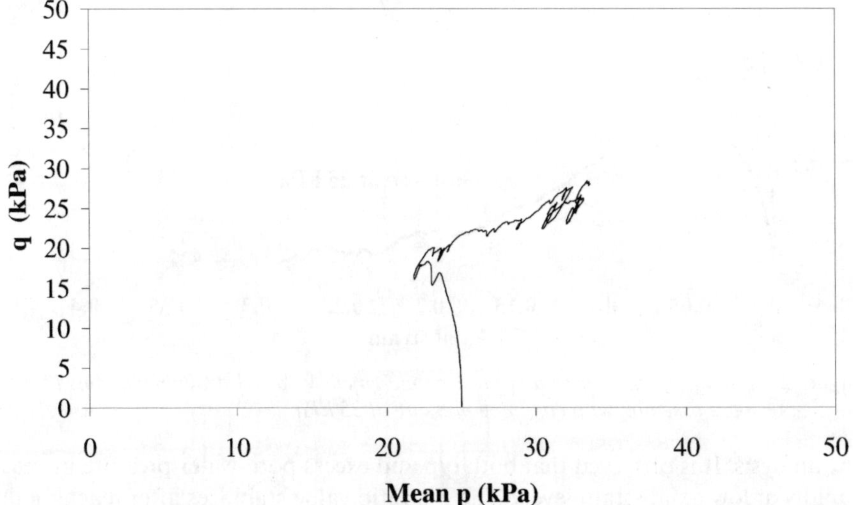

Figure 3   *Plot of deviator stress (q) Vs mean p in undrained test on loose assembly at a confining pressure of 25 kPa*

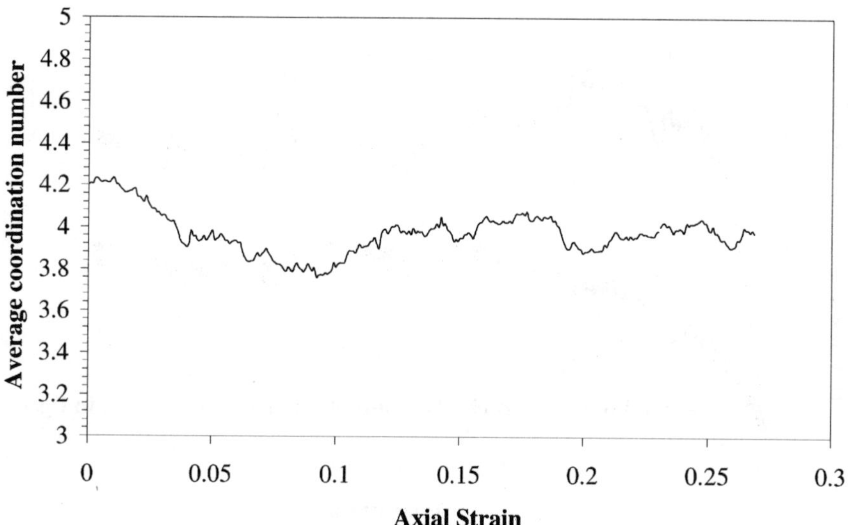

Figure 4   *Plot of average coordination number Vs axial strain in undrained test on loose assembly at a confining pressure of 25 kPa*

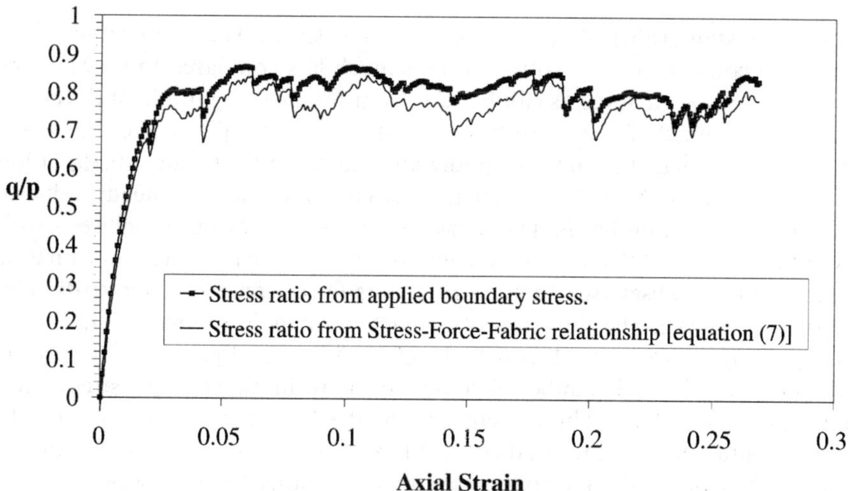

Figure 5   *Plot of stress ratio (q/p) from applied boundary stress and from stress-force-fabric relationship Vs axial strain in undrained test on loose assembly at a confining pressure of 25 kPa*

fabric (arrangement of particles) and (2) normal contact force and tangential contact force distribution in the assembly. Figure 6 shows the plot of deviatoric coefficients of anisotropy of a_r^d, a_n^d, and a_t^d versus axial strain. The deviatoric anisotropy coefficients increase monotonically with strain. The anisotropy in contact force, particularly the normal contact force anisotropy (a_n^d), plays a major role in determining the shear strength of granular media. It reaches a peak value at the same axial strain level at which q/p is maximum, and has

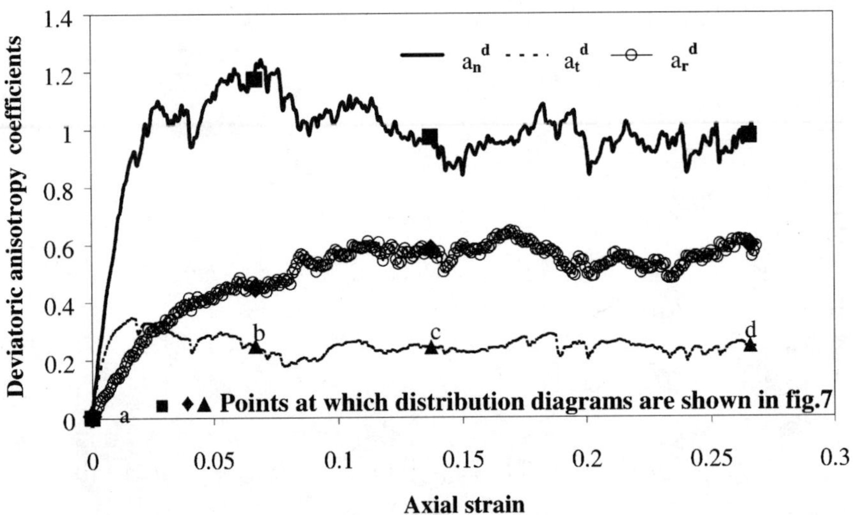

Figure 6   *Plot of deviatoric anisotropic coefficient Vs axial strain in undrained test on loose assembly at a confining pressure of 25 kPa*

a fairly constant value of (0.98) at large strain levels. The contribution from tangential contact force anisotropy (a_t) is much less compared to normal contact force and fabric force as can be seen from the low magnitude of deviatoric tangential contact force anisotropy coefficient. The deviatoric tangential anisotropy coefficient mobilizes rapidly and reaches a peak value (0.35) at low axial strain level (1.8%). Thereafter it decreases and reaches a steady value of 0.25 until large strain levels. The fabric anisotropy coefficient mobilises slowly and gradually. It stabilises at a strain level of (9%) at which average coordination number also stabilises (see Figure 4) remains fairly constant afterwards. This coefficient indicates the variation of contact normals in the major and minor principal stress direction. A constant value of this coefficient at large strain levels indicates that the number of contact normals in the principal stress direction remains constant. This is reflective of the fact that at critical state the normal contact force, tangential contact force and contact normals remain the same in the major and minor principal stress directions. Figure 7 shows plots of contact normals, normal contact force and tangential contact force distribution diagrams using harmonic functions at different stages (points a, b, c, d as indicated in Figure 6). It can be seen from these distribution diagrams that the contact normals distribution and average normal contact force distribution are isotropic at the initial isotropic compacted state, indicating spherical distribution (Figure 7a). There is no preferential directional variation in the magnitude of contact normals and normal contact force. The magnitude of tangential contact force is zero at the isotropic state (Figure 7a). With an increase in deviatoric stress there is an increase in normal contacts orientation and normal contact force anisotropy. The distribution diagram transforms to a peanut shape from an original spherical distribution. It is also evident that the orientation of

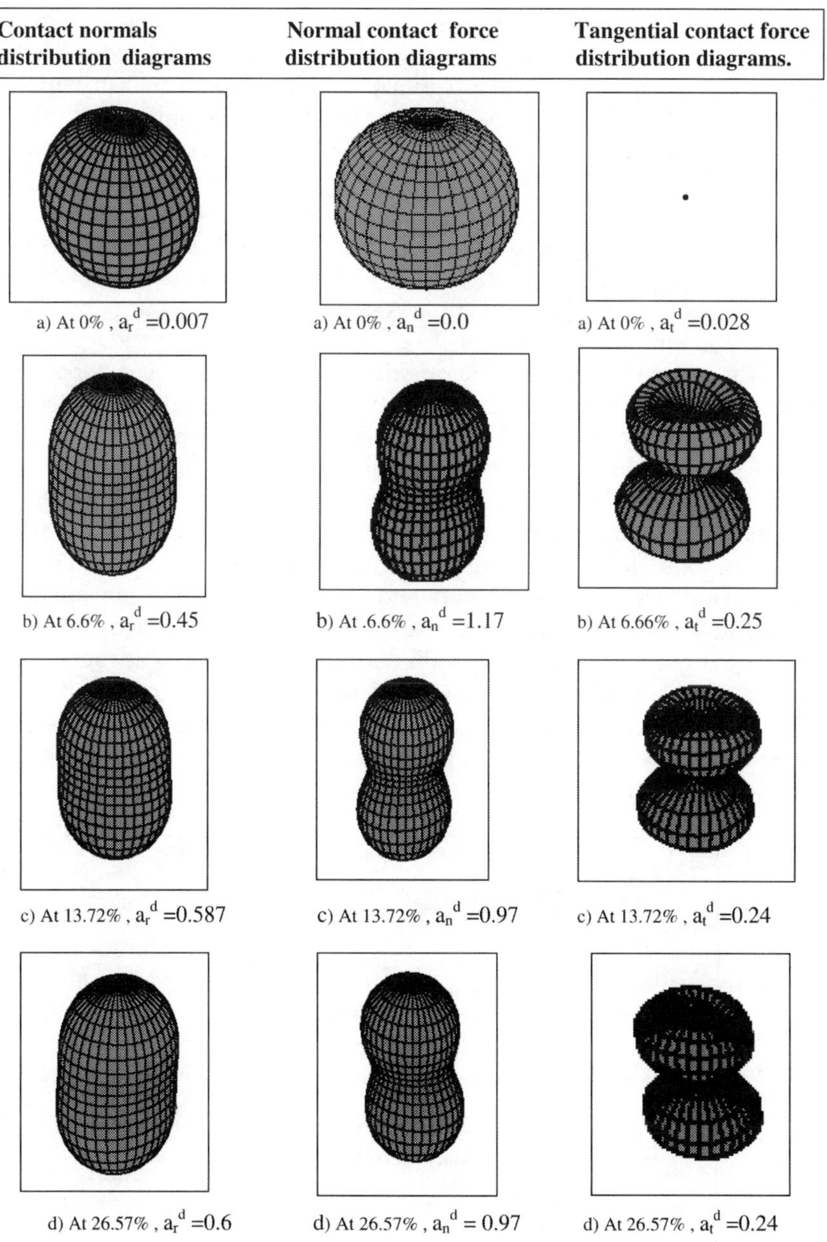

Figure 7 *Plot of contact normal /normal contact force/tangential contact force distribution diagrams at different stages during undrained shear test on loose assembly at a confining pressure of 25 kPa*

the principal vector of contact normals and normal contact force anisotropy coincides with the direction of major principal stress at all stages of deviatoric loading. The tangential contact force develops on a plane inclined at 45° to the direction of the major and minor principal stress plane. The distribution diagram assumes a dumbbell shape. However the magnitude of the tangential contact force anisotropy coefficient is relatively small compared to the contact normal and the normal contact force anisotropy coefficient. This signifies the correspondence between stress tensor, contact forces and fabric (contact normals) at every stage of deviatoric loading. The distribution of contact normals reveals a loss of contacts in the direction of minor principal stress (tensile direction) and an accumulation of contacts in the direction of major principal stress compared to the isotropic condition.

In a similar way, macro and micromechanical analyses are made for all tests to understand the critical state behavior of granular materials under loose and dense conditions. Figure 8a shows the state diagram which is a plot of specific volume (1+e) versus mean stress containing isotropic compression and triaxial shear stress paths. The specific volume is a sum of unit volume of soil particles with the volume of surrounding voids. The slopes of the isotropic compression curves for both loose and dense assemblies are nonlinear. Both curves are approximately parallel to begin with and tend to converge at high stresses. The drained stress paths at confining pressures of 5, 25, 50, 100 and 200 kPa on both loose and dense assemblies along with undrained stress paths at confining pressures of 5 and 25 kPa under loose and dense conditions are shown on the respective isotropic compression curves. The drained stress path for loose and dense conditions at constant confining pressure converge at the same critical

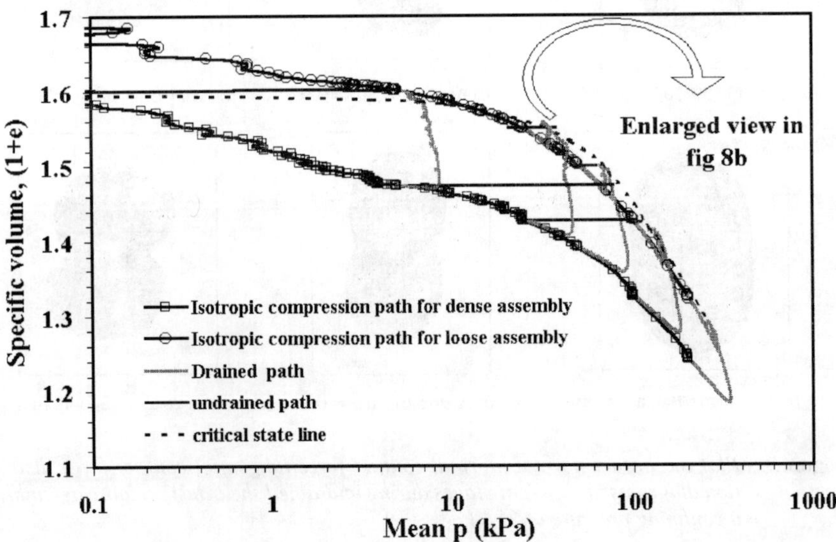

Figure 8a   *Plot of specific volume Vs mean p under different stress paths on loose and dense assemblies at different confining pressures*

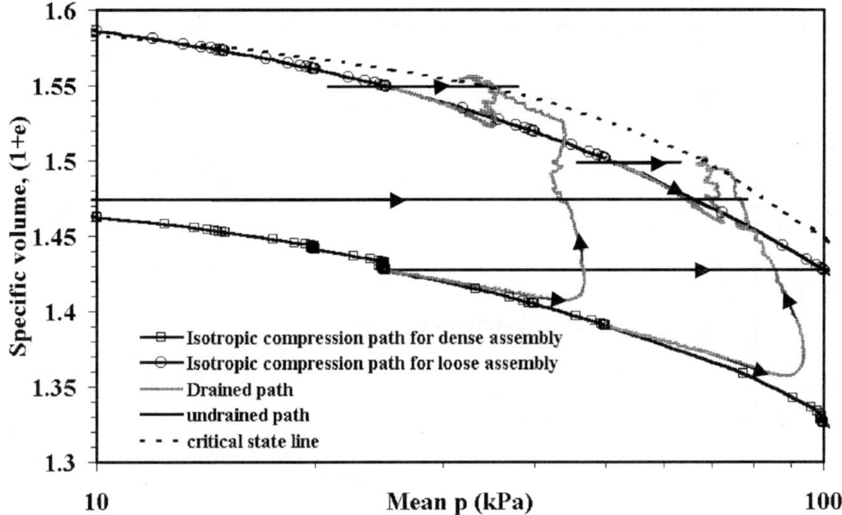

Figure 8b *Plot of specific volume Vs mean p (enlarged view from Figure 8a)*

state as shown in Figure 8b. Critical state is a condition of perfect plasticity at which the material undergoes continuous plastic shearing at constant volume, pore pressure, deviator stress and mean normal stress. This critical state is associated with large strain behavior with excessive deformation. It represents the lower bound residual strength. The critical states will form a 3-D surface in the $q - p - v$ space (v-specific volume). This surface is generally represented by two curves one in the $q - p$ plane and the other in $v - \log p$ plane. Therefore the critical state incorporates both stress state and volume changes in its failure. The proximity of the initial state to critical state has been used as a measure of liquefaction potential. Also if the initial state lies below that of the critical state then such assemblies will show dilation and if the initial state lies above the critical state then such soils show compression upon shearing. Figure 8b clearly highlights that the initial state of the sample does not affect the critical state. All of the stress paths, irrespective of their initial state (location) whether they position below or above the critical state line, finally reach a unique ultimate state at the end of a shear test. The drained or undrained test on loose or dense sample does not have any influence on the location of ultimate critical state. The results are similar to those reported by Poulos[6] and Been et al.[2] The CSL is obtained by joining points of critical state conditions under each stress path. The CSL is a curved line, it is a straight line with a flatter slope at low confining pressires similar to Tatsuoka et al.,[48] and shows a steep increase in slope beyond 25 kPa as shown in Figure 9. The CSL can be approximated as a bilinear line similar to the one suggested by Been et al.[2] for Erksak sand. The isotropic compression paths for loose assemblies and the CSL tend to merge at high stress levels.

Figure 10 shows the plot of stress ratio (q/p) and volumetric strain versus

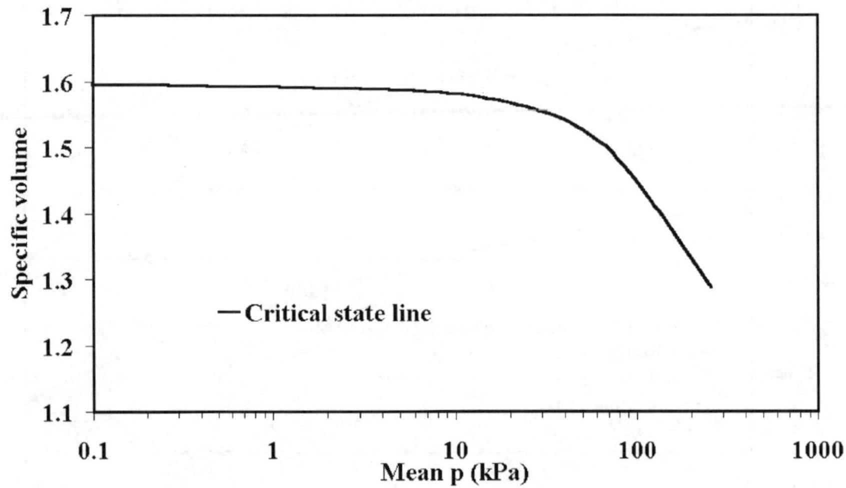

Figure 9   *Plot of specific volume Vs mean p at critical state conditions*

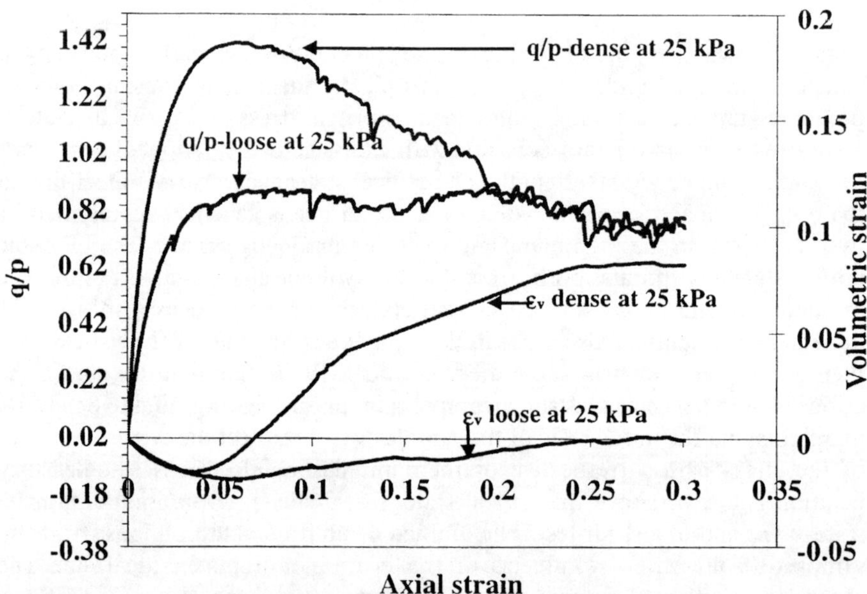

Figure 10   *Plot of q/p and volumetric strain Vs axial strain in drained shear test at a confining pressure of 25 kPa*

axial strain in drained shear tests at 25 kPa under loose and dense conditions. The results indicate that the dense sample undergoes a large amount of dilation, and the loose sample undergoes less dilation. The tests have been carried to large strain levels (beyond 30%) until critical state conditions are met, as indicated in Figure 10. The stress ratio for dense assemblies increases rapidly to a peak value

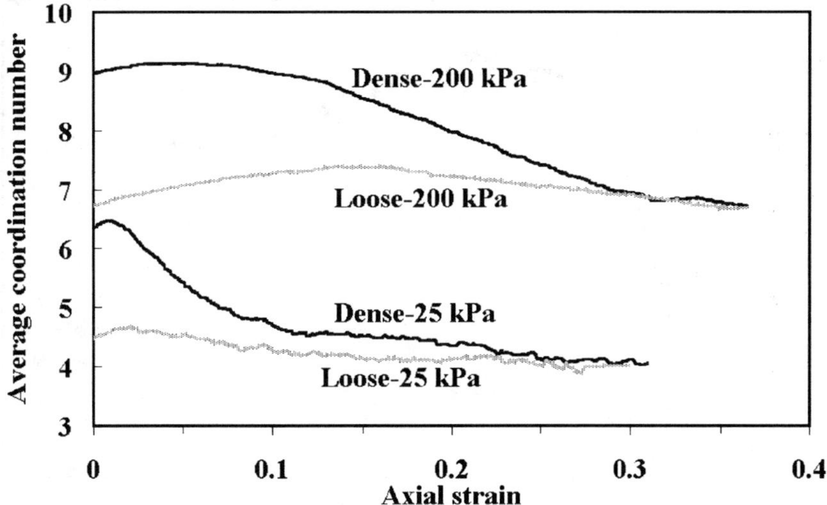

Figure 11　*Plot of average coordination number Vs axial strain in drained shear test at 25 and 200 kPa*

of 1.4 and thereafter dilates (softens) and ultimately reaches a constant value equal to that of a loose assembly at large strain levels. This reinforces the concept that at the critical state both loose and dense assemblies have the same strength irrespective of their initial state. Figure 11 shows the plot of average coordination number versus axial strain in drained shear tests at confining pressures of 25 and 200 kPa under loose and dense states. The dense assemblies at both confining pressures undergo a large reduction in average coordination number. At large strain levels (critical state) both loose and dense assemblies at a given confining pressure have the same average coordination number. Therefore, at critical state, assemblies have the same number of contacts irrespective of their initial density. This reflects that the fabric for both loose and dense assemblies are the same and plastic shearing is taking place at a uniform particle arrangement.

Figures 12, 13 and 14 show the plots of deviatoric anisotropy coefficients of contact forces and contact normal during drained triaxial compression test at 25 kPa confining pressure for loose and dense assemblies. Figure 12 shows the plot of deviatoric coefficient of normal contact force anisotropy versus axial strain. The deviatoric normal contact force anisotropy coefficient increases rapidly for the dense assembly and reaches a peak value of 1.95 at 13.72% axial strain, but it decreases in the post-peak zone and finally attains the same value as that of the loose assembly at large strain levels (30% – critical state). Since this coefficient reflects the magnitude of deviatoric normal contact force in the major and minor principal stress directions, a constant value of this coefficient at critical state indicates plastic shearing at constant deviatoric normal contact force in the major and minor principal stress directions for both loose and dense assemblies.

Figure 13 shows a plot of the deviatoric coefficient of tangential contact force

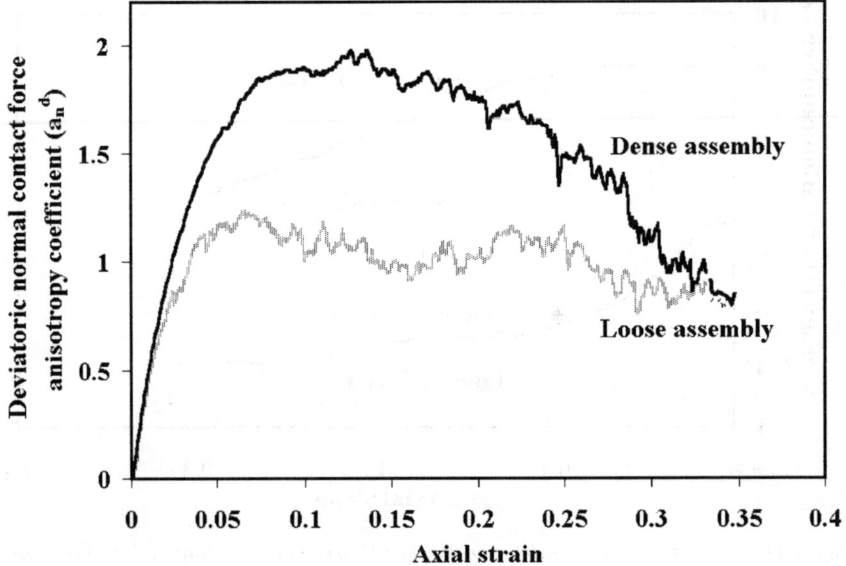

Figure 12   *Plot of deviatoric coefficient of normal contact force anisotropy Vs axial strain in drained test at a confining pressure of 25 kPa*

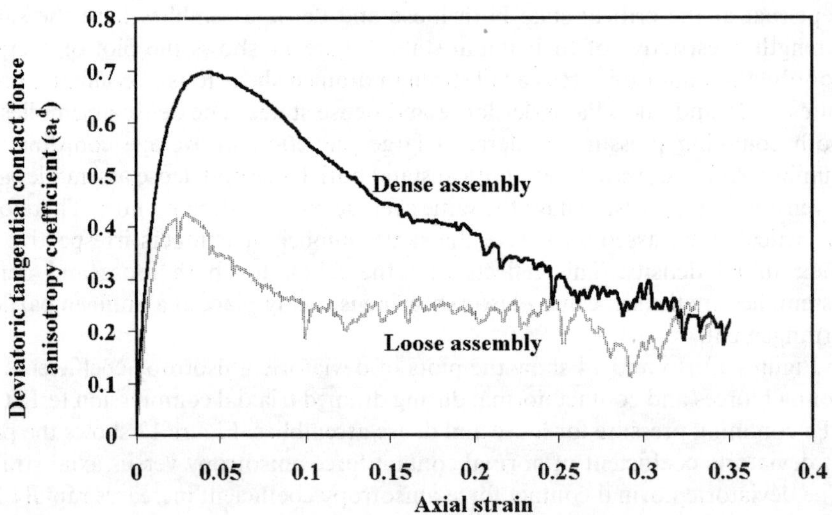

Figure 13   *Plot of deviatoric coefficient of tangential contact force anisotropy Vs axial strain in drained test at a confining pressure of 25 kPa*

anisotropy versus axial strain. Both loose and dense assemblies show a rapid increase in the magnitude, and attain peak values simultaneously at low axial strain levels followed by a gradual decrease. After the peak, the contacts where slip occurs reach some threshold and the particles start dilating which results in a decrease of contact density. As a result, the tangential contact forces in the assembly get released and therefore a decrease in anisotropy coefficient value is observed. Furthermore at large strain levels (critical state), both loose and dense assemblies tend to have the same magnitude of deviatoric tangential contact force anisotropy. This is indicative of the fact that the magnitude of the deviatoric tangential contact force anisotropy coefficient is the same for both loose and dense assemblies at the critical state. Figure 14 shows the plot of the deviatoric contact normals anisotropy coefficient versus axial strain. In both cases the deviatoric fabric force anisotropy coefficient increases slowly when compared to the normal contact force and tangential contact force anisotropy coefficients. The dense assembly reaches a peak value at a fairly large axial strain level (16.23 %). In the small strain region (up to 8.5%) both loose and dense assemblies show a similar variation in the variation of contact normal anisotropy coefficient. But the dense assembly shows an increase up to the peak and later shows a declining trend. At large strain levels both loose and dense assemblies have the same magnitude of deviatoric contact normals anisotropy coefficients. This is indicative of the fact that at critical state the difference in the number of contacts in the major and minor principal stress direction is same for both loose and dense assemblies. This shows that at the critical state both loose and dense assembly has the same configuration of fabric.

5 Conclusions

The numerical simulation results show that qualitatively realistic macroscopic critical state behavior of granular assembly can be generated with DEM

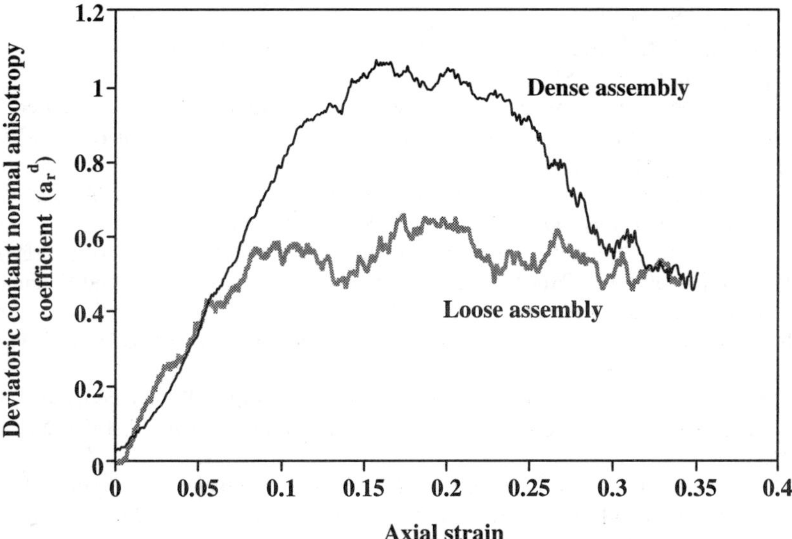

Figure 14 *Plot of deviatoric coefficient of contact normals anisotropy Vs axial strain in drained test at a confining pressure of 25 kPa*

simulations. The isotropic compression paths of loose and dense assembly are nonlinear and pressure dependent. The compression paths are approximately parallel at low stress levels and start converging at high stress levels. The critical state line is a curved line. The behaviour of the sample before the critical state depends on stress path and initial state/fabric. However, the critical state for a given granular material is unique irrespective of its initial state. The strength of the assembly at the critical state is the same for both loose and dense systems. The micromechanical analysis indicates that the average coordination number indicating number of neighboring particles per particle at the critical state is the same for loose and dense system in drained shear tests at constant confining pressure. The deviatoric anisotropy coefficients of normal contact force, tangential contact force and contact normals are the same for loose and dense system at the critical state, which is indicative of the fact that the magnitude of deviatoric normal contact force, tangential contact force and contact normals in the major and minor principal stress directions are the same for loose and dense assemblies.

Acknowledgements

The first author acknowledges the Department of Science and Technology (DST), Govt of India for financial support throughout the project [No. DST/ 23(183)ESS/99].

References

1. D. M. Wood, *Soil Behavior and Critical State Soil Mechanics*, Cambridge University Press, Cambridge, 1990.
2. K. Been, M. G. Jefferies and J. E. Hachey, "The critical state of sands", *Geotechnique*, 1991, **41**, 365–381.
3. A. S. Vesic and G. W. Clough, "Behavior of granular materials under high pressures", *Journal of the soil Mechanics and Foundation Engineering Division, ASCE*, 1968, **94**, No.3, 661–688.
4. K. L. Lee and H. B. Seed, "Drained strength characteristics of sand", *Journal of Soil Mechanics and Foundations Division, ASCE*, 1967, **93**, No.6, 117–141.
5. A. Casagrande, "Liquefaction and cyclic deformations of sands A critical review", Proceedings of the 5th Pan-American Conference on Soil Mechanics and Foundation Engineering, Buenos Aires, 1975, **5**, 80–133.
6. S. J. Poulos, "The steady state of deformation", *Journal of the Geotechnical Engineering Division, ASCE*, 1981, **111**, No.6, 772–792.
7. J. A. Sladen, R. D. D' Hollander, J. Krahn and D. E. Mitchell, "Back analysis of the Nerlerk berm liquefaction slides", *Canadian Geotechnical Journal*, 1985, **22**, No.4, 462–466.
8. G. Castro, R. B. Seed, T. O. Keller and H. B. Seed, "Steady state strength anslysis of lower San Fernando dam slide", *ASCE Journal of Geotechnical Engineering*, 1992, **118**, No.3, 406–427.
9. A. W. Bishop, "Shear strength parameters for undisturbed and remoulded soil specimens", in proceedings of the Roscoe memorial symposium on stress-strain

behavior of soils, edited by R. H. G. Parry, Cambridge University Press, Cambridge, U.K., 1972, 3–58.

10. S. Miura and S. Toki, "Anisotropy in mechanical properties and it's simulation in sand sampled from natural deposits", *Soils and Foundations*, 1984, **24**, No.3, 69–84.

11. H. Hanzawa, "Undrained strength and stability of quick sand", *Soils and Foundations*, 1980, **20**, No.3, 17–29.

12. Y. P. Vaid, E. K. F. Chung and R. H. Kuerbis, "Stress path and steady state", *Canadian Geotechnical Journal*, 1990, **27**, 1–7.

13. A. Alarcon-Guzman, G. A. Leonards and J. L. Chameau, "Undrained monotonic and cyclic strength of sands", *Journal of Geotechnical Engineering Division, ASCE*, 1988, **114**, No.10, 1089–1109.

14. K. Been and M. G. Jefferies, "A state parameter for sands", *Geotechnique*, 1985, **35**, No.2, 99–112.

15. P. A. Cundall and O. D. L. Strack, "A Discrete numerical model for granular assemblies", *Géotechnique*, 1979, **29**, No.1, 47–65.

16. O. D. L. Strack and P. A. Cundall, "Fundamental studies of fabric in granular materials", interim report to National Science Foundation, Department of Civil and Mineral Engineering, University of Minnesota, Minneapolis, Minnesota, 1984, 53 pp.

17. K. Chantawarungal, "Numerical simulations of three dimensional granular assemblies", Ph.D. Thesis, University of Waterloo, Waterloo, Ontario, Canada, 1993.

18. P. Dantu, "Contribution á l'étude mécanique et géometrique des milieux pulvérulents", *Proc. 4th Int. Conf. Soil Mech. Found. Eng.*, London, 1957, **1**, 144–148.

19. G. De Josselin de Jong and A. Verrujit, "Étude photo-élastique d'un empilement de disques", *Cahiers du Groupe Francais de Rhéologie*, 1969, **2**, No.1, 73–86.

20. M. Oda, "Initial fabrics and their relations to mechanical properties of granular materials", *Jap. Soc. Soil Mech. Fdn. Engrg.*, 1972a, **12**, No.1, 17–36.

21. M. Oda, "Mechanism of fabric changes during compressional deformation of sand", *Jap. Soc. Soil Mech. Fdn. Engrg.*, 1972b, **12**, No.2, 1–18.

22. M. Oda, "Deformation mechanism of sand in triaxial compression tests", *Jap. Soc. Soil Mech. Fdn. Engrg.*, 1972c, **12**, No.4, 45–63.

23. A. Drescher and G. De Josselin de Jong, "Photoelastic verification of a mechanical model for the flow of a granular material", *J. Mech. Phy. Solids*, 1972, **20**, 337–351.

24. M. Oda and J. Konishi, "Microscopic deformation mechanism of granular materials in simple shear", *Jap. Soc. Soil Mech. Fdn. Engrg.*, 1974a, **14**, No.4, 25–38.

25. M. Oda and J. Konishi, "Rotation of principal stresses in granular material in simple shear", *Jap. Soc. Soil Mech. Fdn. Engrg.*, 1974b, **14**, No.4, 39–53.

26. O. D. L. Strack and P. A. Cundall, "The Discrete Element Method as a tool for research in granular media, Part I, report to National Science Foundation", Department of Civil and Mineral Engineering, University of Minnesota, Minneapolis, Minnesota, 1978, 97 pp.

27. L. Rothenburg, "Micromechanics of idealized granular systems", Ph.D. Dissertation, Carleton University, Ottawa, Canada, 1980.

28. R. J. Bathurst, "A study of stress and anisotropy in idealized granular assemblies", Ph.D. Dissertation, Department of Civil Engineering, Queen's University, Kingston, Ontario, 1985.

29. P. A. Cundall, BALL–"A Computer Program to Model Granular Media Using the Distinct Element Method", Technical Note TN-LN-13, Advance Technology Group, Dames and Moore, London, 1978, 129–163.

30. C. Thornton and D. J. Barnes, "Computer simulated deformation of compact granular assemblies", *Acta Mechanica*, 1986, **64**, pp. 45–61.

31. L. Rothenburg and R. J. Bathurst, "Analytical study of induced anisotropy in idealized granular materials", *Géotechnique*, 1989, **39**, No. 4, 601–614.

32. T. G. Sitharam, "Numerical simulation of hydraulic fracturing in granular media", Ph.D. Thesis, University of Waterloo, Waterloo, Ontario, Canada 1991.

33. C. Thornton and G. Sun, "Axisymmetric compression of 3D polydisperse systems of spheres", in *Powders and Grains, 93*, (Ed. C. Thoronton), Balkema, Rotterdam, 1993, 129–134.

34. C. Thornton and G. Sun, "Numerical simulation of general three-dimensional quasi-static shear deformation of particulate media", in Numerical Methods in Geotechnical Engineering, Ed. I. M. Smith), Balkema, 1994, 143–148.

35. C. Thornton, in *Solid-Solid Interactions*, M. J. Adams, S. K. Biswas and B. J. Briscoe (Eds.), Imperial College Press, 1996, 250–264.

36. C. Thoronton, M. T. Ciomocos, K. K. Yin and M. J. Adams, "Fracture of particulate solids", in *Powders and Grains 97*, (Eds. R. P. Behringer and J. T. Jenkis), Balkema, Rotterdam, 1997, 131–134.

37. C. Thornton and S. J. Antony, "Quasi-static deformation of particulate media", *Phil. Trans. R. Soc. London A* 356, 1998, 2763–2782.

38. D. Robertson, "Numerical simulation of crushable aggregates", Ph.D. Dissertation, University of Cambridge, 2000.

39. C. Thornton and S. J. Antony, "Quasi-static shear deformation of a soft particle system", *Powder Technology*, 2000, **109**, 179–191.

40. T. G. Sitharam, S. V. Dinesh and N. Shimizu, "Micromechanical modelling of monotonic drained and undrained behaviour of granular media using three dimensional DEM", *Int. J. Numer. Ana. Meth. Geomechanics*, 2002, **26**, 1167–1189.

41. G. R. McDowell and O. Harireche, "Discrete element modelling of soil particle fracture", *Geotechnique*, 2002a, **52**, No.2, 131–135.

42. G. R. McDowell and O. Harireche, "Discrete element modelling of yielding and normal compression of sand", *Geotechnique*, 2002b, **52**, No.4, 299–304.

43. S. V. Dinesh, "Discrete element simulation of static and cyclic behavior of granular media", Ph.D. Thesis, submitted to Indian Institute of Science, Bangalore, India 2002.

44. S. J.Antony, "Evolution of force distribution in a three dimensional granular media", *Physical Review E, American Physical Society*, 2001, **63(1)**, 011302.

45. L. D. Landau and E. M. Lifshitz, *Theory of Elasticity, 3rd Ed.*, Pergamon Press, Oxford, 1986, 1–2.

46. J. Weber, *Recherche concernant les contraintes des Ponts et Chaussées*, 1966, **20**, 3.1–3.20.

47. J. Christofferson, M. M. Mehrabadi and S. Nemat-Nasser, "A micromechanical description of granular material behavior", *Journal of Applied Mechanics*, 1981, **48**, 339–344.

48. F. Tatsuoka, M. Sakamoto, T. Kawamura and S. Fukushima, "Strength and deformation characteristics of sand in plane strain compression at extremely low pressures", *Soils and Foundations*, 1986, **26**, No.1, 65–84.

CHAPTER 6

Key Features of Granular Plasticity

F. RADJAÏ[1], H. TROADEC[1] and S. ROUX[2]

[1]Laboratoire de Mécanique et Génie Civil, CNRS-Université Montpellier II, Place Eugène Bataillon, 34095 Montpellier cedex, France.
[2]Laboratoire Surface du Verre et Interfaces, CNRS-Saint Gobain, 39 Quai Lucien Lefranc, 93303 Aubervilliers Cedex, France
Email: radjai@lmgc.univ-montp2.fr

1   The Mystery of Sand

Sand and other cohesionless or weakly cohesive granular materials share marked *plastic* properties which reflect their common *granular structure*. Intuitively, we understand the flow and irrecoverable deformations of sand as an evident consequence of the relative displacements of the grains (seen as solid particles interacting via contact and friction) caused by an external mechanical action. The mystery is that, this desperately simple (and correct) picture of granular plasticity does not let itself be captured into an equally simple model thoroughly based on the properties of the grains and their organization in space. Recent interdisciplinary research on the matter has even transformed the status of sand as a rather old poetic symbol of simplicity into a paradigm of *complexity*![1] This metamorphose is motivated by the observation that dry granular materials behave very differently both from ordinary molecular fluids when they flow and from ordinary solids when they remain at rest.[2] Obviously, an extraordinarily rich behaviour emerges when simple grains are piled to form a granular structure.

The experimental observation of force inhomogeneity[3] and structural anisotropy[4] in model particle assemblies about four decades ago suggested that a detailed description of granular *microstructure* should provide a key for understanding the quasistatic rheology of granular materials. For two decades, such a description has been offered by *distinct element* numerical simulations.[5,6] The elements are idealized grains (discs, polygons . . . in two dimensions, spheres, polyhedra . . . in three dimensions). The equations of motion of each grain are integrated by taking into account contact interactions and body or boundary forces and displacements. It happens that, even with most basic ingredients, a rich behaviour is observed as in experiments. The simulations provide detailed information about grain motions and contact forces. Fascinating phenomena, such as the *bimodal* transmission of stresses[7] and *collective* particle motions at intermediate scales between the grain size and the system size,[8,9] can be observed. Yet, such often intriguing phenomena have in many ways helped more to reinforce the mystery of sand than to unravel it.

The difficulty here is common to all heterogeneous materials: what is the simplest level of microstructural information, and to what extent does it control the effective properties of the material? How do the effective properties depend on higher-order microstructural information? The phase volume fractions often provide trivial first-order information. In the case of a dry granular material, we distinguish the solid phase from the pore phase. The *solid fraction* ρ (volume fraction occupied by the grains) is known to influence strongly the shear strength and stress-strain behaviour.[10,11] However, the mean coordination number z (the average number of contact neighbours of a grain) can be used, as well, as a descriptor of the average *compactness* of the structure. But, the choice between ρ and z is not a mere matter of taste since the idea is to account properly for *grain scale* mechanisms. The point is that the grains interact via *contact* and this property is an essential ingredient of granular materials. The equilibrium and motion of a grain are thus more related to its *contact neighbours* than to the average free volume accessible to the grain. This implies that the coordination number is more suitable than the solid fraction as state variable.

Recognizing the coordination number z as the *lowest-order* relevant microstructural information, represents already a step beyond the phenomenological approach. This is also a rewarding choice as it naturally points to higher order microstructural information. In fact, the coordination number, as an average over all grains in a control volume, does not make much sense to the grains which always have an *integer* number of contact neighbours. Hence, a more detailed description of the microstructure requires the *connectivity* function $p_c(c)$, defined as the fraction of grains with exactly c contact neighbours.

In order to describe the grain equilibrium states, which underlie the yield properties of a granular medium, further information is required about the angular positions of the contact neighbours. For this purpose, one may rely on the *fabric tensors* of increasing order ($\vec{n} \otimes \vec{n}$, $\vec{n} \otimes \vec{n} \otimes \vec{n} \otimes \vec{n}$, etc) constructed from contact normals $\vec{n}$, or resort to the 1-contact probability density function $p(\vec{n})$ defined as the probability that a contact normal is oriented along $\vec{n}$.[12] An exact description of the environment of a grain requires, however, *multi-contact* probability density functions $g_c(\vec{n}_1, \vec{n}_2, \ldots, \vec{n}_c)$ corresponding to the probability that the contact neighbours (for a grain with c contact neighbours) occupy the angular positions ($\vec{n}_1, \vec{n}_2, \ldots, \vec{n}_c$) around the grain.[13,14] Both p_c and g_c are controlled by steric exclusions of the grains and they depend on the *composition* (grain shapes and size distribution). More importantly, they evolve with plastic strain.

We see that the *steric constraints* (mutual exclusions together with excluded-volume effects that impose an upper bound on the number of contact neighbours) result in a dependence of the yield properties of a granular medium on nontrivial features of the microstructure (beyond the solid fraction). This is also true for the flow properties (dilatancy) which require the compatibility of the global imposed strain with local steric constraints. In the same way, the hardening properties reflect the evolution of the microstructure which involves both discontinuous changes, due to creation and loss of contacts between the grains, and distortions as a result of the rotations of persisting contact normals.

The apparent mystery of granular materials reflects not only the complexity of their microstructure but also the fact that the effective plastic properties involve nontrivial details of this microstructure and its evolution. The ambition of this contribution is not to present a model of granular plasticity based on the microstructure though this is obviously the long term scope of this work. Instead, we discuss a number of basic behaviours in the light of simple microstructural considerations by indicating at each step the route to a more fundamental approach. In particular, we emphasize the role of steric constraints, compactness and structural anisotropy, retained as the most salient microstructural information, with respect to shear strength, stress-strain and volume-change properties in model granular media composed of rigid (two-dimensional) disks interacting via contact and friction.

2 Granular Friction

The granular microstructure can be seen as the disordered grain configuration that spans the space from the contact scale to the system boundaries. The query is how this microstructure controls the scale-up of the contact rheology (local behaviour) to the macroscopic scale (global behaviour). Within the hard-particle approximation, assumed here, the contact rheology is approximated by two "contact laws": the Signorini condition (mutual exclusion of the grains) and the Coulomb friction law. The Signorini condition simply relates the distance δ between two particles and the corresponding normal reaction force N: For $\delta \neq 0$ (no contact), the normal force is zero, $N = 0$, whereas for $\delta = 0$ (contact), N can take arbitrarily large positive (compressive) values. This relation is shown as a graph in Figure 1(a). This is a *nonsmooth* relation in the sense that the set of admissible pairs (δ, N) can not be reduced to a (mono-valued) function.[15–17] On the other hand, the Coulomb friction law relates the tangential force T and the sliding velocity v_s at contacts between grains. Again, the set of admissible pairs (v_s, T), shown as a graph in Figure 1(b), can not be reduced to a function: at $v_s = 0$, the friction force T can take any value in the range $[-T_s, T_s]$, with $T_s = \mu N$, where μ is the coefficient of friction. The Coulomb friction law and the Signorini condition have the remarkable property that they involve *no force scale*. The range of admissible contact forces $\vec{f} = (N, T)$ is inside a cone $T_s = \mu N$ with $N > 0$.

As a result of the absence of intergrain forces of cohesion ($N > 0$), the stress state is, for the most part, controlled by the boundary stresses and bulk forces. Classically, the Cauchy stress tensor $\overline{\overline{\sigma}}$ in a static granular medium is an average

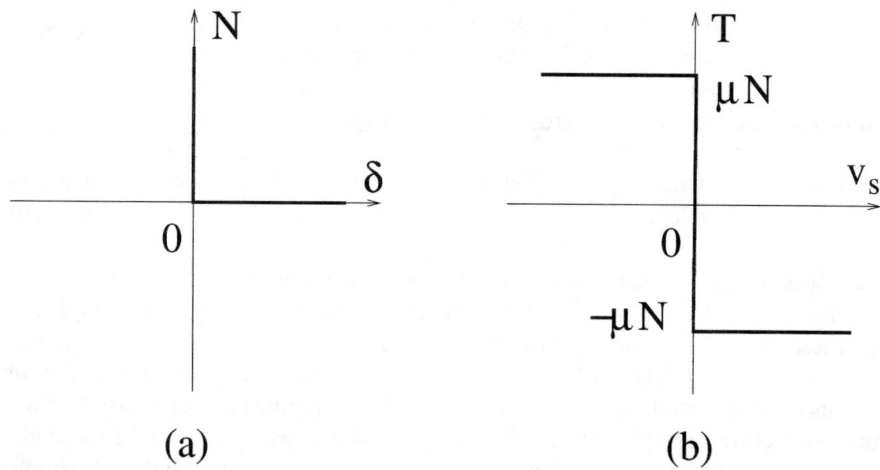

Figure 1 *Nonsmooth contact laws: (a) Signorini condition, (b) Coulomb friction law*

quantity involving contact forces $\vec{f} = (N,T)$ and intercentre vectors $\leftrightarrow$ joining the centres of contact neighbours:[18-20]

$$\overline{\overline{\sigma}} = n_c \langle \vec{f} \otimes \leftrightarrow \rangle \tag{1}$$

The symbol $\otimes$ stands for dyadic product and n_c is the number density of contacts. The averaging $\langle \ldots \rangle$ runs over the whole contact set in a control volume. Since $\overline{\overline{\sigma}}$ is linear in $\vec{f}$, and the forces belong to a cone, the domain of admissible stresses is also necessarily inside a cone $\tau_s = Mv$ with $v > 0$ (compressive) in the stress space, where $\tau_s \equiv \vec{n}\overline{\overline{\sigma}}\,\vec{t}$ and $v \equiv \vec{n}\overline{\overline{\sigma}}\,\vec{n}$ are respectively the shear and normal stresses on the yield plane (line, in two dimensions) characterized by its normal and tangential unit vectors $\vec{n}$ and $\vec{t}$, and M is a *global* (or effective) coefficient of friction.

The above argument recovers the Coulomb yield criterion, with $\varphi = \tan^{-1}(M)$ corresponding to the *internal angle of friction*. Equivalently, the stress ratio q/p, where p and q are the mean and deviator stresses respectively, is given by $q/p = \sin\varphi$. The important point here is that, this argument does not refer to the microstructure, so that M is defined *whatever* the mechanical state of the medium. The usual Coulomb yield criterion corresponds only to the steady state (the "critical state" of soil mechanics[21]) which is a particular state of the material (reached upon sufficient monotonic shearing and in which no volume change occurs). The Coulomb model is thus a *rigid-plastic* behaviour without state variables (no evolution of the microstructure). This model can be considered as a simple scale-up of contact friction with the yield plane (determined *a posteriori*) playing the same role as the contact plane between two grains (fixed *a priori*). Since the material is assumed to possess a unique state, corresponding to a unique microstructure that does not evolve, *the* internal angle of friction φ in the Coulomb model is an intrinsic property of the material.

However, since the internal angle of friction φ is defined whatever the microstructure, the Coulomb model can readily be generalized to arbitrary states through the dependence of φ on the microstructure.[13,22] In other words, assuming that p_c and g_c contain the relevant microstructural information, the yield properties of a granular material are characterized by the function $\varphi = \varphi(p_c, g_c)$. Upon shearing, the distributions p_c and g_c evolve incrementally with plastic strain $\dot{\varepsilon}_s$: $\dot{p}_c = \dot{p}_c(\dot{\varepsilon}_s, p_c, g_c)$ and $\dot{g}_c = \dot{g}_c(\dot{\varepsilon}_s, p_c, g_c)$. The steady state corresponds to $\dot{p}_c(\tilde{p}_c, \tilde{g}_c) = \dot{g}_c(\tilde{p}_c, \tilde{g}_c) = 0$, where $\tilde{p}_c$ and $\tilde{g}_c$ are steady-state distribution functions. Hence, in this framework the internal angle of friction is *state dependent*, only the steady-state value $\tilde{\varphi}(\tilde{p}_c, \tilde{g}_c)$ being a material property. In the following, we will discuss the potential predictions of this generalization of the Coulomb model to include state variables pertaining to the microstructure.

3   The Fabric

The distributions g_c and p_c contain rich information about local structures characterized by the positions of contact neighbours in the angular interval $[0, 2\pi[$ around a typical grain with c contact neighbours. The drawback with an approach based on too detailed microstructural information is that it requires simplifications at the level of the composition and resort to numerical procedures in order to obtain quantitative results. In fact, discrete element simulations can be seen as an extreme approach totally based on *all* details of the microstructure (positions and velocities of all grains). This approach provides quantitative results and, due to the fast increase of the available computer power and memory, it can be used as a precious investigation tool in the domain of granular materials. But because of the profusion of information, it does not readily help to get insight into the behaviour. In a microscopic approach, a useful strategy is to begin with the lowest-order microstructural information and to enrich the description by including higher order information in a progressive manner.[13]

Key information contained in g_c is the *angular exclusions*: the angular positions of contact neighbours around a grain are such that, due to their mutual hindrances, there is a finite difference between them always larger than an *exclusion angle* $\delta\theta_{min}$ (see Figure 2), so that[13,14]

$$g_c = 0 \Leftrightarrow \vec{n}^\alpha \cdot \vec{n}^\beta < \cos(\delta\theta_{min}) \tag{2}$$

The exclusion angle is about $\pi/3$ for grains of nearly the same size. The angular exclusions, as a local property of the granular structure, obviously play a key role in yield, flow and hardening properties of a granular medium by virtue of the constraints they impose on grain motions and accessible equilibrium states.

The coordination number $z = \sum_c c \, p_c$ describes the average connectivity of granular structure. An equivalent parameter is the valence number v, defined as the average number of edges per void cell (a loop of contiguous grains surrounding a pore).[8] The Euler formula for the grain network allows us to relate the

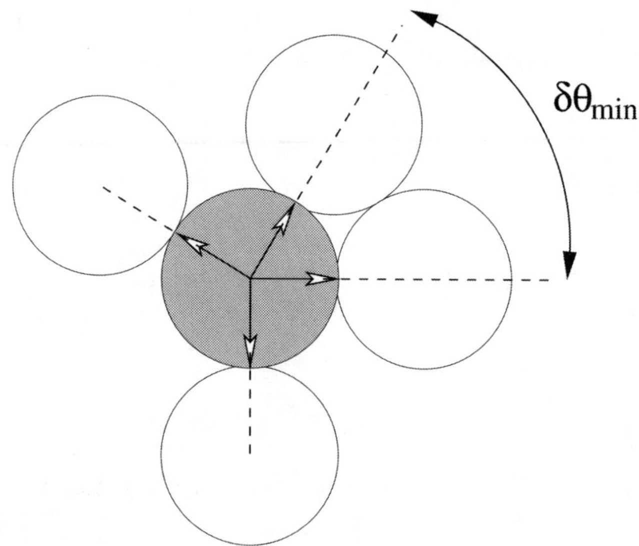

Figure 2   *Illustration of angular exclusions*

valence number to the coordination number: $v = 2z/(z - 2)$. The coordination number is limited by an upper bound z_{max} as a result of angular exclusions. z_{max} is equal to 6 for grains of tight polydispersity (weak size dispersion) in two dimensions in the presence of long-range (crystalline) order and equal to 4 in the presence of disorder.[23] There is also a lower bound z_{min} imposed by the requirement of grain equilibrium. Numerical simulations show that, although a grain can in principle be equilibrated under the action of *two* frictional contact forces, the coordination number never decreases below 3.

Other basic information included in the function g_c is the 1-contact distribution function $p_\theta(\theta)$ of contact directions θ which is obtained by integrating $\sum_c p_c g_c$ over all angles except one. Notice that, due to angular exclusions, the multi-contact distribution g_c is not a product of 1-contact distribution functions: $g_c(\theta_1, \ldots, \theta_c) \neq p_\theta(\theta_1) \ldots p_\theta(\theta_c)$. The distribution p_θ is easily accessible from (two-dimensional) experiments and simulations, and it has been extensively investigated in the past.[4,7,19,20,24] A fundamental observation is that p_θ is not uniform: there are preferred contact directions. In other words, the granular structure is generically *anisotropic*. This bias in contact directions is induced by the relative motions of grains: contacts are gained along the directions of compression and lost along the directions of extension. This means that, starting from an isotropic packing and applying a monotonic homogeneous shearing, the function p_θ is expected to tend to a simple form reflecting the elongational deformation in different directions, namely $p_\theta = (1/2\pi)(1 + a \cos 2(\theta - \theta_c))$, where a represents the *anisotropy* of the packing, and θ_c is the average contact direction. Indeed, this simple form, with two parameters a and θ_c characterizing the anisotropy and its direction, provides a reasonable fit for a number of numerical

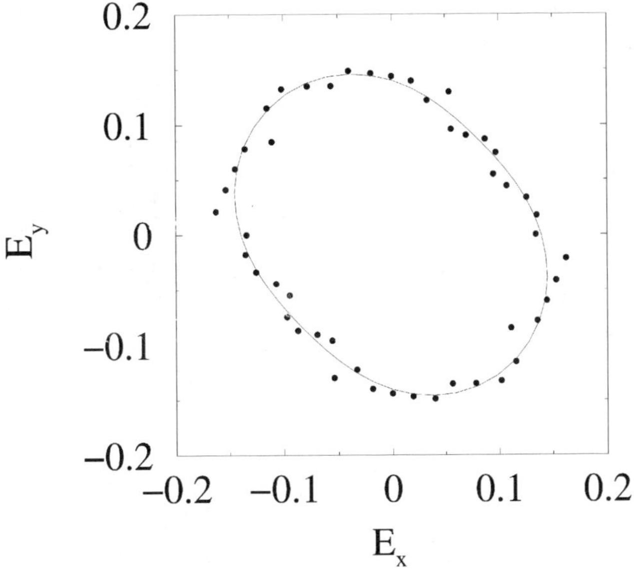

Figure 3 *Polar representation of the distribution function of contact normal orientations in a simple shear simulation. The function is normalized such that its integral is equal to the coordination number (see the definition of function E(θ) defined in the text)*

and experimental data;[19,24] see Figure 3. The values of a and θ_c can be more conveniently extracted from the fabric tensor $\overline{\overline{F}} = \langle \vec{n} \otimes \vec{n} \rangle$.[12] It is easy to show that $a = 2(F_1 - F_2)$, where $F_{1,2}$ are the eigenvalues, and θ_c is the major principal direction of the fabric tensor.

Retaining a, θ_c and z as the only relevant state variables, the functions g_c and p_c can be constructed by requiring that their informational content is strictly equivalent to the angular exclusions, on one hand, and to the knowledge of a, θ_c and z, on the other hand.[13,14] In other words, the neglected information is simply replaced by *disorder*. In practice, this amounts to maximizing the statistical entropy associated with g_c and p_c constrained by the available information. The estimated distributions g_c by means of this method have been shown to compare well with numerical simulations in the steady state for $c = 4$ and $c = 5$ in two dimensions.[14]

4 Shear Strength

The anisotropy of the structure has a fundamental implication: the failure criterion φ can not be isotropic as in the basic Coulomb model. This means that the shear strength for an anisotropic microstructure varies with the direction of loading. In practice, this effect is more visible on the stress-strain behaviour in different directions and it has been observed in experiments.[10] This effect is particularly important when the direction of shear is reversed (unloading) in which case a long transient shear occurs before the critical state in a new direction is reached.[24] Let us recall that the basic Coulomb model (with an isotropic failure criterion) simply predicts that the system remains in the critical state for all shear directions.

The dependence of the internal angle of friction on structural anisotropy a is a direct consequence of the fact that the distribution p_θ enters the micromechanical expression of the stress tensor (Equation 1). Setting $\vec{f} = N\vec{n} + T\vec{t}$, $\leftrightarrow = l\vec{n}$ and $\vec{n} = (\cos\theta, \sin\theta)$ for a two-dimensional packing of disks and neglecting the correlations between l and N (nearly absent in two dimensions), the expression of the stress tensor yields the following equations for the mean stress p and the deviatoric stress q:

$$p = \frac{n_c l_m}{2}\langle N\rangle \tag{3}$$

$$q = \frac{n_c l_m}{2}\left(\langle N\cos 2(\theta - \theta_\sigma)\rangle + \langle T\sin 2(\theta - \theta_\sigma)\rangle\right) \tag{4}$$

where l_m is the mean intercentre distance. Notice that the two terms of the expression of q in Equation 4 correspond to the *correlation* between the force amplitude (N or T) and the contact direction (θ).

The measure for the averaging operation $\langle\ldots\rangle$ in its integral form is $p_e(\theta, N, T)$ $d\theta\, dN\, dT$ (the average of a quantity A is given by $\langle A\rangle = \int A\, p_e d\theta\, dN\, dT$). The probability distribution function $p_e(\theta, N, T)$ is the probability for a contact to have its direction along θ and to carry normal and tangential forces N and T, respectively. By definition, $p_\theta(\theta) = \int p_e\, dN\, dT$. The distribution p_e allows also to define the average forces $N_m(\theta)$ and $T_m(\theta)$ by

$$N_m(\theta)\, p_\theta(\theta) \equiv \int N\, p_e\, dN\, dT \tag{5}$$

$$T_m(\theta)\, p_\theta(\theta) \equiv \int T\, p_e\, dN\, dT \tag{6}$$

In the absence of correlations, Equation 4 implies $q = \frac{n_c l_m}{2}\left(\langle N\rangle\langle\cos 2(\theta - \theta_\sigma)\rangle + \langle T\rangle\langle\sin 2(\theta - \theta_\sigma)\rangle\right)$. But, due to equilibrium, we have $\langle T\rangle = 0$, so that $q/p = \langle\cos 2(\theta - \theta_\sigma)\rangle$. For an isotropic structure ($p_\theta(\theta) = 1/2\pi$), this implies $q/p = 0$. For an anisotropic structure, the largest value of the stress ratio q/p is obtained if the distribution $p_\theta(\theta)$ of contact directions varies as $\cos 2(\theta - \theta_c)$ with $\theta_c = \theta_\sigma$. This implies *coaxiality* between the fabric tensor $\overline{\overline{F}}$ and the stress tensor. Then, the Fourier expansion

$$p_\theta(\theta) = \frac{1}{2\pi}[1 + a\cos 2(\theta - \theta_c) + \ldots] \tag{7}$$

where a is the amplitude of anisotropy, yields $q/p = \int_0^{2\pi} p_\theta(\theta)\cos 2(\theta - \theta_c)d\theta = a/2$.

Note that for the largest anisotropy $a = 1$ (which can never happen due to structural disorder), the largest stress ratio is $q/p = 1/2$, corresponding to $\varphi = 30°$.

The simulations show, however, a strong correlation between the forces and the contact directions.[19] According to Equation 4, the stress ratio is maximized with respect to θ if N varies as $\cos2(\theta - \theta_\sigma)$, T varies as $\sin2(\theta - \theta_\sigma)$ and $p_\theta(\theta)$ varies as $\cos2(\theta - \theta_\sigma)$. Of course, it is highly improbable that this should happen individually for each contact, but the average forces $N_m(\theta)$ and $T_m(\theta)$ as a function of contact direction do show this behaviour in a monotonic deformation.[19] More generally, starting with a given fabric, the stresses can be applied in *arbitrary* directions, and fabric changes are not immediate. From Equation 4 and using the Fourier expansions

$$N_m(\theta) = \langle N \rangle [1 + a_n \cos 2(\theta - \theta_n) + \ldots] \tag{8}$$

$$T_m(\theta) = \langle N \rangle [a_t \sin 2(\theta - \theta_t) + \ldots] \tag{9}$$

where θ_n and θ_t are the preferred directions of N_m and T_m, we get

$$\frac{q}{p} = \frac{a_n}{2} \cos 2(\theta_\sigma - \theta_n) + \frac{a_t}{2} \cos 2(\theta_\sigma - \theta_t) + \frac{a}{2} \cos 2(\theta_\sigma - \theta_c) \tag{10}$$

where cross products among the anisotropies a, a_n and a_t have been neglected. This is an interesting relation as it shows that the shear strength depends on the direction of loading (as expected from the anisotropy of the structure). It is important to remark that the presence of an anisotropic fabric that is not coaxial with the stress tensor breaks the chiral (left-right) symmetry of the problem with respect to the principal stress directions. This means that, in order to interpret correctly the phase differences in Equation 10, one should ensure that the amplitudes a, a_n and a_t, as well as q, all have the same sign, e.g. positive.

Equation 10 separates two microstructural origins of the shear strength in a granular packing: 1) structural anisotropy, represented by the parameter a; 2) force anisotropy captured into the parameters a_n and a_t. The parameters a_n and a_t can be considered also as structural properties that, in the last analysis, are linked with the anisotropy a. This link is complex, involving the microstructure beyond the first neighbours due to force correlations. Since the anisotropy and phase differences are internal (hardening) parameters from a micromechanical point of view, the expression of the stress ratio in Equation 10 can be interpreted as that of the *yield* stress ratio $\sin\varphi$ for a given set of internal parameters. Then, θ_σ should be replaced by space direction θ. On the other hand, we have to set $\varphi = 0$ for the directions that lead to a negative value of the deviator stress. Starting with an arbitrary initial state, *monotonic* shearing tends to bring the structure to a force-fabric coaxial state $\theta_c = \theta_n = \theta_t$. In this case, we have $\varphi = 0$ for $\theta_c + \pi/4 < \theta < \theta_c + 3\pi/4$ since q in Equation 10 is negative for these directions. The expression of the internal angle of friction becomes

$$\sin \varphi = \begin{cases} \dfrac{1}{2}(a_n + a_t + a)\cos 2(\theta - \theta_c) & \text{if } \theta < \theta_c + \pi/4 \text{ or } \theta_c + 3\pi/4 < \theta \\ 0 & \text{if } \theta_c + \pi/4 \le \theta \le \theta_c + 3\pi/4 \end{cases} \qquad (11)$$

The largest value of the shear strength is then $q/p = (a + a_n + a_t)/2$ occurring in the direction $\theta = \theta_c$.[19] When the stress principal axes rotate, phase differences may still persist even in monotonic shearing. Then, according to Equation 10, the shear strength is lower than $(a + a_n + a_t)/2$ due to those phase differences.

5 Stress-Strain Behaviour

The compactness, in terms of the solid fraction ρ or the coordination number z, does not enter the expression of the stress ratio in Equation 10. At first sight, this might appear in contradiction with the observation that the stress-strain behaviour is crucially dependent on the initial compactness of a granular sample. We know also from experiments that when an initially isotropic assembly is subjected to shearing, the stress ratio always increases with shear strain whatever the initial density, whereas the density either increases, if the initial state is loose, or decreases, if the initial state is dense.[10] Here, we would like to show that these behaviours are reconciled if the shear strength depends only *implicitly* on the compactness through the anisotropy, the *accessible* anisotropies to the microstructure being constrained by the compactness.

There are two ways to evaluate these constraints: from the grain-scale environments or from a global point of view. Let us first consider one grain with its c contact neighbours, all having nearly the same size so that the exclusion angle is $\pi/3$ (see Figure 2). The components of the local fabric tensor associated with the grain are $F_{ij} = (1/c)\sum_{a=1}^{c} n_i^a n_j^a$, where $\vec{n}^a$ is the normal unit vector of the contact a. Maximizing the anisotropy $a = 2(F_1 - F_2)$ in the presence of steric constraints Equation 2 yields:[14]

$$a_{\max}(c) = \frac{2\pi}{3\sqrt{3}}\left(\frac{c_{\max}}{c} - 1\right) \qquad (12)$$

where $c_{\max}$ is the largest possible number of contact neighbours. Equation 12 shows that $a_{\max}$ increases nonlinearly from 0 for $c = 6$ to ≈ 2.4 for $c = 2$. Of course, the largest local anisotropy, by definition, is 2. For $c = 4$, we get $a_{\max} \approx 0.6$. The corresponding configuration is shown in Figure 4. Due to disorder, the largest anisotropies observed in two dimensions for slightly polydisperse systems are much lower (less than 0.2).

Alternatively, we may arrive at an expression for $a_{\max}$ by considering that all microstructural states of a granular assembly are enclosed between two *isotropic limit states*: (1) the *densest* isotropic state, characterized by $z_{\max}$ or $\rho_{\max}$, and (2) the *loosest* isotropic state, characterized by $z_{\min}$ or $\rho_{\min}$. These limit states are

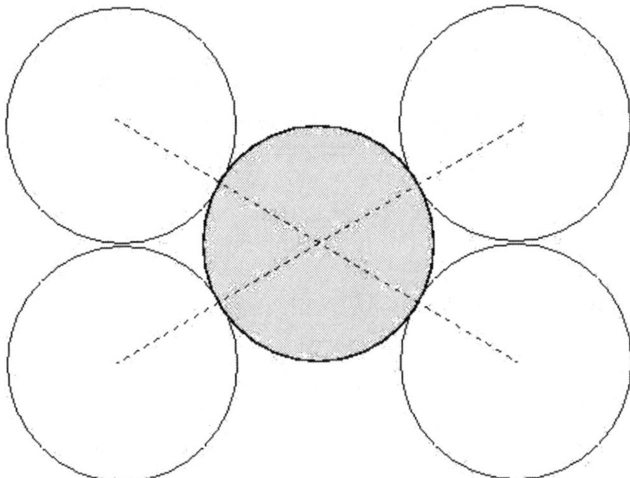

Figure 4 *A local configuration with four contact neighbours presenting the largest local anisotropy*

more difficult to reach than the steady state. In particular, it is generally difficult to bring a granular system towards a dense isotropic state by means of isotropic compaction. The reason is that the rearrangements occur mainly in the presence of shearing, and the latter induces structural anisotropy. Let us also mention here the concept of *random close packing* which refers to the densest disordered packing. This concept is not well-defined since the "randomness" to which its definition refers has not been specified in terms of the microstructure.[25] For example, relying only on the compactness and anisotropy as the relevant microstructural information, there is a continuous set of randomly close-packed systems for the anisotropy varying from zero to its largest value. The densest isotropic state does not need to be disordered. In two dimensions and for a monodisperse system, it corresponds to a triangular network with $z_{max} = 6$ and $\rho_{max} = 0.91$. In the absence of long-range correlations in contact directions, both z and ρ are always below these values. If the requirement of randomness in contact directions (no short-range correlations) is added, then $z_{max} \approx 4$.[23] In practice, partial crystallisation may occur spontaneously in the presence of gravity or flat walls, allowing z to exceed 4.

In general, z_{max} and ρ_{max} are controlled by steric constraints, whereas z_{min} or ρ_{min} are clearly related to the condition of grain equilibrium. The latter (in the absence of rolling strength between grains) implies that the fraction of grains with $c = 2$ can not be large. As a result, a loose isotropic packing should be constructed with $c = 3$. To reach the loosest structure, the pore volume fraction should be maximized. At the grain scale, it amounts to maximizing the area of a void cell. The valence number in a packing with $z = 3$, is $v = 6$, and simple geometry shows that *regular* hexagons (assuming that the distances between grain centres are all equal) have the largest area for a fixed circumference. As a result, the loosest local structure with $c = 3$ has also the property to be isotropic (the regularity of a polygon implies isotropy of the corresponding fabric). Hence, the loosest isotropic state may be identified with a hexagonal (honeycomb) packing with $z_{min} \approx 3$ and $\rho_{min} = \pi/3\sqrt{3} \approx 0.6$. Obviously, still looser structures with lower coordination numbers may be constructed and similar arguments may be used, but such structures are unstable due to the presence of grains with two contacts (contact chains). In fact, none of the two isotropic limits with $z = 3$ and $z = 6$ can be reached in practice. But, as *reference* states, they provide an intuitive representation of the limit isotropic states and their properties.

In order to characterize the geometrical states by a single function, let us introduce the function $E(\theta) = zp_\theta(\theta)$. This is the distribution of contact directions normalized by the coordination number, so that $\int_0^{2\pi} E(\theta)d\theta = z$. The two limit isotropic states are then represented by $E_{max}(\theta) = z_{max}/2\pi$ and $E_{min}(\theta) = z_{min}/2\pi$. Equivalently, we introduce the fabric tensor $\overline{\overline{G}} \cong z\overline{\overline{F}}$ with eigenvalues $G_1 = zF_1$ and $G_2 = zF_2$. The limit isotropic states correspond to $G_1 = G_2 = G_{min} = z_{min}/2$ and $G_1 = G_2 = G_{max} = z_{max}/2$. Let z ($\in [z_{min}, z_{max}]$) be the current coordination number. Starting with an isotropic distribution $E(\theta) = z/2\pi$, we may construct an anisotropic distribution with a fixed value of z by adding contacts in one direction, θ_c, and removing as many contacts in the perpendicular direction $\theta_c + \pi/2$. The condition that z remains constant implies

$$G_1 + G_2 = z \tag{13}$$

and the anisotropy is given by

$$2(G_1 - G_2) = az \tag{14}$$

This procedure can continue until either $G_1 = G_{max}$ (*gain saturation*) or $G_2 = G_{min}$ (*loss saturation*). By virtue of Equation 13, when one of these two extremes is reached, the anisotropy can be increased no more with a fixed value of z. Hence, the largest anisotropy $a_{max}(z)$ for a specified value of z is fixed either by Gain Saturation (GS) or by Loss Saturation (LS). Using Equations 13 and 14 we get

$$a_{max}(z) = \min\left\{2\left(1 - \frac{z_{min}}{z}\right), 2\left(\frac{z_{max}}{z} - 1\right)\right\} \tag{15}$$

This function is shown in Figure 5. By construction, $a_{max}(z_{min}) = a_{max}(z_{max}) = 0$. The largest possible anisotropy a_{Max} is

$$a_{Max} = a_{max}(z_{mean}) = 2\frac{z_{max} - z_{min}}{z_{max} + z_{min}} \tag{16}$$

for $z = z_{mean} = (z_{max} + z_{min})/2$. Using $z_{min} = 3$ and $z_{max} = 4$, we get $a_{Max} \approx 0.28$. According to Equation 15, a_{max} increases with z for $z < z_{mean}$, and it declines with

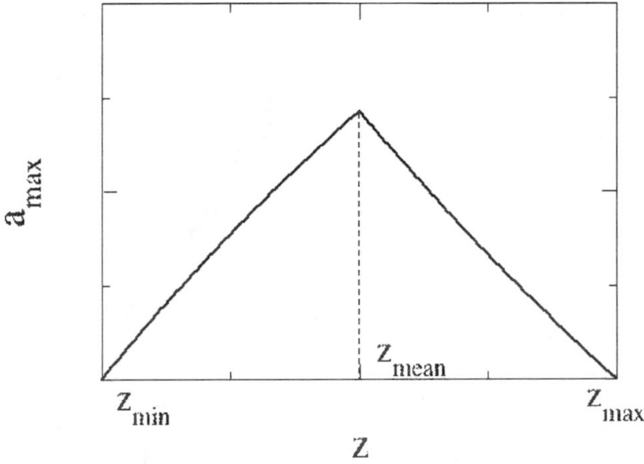

Figure 5 *Maximal anisotropy a_{max} as a function of coordination number z*

z for $z > z_{mean}$. When the anisotropy $a = a_{Max}$ is reached along a monotonic path, the anisotropy and the coordination can evolve no more since both contact gain and contact loss are saturated. This suggests that the steady state corresponds to the intersection between the two regimes, so that $\tilde{z} = z_{mean}$ and $\tilde{a} = a_{Max}$. Note that the expression of $a_{max}(c)$ in Equation 12 has the same functional dependence on c as $a_{max}(z)$ in Equation 15 on z in the gain saturation regime.

In Figure 6 we have shown an example of the evolution of the anisotropy a with the coordination number z in a biaxial compression test simulated by the contact dynamics method for two initially quasi-isotropic samples with different initial coordination numbers $z_0 = 3.1$ and $z_0 = 3.7$. The composition and the coefficient of friction ($\mu = 0.8$) are similar in the two samples. In both cases z tends to the same steady-state value $\tilde{z} \approx 3.35$ for which $\tilde{a} \approx 0.24$. From these values and using Equation 16 together with the assumption that $\tilde{a} = a_{mean} = (z_{min} + z_{max})/2$, we get $z_{min} = 2.94$ and $z_{max} = 3.75$. The saturation curve (Equation 15) is shown for these values in Figure 6. We see that in the gain saturation regime (as z decreases from $z_0 = 3.7$) the anisotropy reaches and then follows closely the saturation curve up to the steady state. Hence, in this regime, Equation 15 provides an excellent fit to the data with only one fitting parameter z_{max}. In the loss saturation regime (as z increases from $z_0 = 3.1$), the data remain below the saturation curve, reached only at the steady state. Even with very slow shearing, the evolution of a loose sample from an initially isotropic state is strongly unstable and dynamic. As a result, the rearrangements at the initial stages of evolution (as long as the coordination number is low) do not give rise to an oriented gain of contacts. This leads to a slow increase of the anisotropy compared to the coordination number. This effect disappears as the coordination number becomes larger.

Coming back to the monotonic stress-strain behaviour, we see that the simple arguments developed above and partially corroborated by numerical simulations, allow us to characterize the steady state in a simple way and to distinguish

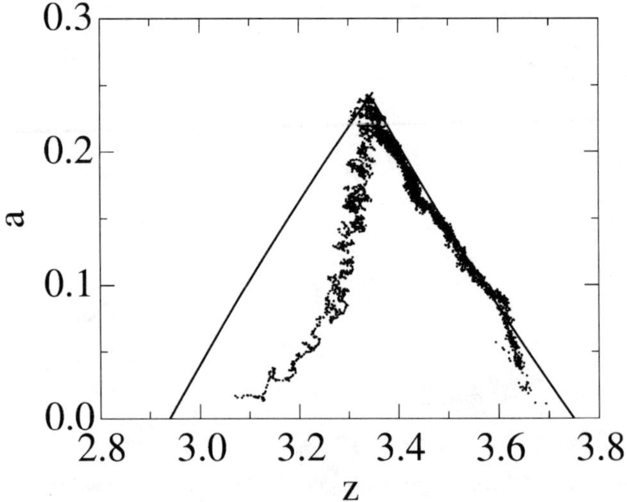

Figure 6   *Evolution of the anisotropy a with the coordination number z in a biaxially compressed assembly of rigid grains simulated by the contact dynamics method for an initially dense (right curve) and an initially loose (left curve) sample. The plain curve corresponds to the theoretical prediction (Equation 15)*

two regimes: (a) $z_0 < \tilde{z}$ (initially loose system): since $z_0 < \tilde{z}$, z increases towards $\tilde{z}$ by gain of contacts. The system remains in the LS regime during its evolution towards the steady state. Hence, $a_{\max}$ and the shear strength, as a result, increase monotonically with z to saturate at their steady-state value; (b) $z_0 > \tilde{z}$ (initially dense system): since $z_0 > \tilde{z}$, z decreases towards $\tilde{z}$ by loss of contacts. The system remains in the GS regime. As a result, $a_{\max}$ and the shear strength increase until z reaches its steady-state value. It is also worth noting that shear localization need not worry us about the application of Equation 15 since the same mechanisms continue to be active *inside* the shear zones.[26,27]

6   Equilibrium States

The equilibrium states of a granular packing subjected to external forces are globally characterized by the probability distribution $p_e(\theta, N, T)$. This function contains (by integration) the distributions of contact directions $(p_\theta(\theta))$, normal forces $(p_N(N))$, friction forces $(p_T(T))$ and friction mobilization $(p_\eta(\eta)$, with $\eta \cong T/\mu N)$, as well as force-fabric correlations. A major task in micromechanical modelling of granular plasticity is to deduce the distribution $p_e(\theta, N, T)$ from *purely geometrical* descriptors of the microstructure such as g_c and p_c. This is a difficult task since numerical investigations reveal broad force distributions and ingenious force-fabric correlations. Often qualitative concepts such as "force chains" or "arching" are used to describe the patterns observed in experiments (optical visualization of stresses by means of the photoelastic effect, for example).[3] Here, we present a brief review of some outstanding properties of

$p_e(\theta, N, T)$ with the motivation of understanding how a grain is equilibrated on average.

A peculiar aspect of force distribution, which we analyze in more detail here by focussing on normal forces, is the occurrence of numerous weak forces together with a number of strong forces appearing often sequentially (force chains).[3,6,28–30] The distribution p_N of normal forces in a macroscopically homogeneous system shows that about 57% of forces are below the mean force $\langle N \rangle$ and they have a *nearly* uniform distribution. They contribute only about 29% to the average pressure.[28] The number of forces larger than the mean decays almost exponentially.

The huge number of weak forces, as a consequence of grain frustrations or arching, with a nearly uniform distribution is a source of weakness for the system. Weak-force regions inside a packing correspond to locally weak pressures. Such regions are naturally more susceptible to fail. A quantitative analysis of grain rearrangements shows that during quasistatic evolution of the system, these weak regions undergo *local rearrangements*,[31] and nearly all sliding contacts (where the friction force is fully mobilized) are localized in weak regions.[7] Let $A(\xi)$ be the set of all contacts where the magnitude of the force is below a cutoff force ξ normalized by the mean force $\langle N \rangle$ in a macroscopically homogeneous packing. The set $A(\xi = \infty)$ corresponds then to the whole contact set. A simple analysis of simulation data shows that the fraction $\rho_s(\xi)$ of sliding contacts in the set $A(\xi)$ is not independent of ξ. Indeed, ρ_s increases with ξ from zero and tends to its largest value (8% for $\mu = 0.5$) at $\xi \cong 1$, corresponding to the average force. This means that, sliding occurs almost exclusively at contacts carrying a force below the average force.

Another interesting property of weak contacts in a granular medium is that, in monotonic shearing, the weak contacts are on average perpendicular to the direction of strong contacts.[7] To show this, we consider the partial fabric tensor $\overline{\overline{F}}(\xi)$ defined as the restriction of the expression of $\overline{\overline{F}}$ (defined above) to the contacts belonging to the set $A(\xi)$. The corresponding anisotropy $a(\xi)$ is a function of ξ. A typical example of $a(\xi)$ is shown in Figure 7. In this figure, a positive value of a indicates that $\theta_c(\xi)$ (average contact direction in the set $A(\xi)$) is along the compression axis, i.e. $\theta_c(\xi) \approx \theta_\sigma$, whereas a negative value of a corresponds to the direction $\theta_\sigma + \pi/2$. We see that the direction $\theta_c(\xi)$ is *perpendicular* to the axis of compression ($a < 0$) for weak forces ($\xi < 1$). The anisotropy increases (in absolute value) as ξ increases, and reaches a maximum at $\xi = 1$ corresponding to the average force. As ξ is further increased, a becomes less negative and finally changes sign at $\xi \approx 2$. This shows that the contacts which carry a force above the mean are preferentially oriented parallel to the axis of compression, and their positive contribution to a outweighs the negative contribution of the forces below the mean. Distinguishing in this way *strong contacts* (contacts where $N > \langle N \rangle$) from *weak contacts* (contacts where $N < \langle N \rangle$), we decompose the fabric tensor and the anisotropy a as a sum of two terms with opposite phases:

$$a = a_s - a_w \tag{17}$$

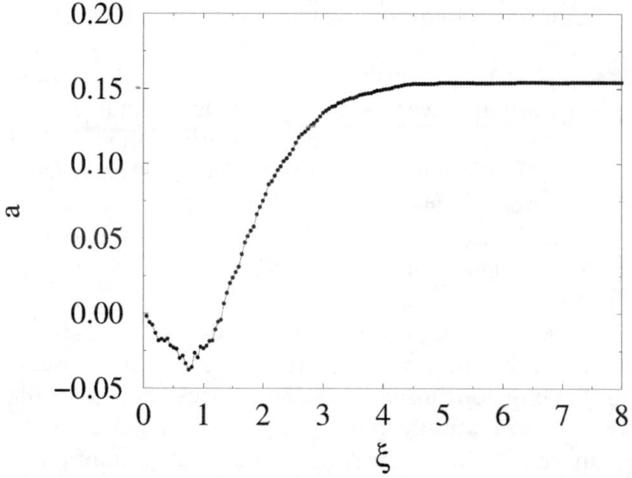

Figure 7 *Partial anisotropy a as a function of the cutoff force ξ in a sheared grain assembly*

where a_s and a_w are the (positive) anisotropies of the strong and weak contact sets, respectively. The weak fabric tensor $\overline{\overline{F}}_w$ is *orthoaxial* with the strong fabric tensor $\overline{\overline{F}}_s$.

A similar analysis can be applied to the stress tensor whose micromechanical expression is given by Equation 1. By restricting the averaging operation to the set $A(\xi)$, we get partial stress tensors $\overline{\overline{\sigma}}(\xi)$. In particular, let us consider the partial stress ratio $q(\xi)/p$ as a function of ξ; see Figure 8. Amazingly, for all $\xi <$ 1, we have $q(\xi)/p \approx 0$! This means that nearly the whole *deviator* stress is carried by the strong contact set. The weak contact set contributes only to the average

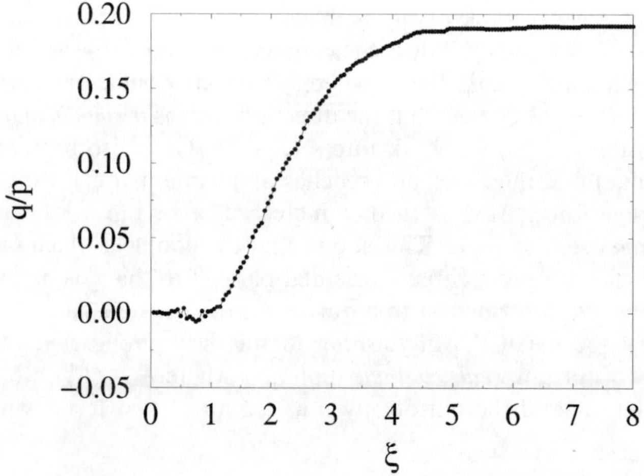

Figure 8 *Partial stress ratio q/p as a function of ξ in a sheared grain assembly*

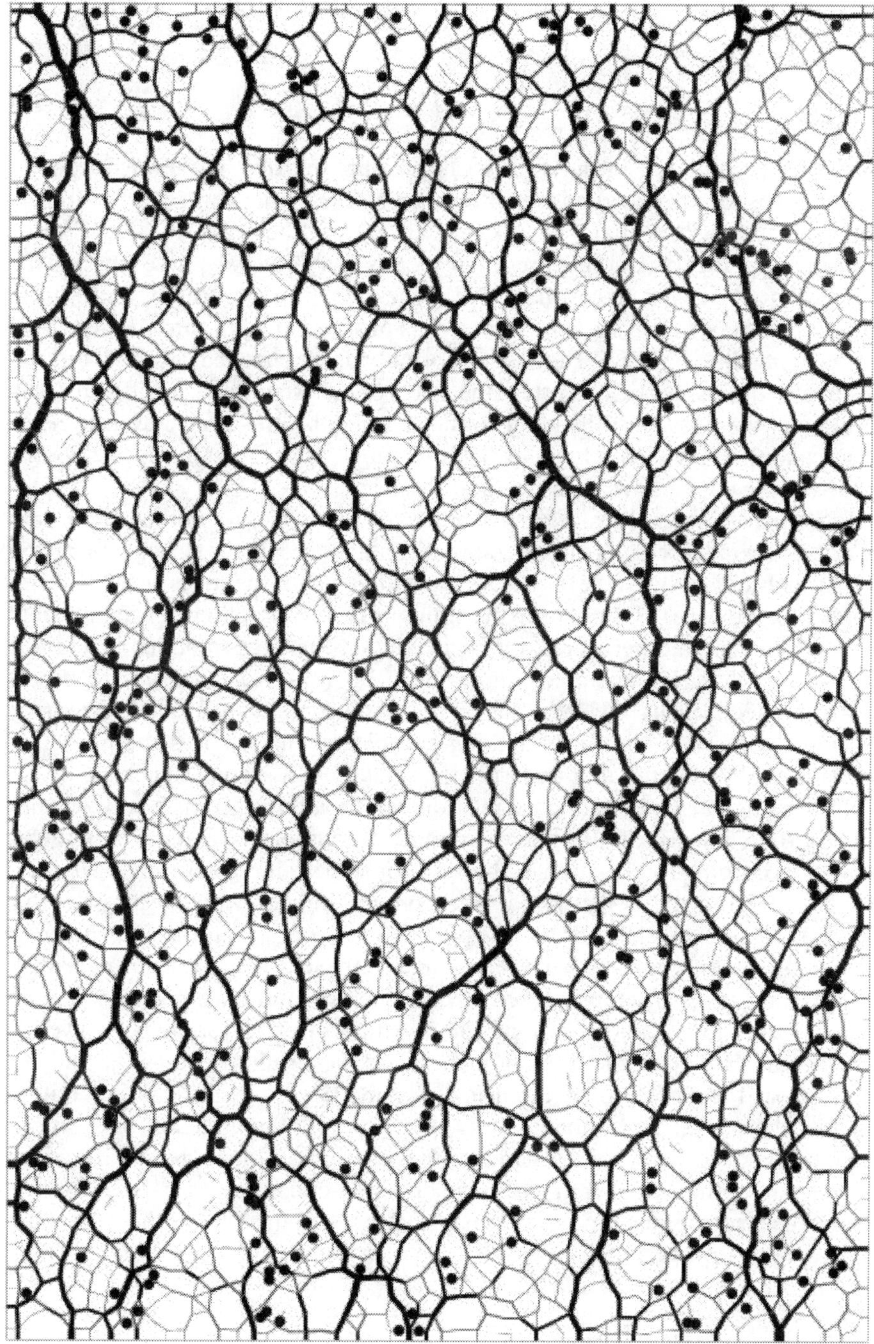

Figure 9   *A map of normal forces in a biaxially sheared granular assembly. The width of segments joining particle centres are proportional to the normal force. The strong and weak contacts are shown in black and grey, respectively. Full circles show the positions of sliding contacts*

pressure. This suggests an additive decomposition of the stress tensor as following:[7]

$$\bar{\bar{\sigma}} = p_w \bar{\bar{I}} + \bar{\bar{\sigma}}_s \tag{18}$$

where $\bar{\bar{I}}$ is the unit tensor, p_w is the weak pressure (carried by the weak contact set) and $\bar{\bar{\sigma}}_s$ is the strong stress tensor (carried by the strong contact set). Numerical simulations show that $p_w/p \approx 0.29$. Hence, from the point of view of stress transmission, the weak contact set corresponds to a *liquidlike* phase whereas the strong contact set appears as the *solidlike* backbone of the medium transmitting deviatoric stresses. The same behaviour was observed also in three-dimensional granular media. Figure 9 shows the normal forces encoded as the width of segments joining particle centres, the two contact sets by two different grey levels, and the positions of sliding contacts, which belong almost exclusively to the weak set, in an assembly of 4000 grains subjected to biaxial compression by means of the contact dynamics method.

From the above analysis, we arrive at the following picture of the equilibrium of grains: strong forces act on the grains at contacts (inside the strong fabric) that are strongly *aligned* with the major principal direction of the stress tensor. Lateral weak forces (the weak fabric) prop the grains against *deviations* from alignment at strong contacts. In other words, the weak contacts play the same stabilizing role with respect to the grains supporting strong forces as the *counterforts* with respect to an architectural arch. Hence, this bimodal transmission of shear stresses may be understood as a "tensorial arching effect" that materializes the particular stress-fabric correlation that underlies the stability of a granular packing.

This particular stress-fabric correlation can be interpreted as a way of optimizing the capacity of the microstructure to support largest shear stresses. Indeed, according to Equation 4, the deviator stress q increases if a larger number of strong normal forces N occur at contacts aligned with the stress tensor ($\theta \approx \theta_\sigma$). This obviously leads to a surplus of weak contacts in the perpendicular direction. This mechanism can be compared to the loss-gain mechanism leading to the increase of structural anisotropy. In the loss saturation regime where no more contact loss is allowed along the direction of extension, either z evolves (together with a_{max}) or the stress ratio increases as a result of *extra* weakening of contacts in the direction of extension. Symmetrically, in the gain saturation regime where no more contact gain is allowed along the direction of compression, either z evolves (together with a_{max}) or the stress ratio increases as a result of extra strengthening of contacts in the direction of compression. In both cases the force anisotropies a_n and a_t increase, leading to an increase of the stress ratio according to Equation 10.

7   Fabric Evolution

The incremental evolution of state variables with plastic strain determines the hardening properties of a granular material. Let $E[\theta - \theta_c(t)] \cong z(t)p_\theta[\theta - \theta_c(t)]$ be the distribution of contact directions at "time" t. The incremental evolution of the fabric is then given by the total derivative $\dot{E}(\mathrm{E}, \overline{\overline{G}})$ as a function of E and the velocity gradient tensor $\overline{\overline{G}}$. Let θ_ε be the major principal direction of $\overline{\overline{G}}$. The rotational invariance of the system implies that $\dot{E}$ is a function of $\theta_\varepsilon - \theta_c$. In other words, the fabric change at any moment depends on the direction θ_ε of loading with respect to the fabric.

Assuming that z and a follow the saturation curve $a_{\max}(z)$ (Equation 15), the evolution of the fabric is simply reduced to that of $z : \dot{z} = \dot{z}\,(z, \overline{\overline{G}})$. Since z is an average over all directions, the deviator strain rate and rigid rotations are not expected to influence $\dot{z}$. Hence, only the volumetric strain rate $\dot{\varepsilon}_p$ should be taken into account, i.e. $\dot{z} = \dot{z}(z, \dot{\varepsilon}_p)$. The assumption of rate independence in quasistatic loading implies further that $\dot{z}$ depends on $\dot{\varepsilon}_p$ through a positively homogeneous function of degree 1. As a result,

$$\dot{z} = f(z)\,\dot{\varepsilon}_p \tag{19}$$

where $f(z)$ is an unknown function of z. Since the grains are assumed to be infinitely rigid, the volumetric strain rate $\dot{\varepsilon}_p$ represents only plastic volume increments (due to grain displacements). Figure 10 displays z as a function of the total volume change ε_p normalized by the initial volume in a numerical simulation with 4000 (circular) grains. The data is well fitted by the following form:

$$z = z_0 + (\tilde{z} - z_0)(1 - e^{-k\varepsilon_p}) \tag{20}$$

where $k \approx 1500$ is a numerical factor, and z_0 and $\tilde{z}$ are the initial and steady-state values of the coordination number, respectively. Such a large value of k indicates that tiny volume changes induce indeed large structural and stress changes in the system. We also observe that z reaches its steady-state value well before the solid fraction (ε_p continues to increase). Equation 20 implies $f(z) = k\,(\tilde{z} - z)$. Note also

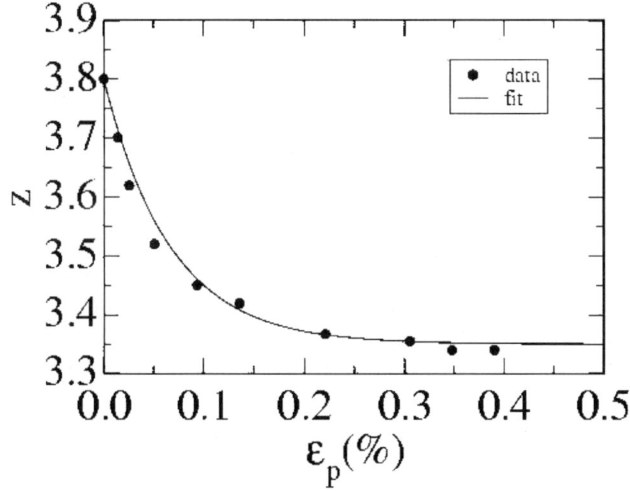

Figure 10 *Evolution of the coordination number z as a function of the volumetric strain ε_p*

that, according to Equation 20, z increases if $z < \tilde{z}$, and it decreases in the opposite case. Equation 20 together with Equation 15 determines the evolution of the fabric along a monotonic path.

When the strain is reversed ($\theta_\varepsilon = \theta_c + \pi/2$ assuming $\theta_\varepsilon \approx \theta_c$ along a monotonic path) loss and gain directions, corresponding respectively to the directions of extension and compression, become orthogonal to the principal fabric directions. As a result, the fabric anisotropy, and the deviator stress, decreases and the principal fabric directions rotate. The path in the (z, a) space is then below the saturation curve $a_{\max}(z)$. The fabric evolves until a new point on the saturation curve is reached. Along this path the fabric anisotropy a, its direction θ_c and the coordination number z are not subjected to gain or loss saturation constraints.

The evolution of E can be considered at the grain scale. Three elementary processes contribute to fabric change: contact *gain*, contact *loss* and contact *advection*. The latter refers to the rotation of *persisting* contact normals, corresponding thus to a *continuous* change in the contact direction, whereas the contact gain and loss are *discontinuous* (topological) changes in the fabric. Let $I^+(\theta)$ and $I^-(\theta)$ be respectively the gain and loss rates per grain along the direction θ. The contact *induction* function $I(\theta) = I^+(\theta) - I^-(\theta)$ represents the *net* change in the number of contacts due to gain and loss. Along a given direction, I is positive if there is more gain than loss, and it is negative if there is more loss than gain. We introduce also an *advection current* $J(\theta)$ corresponding to the mean current of persisting contacts at θ in the space of angles. Then, $\partial J/\partial\theta(\theta)$ is the mean change in the number of persisting contacts along θ due to contact rotations. The difference $I - \partial J/\partial\theta$ in each direction gives the variation $\partial E/\partial t$ of the number of contacts in that direction. Hence, E satisfies the following *balance equation*:[13,32]

$$\frac{\partial E}{\partial t} + \frac{\partial J}{\partial \theta} = I \tag{21}$$

The two functions I and J depend both on the fabric E and the velocity gradient tensor $\overline{\overline{G}}$. The induction function I is naturally proportional to the relative normal velocity $v(\theta)$. In the same way, the advection function J is proportional to the relative tangential velocity $u(\theta)$ of grain centres. Since contact gain takes place between grains that are *not* in contact, we extend the contact network and the fabric, represented here by the function E, to such pairs of grains. In fact, simulations show that the fabric remains almost unchanged upon such an extension provided the "gap" is sufficiently small compared to the mean grain size. Then, $v(\theta)$ can take both positive values (when two grains approach to one another) and negative values (when two grains separate from one another) at potential contacts. By considering potential contacts, we simply assume a *prospective* point of view, i.e. we consider the *oncoming contacts* instead of outgoing ones.

As a result of steric exclusions the local velocities $v(\theta)$ and $u(\theta)$ are different from global (far-field) velocities which are dictated by the macroscopic velocity gradient tensor (in a homogeneously sheared assembly). Indeed, the application of an affine velocity field to a granular assembly leads to the interpenetration of touching grains. The local velocities thus provide a "correction" to the imposed velocity field, restoring the mutual exclusions of the grains. The local velocity field can be characterized by the probability density function $p_f(\theta, v, u)$ defined as the probability density for an effective or potential contact to have its direction along θ and where the relative normal and tangential velocities are v and u, respectively. By definition, $p_\theta(\theta) = \int p_f\, dv\, du$. The distribution p_f allows the definition of average velocities $v_m(\theta)$ and $u_m(\theta)$ by

$$v_m(\theta)p_\theta(\theta) \equiv \int v\, p_f\, dv\, du \tag{22}$$

$$u_m(\theta)p_\theta(\theta) \equiv \int u\, p_f\, dv\, du \tag{23}$$

Figure 11 shows the mean steady-state velocity field (v_m, u_m) in the frame of a typical grain for different gaps in a simple shear simulation. We see that the near field (below a gap nearly equal to l_m) is considerably different from the far field. While the far field is a simple affine velocity field, the local field is inhomogeneous. The normal velocities are enhanced along the principal strain rate

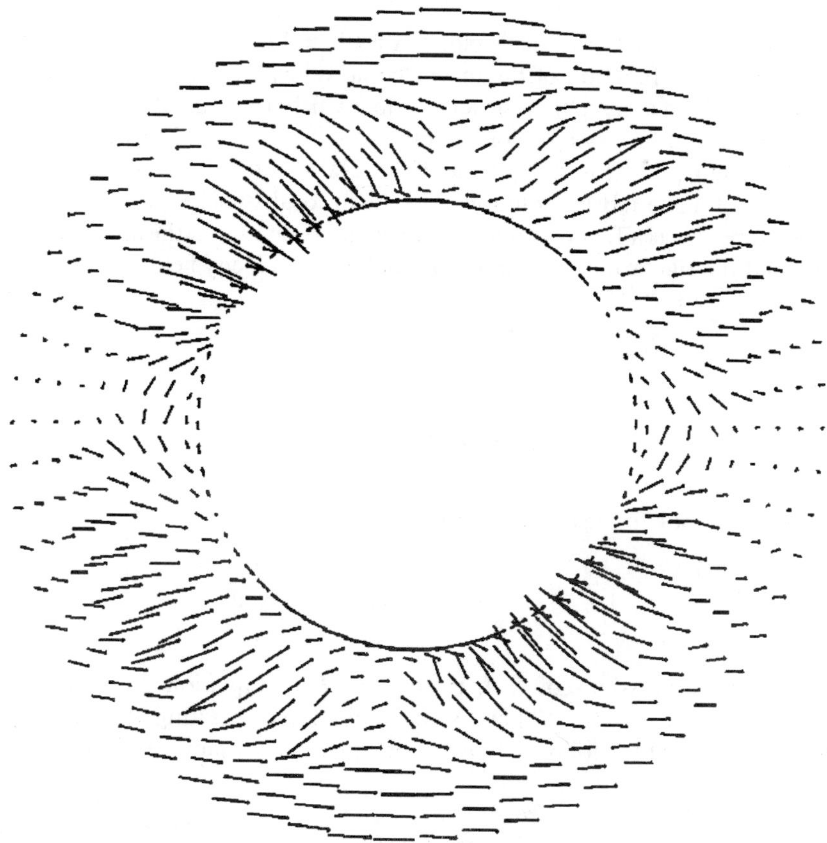

Figure 11 *The mean velocity field in the vicinity of a typical grain in a steady shear flow*

directions (at $\pm \pi/4$ to the vertical) and recirculation currents occur in front of the particle along the flow direction.

The detailed balance of contact events can be obtained by expressing I and J as a function of the local velocities $v_m(\theta)$ and $u_m(\theta)$. The advection current is the product of $u_m(\theta)$ and the density $\delta_J(\theta)$ defined as the mean number of contacts along the grain *circumference* ("surface" density in three dimensions) as a function of the contact direction θ. Considering an angular interval $\delta\theta$ at a distance l_m, this density is $\delta_J(\theta) = E(\theta)\delta\theta/(l_m\delta\theta) = E(\theta)/l_m$. The advection current is then given by

$$J(\theta) = \delta_J(\theta)u_m(\theta) = E(\theta)\, u_m(\theta)/l_m \qquad (24)$$

The contact induction is the product of $v_m(\theta)$ and the *cross section* $\delta_I(\theta)$ given by the length element ("surface" element in three dimensions) $l_m\delta\theta$ times the mean number of contact neighbours per unit *area*, $E(\theta)\delta\theta/(1/2l_m^2\delta\theta)$, along the direction θ. Thus, the induction function is

$$I(\theta) = \delta_I(\theta)v_m(\theta) = 2E(\theta)v_m(\theta)/l_m \qquad (25)$$

We need also to specify the local velocities $v_m(\theta)$ and $u_m(\theta)$ at effective or potential contacts as a function of the macroscopic velocity gradient tensor $\overline{G}$. The simplest idea is to assume that the local velocity field is identical to the far field. Although this "affine assumption" contradicts the numerical observation displayed in Figure 13, it is still instructive to evaluate the outcomes of such an assumption for the fabric evolution. The mean local velocities at a distance l_m can be expressed as a function of the volumetric strain rate $\dot{\varepsilon}_p$, the deviator strain rate $\dot{\varepsilon}_q$ and the rigid rotation (semi-rotational of the velocity field) ω:

$$\frac{v_m(\theta)}{l_m} = \frac{\dot{\varepsilon}_p}{2} + \frac{\dot{\varepsilon}_q}{2}\cos 2(\theta - \theta_\varepsilon) \qquad (26)$$

$$\frac{u_m(\theta)}{l_m} = \omega + \frac{\dot{\varepsilon}_q}{2}\sin 2(\theta - \theta_\varepsilon) \qquad (27)$$

Now, we may look for a solution of the following form:

$$E(\theta, t) = \frac{z(t)}{2\pi}\{1 + a(t)\cos 2[\theta - \theta_c(t)]\} \qquad (28)$$

Inserting the expressions of I, J, v_m, u_m and E in Equation 21, multiplying alternatively by 1, $\cos 2(\theta - \theta_c)$ and $\sin 2(\theta - \theta_c)$ and integrating each time with respect to θ, we get the following system of equations for fabric evolution:

$$\begin{cases} \dot{z} = z\dot{\varepsilon}_p \\ \dot{a} = \sqrt{2}\,\dot{\varepsilon}_q \cos 2(\theta_c - \theta_\varepsilon + \pi/8) \\ 2a(\dot{\theta}_c - \omega) = \sqrt{2}\,\dot{\varepsilon}_q \sin 2(\theta_c - \theta_\varepsilon - \pi/8) \end{cases} \qquad (29)$$

The first equation is to be compared with Equation 19 for the monotonic case. The factor z is obviously well below $k(\tilde{z} - z)$ (with a large value of k) obtained for monotonic deformation. The second equation shows that the variation of the anisotropy depends on the direction θ_ε of loading with respect to the fabric direction θ_c. In particular, when $\theta_c - \theta_\varepsilon > \pi/4$, the anisotropy falls off ($\dot{a} < 0$). If $\omega \neq 0$, there is no steady-state solution. With a constant deviator $\dot{\varepsilon}_q$, the anisotropy grows and oscillates at the same time. However, in deriving Equation 29, we have assumed that a and z are independent parameters while in the steady state, or more generally along a monotonic path, they are linked together through the gain or loss saturation condition Equation 15. If $\omega = 0$, the steady-state solution ($\dot{a} = \dot{\theta}_c = 0$) is characterized by $\theta_c - \theta_\varepsilon = \pi/8$. Since θ_c tends to coincide with θ_σ (major principal direction of the stress tensor), this phase difference may be interpreted as *non-coaxiality* between stress and strain-rate tensors.

We get basically the same equations as Equation 29 if we use the *local* velocity gradient tensor. Then, $\dot{\varepsilon}_p$ and $\dot{\varepsilon}_q$ should be interpreted as the local spherical and deviator strain rates. The problem, however, is not in the distinction between the local and macroscopic strain rate tensors. The point is that the local velocities v_m and u_m involved in the balance equation are not related in a simple way as that assumed in Equations 26 and 27 (affine assumption) to the velocity gradient tensor. In fact, the local velocity field is inhomogeneous (see Figure 11) and it evolves with the fabric.

A simple way to circumvent this difficulty is to rely on Fourier expansions of $v_m(\theta)$ and $u_m(\theta)$ to the desired order. A similar approach was used for contact forces in Equations 8 and 9. The properties of the local velocity field are then reflected in the Fourier coefficients. To keep to the lowest nontrivial order (for E, v_m and u_m), let us set

$$\frac{v_m(\theta)}{l_m} = \frac{c_n\dot{\varepsilon}_p}{2} + \frac{b_n\dot{\varepsilon}_q}{2}\cos 2(\theta - \theta_\varepsilon) \tag{30}$$

$$\frac{u_m(\theta)}{l_m} = \omega + \frac{b_t\,\dot{\varepsilon}_q}{2}\sin 2(\theta - \theta_\varepsilon) \tag{31}$$

where the coefficients c_n, b_n and b_t may evolve with time or with the anisotropy. These expressions are more general than Equations 26 and 27, but preserve the same symmetries. Equations 30 and 31, together with Equations 21, 24, 25 and 28, yield the following system of equations for the evolution of the fabric parameters:

$$\begin{cases} \dot{z} = c_n z\dot{\varepsilon}_p \\ \dot{a} = b\dot{\varepsilon}_q \cos 2(\theta_c - \theta_\varepsilon + \lambda) \\ 2a(\dot{\theta}_c - \omega) = b\dot{\varepsilon}_q \sin 2(\theta_c - \theta_\varepsilon - \lambda) \end{cases} \tag{32}$$

with $b = \sqrt{b_n^2 + b_t^2}$ and $\tan(2\lambda) = b_t/b_n$. The predictions are essentially the same as before, but there are now three parameters which can be adjusted. The parameter c_n allows for a more realistic estimation of the variations of z as a result of volume changes. The phase λ is zero only if $b_t = 0$, i.e. when the distortions of the contact network due to contact advection can be neglected compared to discontinuous changes resulting from loss and gain. This probably is the case in sufficiently dense and loose states and when $\omega = 0$. Then, the system tends to a state where the anisotropy increases linearly with deviator strain (until the saturation limit is reached) and $\theta_c = \theta_\varepsilon$. The case where $b_n = 0$ corresponds to the states where the coordination number does not evolve any longer, but the volume continues to change as a result of contact network distortions. Then, $\lambda = \pi/4$. If $\omega = 0$, the system tends to a state where the anisotropy increases linearly (if b remains constant) with deviator strain and $\theta_c = \theta_\varepsilon + \pi/4$. A less phenomenological description of fabric evolution requires a proper modelling of local velocities by accounting for steric constraints.

8   Dilatancy

When a granular material composed of rigid grains is deformed, irreversible changes in solid fraction ρ occur as a result of collective grain rearrangements. The solid fraction, as a geometrical property determined by the microstructure, may be considered as a descriptor of the microstructure and used as a *state* variable. When *free* volume changes are allowed, the solid fraction evolves with the applied deviatoric strain rate $\dot{\varepsilon}_q$. Reynolds' *dilatancy* refers to this incremental volume change produced by shearing and characterized generally by a dilation angle ψ: $\tan\psi \cong - \dot{\varepsilon}_p / \dot{\varepsilon}_q$ (The negative sign here ensures that negative values of the dilation angle correspond to contraction according to our sign conventions). The dilation angle ψ is thus a basic plastic flow property of a granular material, and like the internal angle of friction φ, it has to be specified as a function of the fabric.[13] The assumption of rate independence in quasi-static loading implies that all rates, including $\dot{\varepsilon}_p$, should depend on $\dot{\varepsilon}_q$ through a positively homogeneous function of degree 1. As a result, ψ is independent of $\dot{\varepsilon}_q$.

While the internal angle of friction φ basically reflects the Coulomb friction law (at the contact scale), the dilation angle ψ is a purely *structural* property which has no counterpart at the contact scale. In fact, the Coulomb friction law can be seen as a non-associated rigid-plastic behaviour (no relative displacement normal to the slip plane).[13] Therefore, the normality rule is not expected to hold at the macroscopic scale. Experiments show, indeed, that ψ is generally different from φ.[10,11] If the fabric did not evolve with plastic strain (no state variable), then the volume would not change and we would have $\psi = 0$ as in the basic Coulomb model. Since the solid fraction is a function of the fabric, and fabric changes depend on the direction of loading, we expect that ψ is different for different principal strain rate directions. Formally, let us consider E as the lowest-order representation of the fabric. Then, the solid fraction $\rho(E)$ is a function of E. Deriving with respect to "time", we get $\dot{\rho} = -\rho\dot{\varepsilon}_q\tan\psi = \dot{\rho}\,[E, \dot{E}(E, \overline{\overline{G}})]$. This shows that the dilation angle depends on $\theta_c - \theta_\varepsilon$, the coordination number z and the anisotropy a. The dependence on the loading direction has been observed in experiments.[10] For instance, it has been shown that the dilation angle for dense sand tested at $\theta_c - \theta_\varepsilon = \pi/2$ is comparable to that for loose samples at $\theta_c = \theta_\varepsilon$.[4]

Since the fabric evolution in Equation 32 is expressed in terms of strain rates, it is straightforward to get the following expression for the dilation angle:

$$\tan\psi = -\frac{\dot{z}}{z}\frac{b}{c_n\dot{a}}\cos 2\,(\theta_c - \theta_\varepsilon + \lambda) \tag{33}$$

As required, this expression depends on the loading direction and evolves with the coordination number and the anisotropy. Along a monotonic path, from Equation 15, we have

$$\dot{a} = \begin{cases} (2-a)\dfrac{\dot{z}}{z} & \text{loss saturation} \\[2ex] -(2+a)\dfrac{\dot{z}}{z} & \text{gain saturation} \end{cases} \tag{34}$$

Combining Equations 33 and 34, we get

$$\tan\psi = \begin{cases} \dfrac{-b}{c_n(2-a)}\cos 2(\theta_c - \theta_\varepsilon + \lambda) & \text{loss saturation} \\[3ex] \dfrac{b}{c_n(2+a)}\cos 2(\theta_c - \theta_\varepsilon + \lambda) & \text{gain saturation} \end{cases} \tag{35}$$

The largest value of ψ predicted by Equation 35 is fixed by the ratio b/c_n. Since c_n is large, ψ should be quite small unless we assume very large values of b. In fact, c_n (for z) and b (for a) should be of the same order of magnitude when discontinuous structural changes dominate. This is the case in very dense and loose states. On the other hand, the steady state corresponds to $\psi = 0$ which implies $\theta_c = \theta_\varepsilon + \pi/4 - \lambda$.

The focus of this section was put on general trends and the relation between fabric evolution and dilatancy. The equations of fabric evolution together with the function $\rho(E)$ are sufficient for the estimation of dilatancy. The function $\rho(E)$ can be evaluated by a tessellation of space into void cells. For example, in the simple case of a regular packing with $z = 4$, it is easy to show that $\rho = \pi/2\sqrt{4 - a^2}$, where a is the anisotropy of the cells. The statistics of valence numbers allows us to obtain an estimation of the mean value of ρ over the cells as a function of z, a and θ_c. However, both for fabric evolution (local velocities) and for the evaluation of ρ it seems necessary to introduce higher order microstructural information. But, one might still keep with z, a and θ_c as the most basic state variables provided phenomenological parameters are introduced.

9 Concluding Remarks

We discussed basic plastic properties of granular materials as they may be understood or modelled from the lowest order description of granular fabric.

The average connectivity or compactness is characterized by the coordination number whereas the preferred orientations of contact normals are represented by the (lowest-order) fabric anisotropy and its direction. It was argued that in a material composed of rigid grains, the macroscopic behaviour is essentially rigid-plastic characterized by a single angle of internal friction as yield criterion. The heart of a microscopic approach is the dependence of this criterion and the solid fraction on the fabric, as *physical* state variables, and the evolution of the latter with plastic strain.

A model based on this approach is all the more successful as it meets the steric constraints and the requirement of grain equilibrium. This can be achieved by considering the local environments through the multicontact distribution functions that may contain the desired level of microstructural information. We introduced a particularly simple approach which is more global in nature but incorporates steric constraints and grain equilibrium through "limit states". The presence of these limit states, which without loosing generality can be isotropic or not, leads to "gain" and "loss" saturation regimes for the fabric. It was shown that the stress-strain behaviour along known stress paths matches this microscopic picture quite well provided the shear strength is an increasing function of the fabric anisotropy along a monotonic path. This point was studied in more detail by considering the micromechanical expression of the stress tensor, and it was shown that the shear strength depends linearly on the fabric anisotropy but also harmonically on the principal stress directions with respect to the fabric. Using numerical results, we also emphasized the generic "bimodal" nature of force transmission in a granular medium that suggests a mechanism (similar to loss and gain for contact anisotropy) that allows the system to optimize the force anisotropy and hence the shear strength.

The fabric and volume changes (along and below the saturation limits) were discussed in the light of a balance equation for gain, loss and advection of contact neighbours. A key point in the evaluation of gain, loss and advection rates is the local velocity field which was shown to be non-affine and inhomogeneous due to steric constraints. We studied the outcome of a simple non-affine assumption with three parameters which leads to equations for the evolution of the coordination number, the fabric anisotropy and its direction as a function of the deviator strain and direction of loading. These equations predict an exponential evolution of the coordination number with volumetric strain, a linear evolution of the anisotropy with deviatoric strain and a generic non-coaxiality between the fabric and strain principal directions. Interpreted in terms of the dilation angle, they show that the dilation angle is harmonically dependent on the loading direction.

The granular plasticity is a vast domain mostly based on phenomenological interpretation of experimental tests. Before a microscopic approach can reach the same predictive level, the main objective is to *understand* the trends from the lowest-level description of the microstructure. This was the main goal of this contribution, but not the only one. It is also important to note that even basic understanding of shear strength and fabric evolution properties requires further investigation of the fabric behaviour, local velocities and space correlations along *complex* loading paths. This is a major challenge due to demanding statistics both for experiments, which provide only limited information about the fabric, and for numerical discrete element methods, which do not allow for a large number of grains to be simulated at the same time. In this context, a model based on a hierarchical description of the fabric can provide a useful strategy allowing us to design genuine simulations or experiments in order to establish the relevant parameters and the level of description to be used in a particular application.

References

1. P. G. de Gennes, *Rev. Mod. Phys.*, 1999, **71**, S374.
2. H. M. Jaeger and S. R. Nagel, *Rev. Mod. Phys.*, 1996, **68**, 1259.
3. P. Dantu, Proceedings of the 4th International Conference on Soil Mechanics and Foundation Engineering, Butterworths Scientific Publications, London, 1957, **1**, 144.
4. M. Oda, *Soils and Foundations*, 1972a, **12**, 17; 1972b, **12**, 1.
5. P. A. Cundall and O. D. L. Stack, *Geotechnique*, 1979, **29**, 47.
6. P. A. Cundall, A. Drescher and O. D. L. Strack, IUTAM Conference on Deformation and Failure of Granular Materials, Delft, 1982, 355.
7. F. Radjai, D. E. Wolf, M. Jean and J. J. Moreau, *Phys. Rev. Lett.*, 1998, **80**, 61.
8. M. R. Kuhn, *Mechanics of Materials*, 1999, **31**, 407.
9. F. Radjai and S. Roux, *Phys. Rev. Lett.*, 2002, **89**, 064302.
10. J. K. Mitchell, *Fundamentals of Soil Behaviour*, Wiley, New York, 1993.
11. D. M. Wood, *Soil Behaviour and Critical State Soil Mechanics*, Cambridge University Press, Cambridge, 1990.
12. M. Satake in Proceedings of the IUTAM Symposium on Deformation and Failure of Granular Materials, Delft, ed. P. A. Vermeer and H. J. Luger, Balkema, Amsterdam, 1982, 63.
13 S. Roux and F. Radjai in "Mechanics for a New Millennium", ed. H. Aref and J. Philips, Kluwer, Netherlands, 2001, 181.
14. H. Troadec, F. Radjai, S. Roux and J. C. Charmet, *Phys. Rev. E*, 2002, **66**, 041305.
15. J. J. Moreau, in "Nonsmooth mechanics and applications", CISM Courses and Lectures, 1988, **302**, 1.
16. J. J. Moreau, *Eur. J. Mech. A*, 1994, **13**, 93.
17. M. Jean and J. J. Moreau, Proceedings of Contact Mechanics International Symposium, Presses Polytechniques et Universitaires Romandes, Lausanne, Switzerland, 1992, 31.
18. J. Christoffersen, M. M. Mehrabadi and S. Nemat-Nasser, *J. Appl. Mech.*, 1981, **48**, 339.
19. L. Rothenburg and R. J. Bathurst R. J., *Geotechnique*, 1989, **39**, 601, R. J. Bathurst and L. Rothenburg, *Mechanics of Materials*, 1990, **9**, 65.
20. B. Cambou, in "Powders and Grains 93", ed. C. Thornton, Balkema, Amsterdam, 1993, 73.
21. A. Schofield and P. Wroth, *Critical State Soil Mechanics*, McGraw-Hill, London, 1967.
22. S. Roux and F. Radjai, in "Physics of Dry Granular Media", ed. H. Herrmann *et al.*, Kluwer, Dordrecht, 1999, 229.
23. J. P. Troadec and J. A. Dodds in "Disorder and Granular Media", ed. D. Bideau and A. Hansen, Elsevier, Amsterdam, 1993.

24. F. Calvetti, G. Combe and J. Lanier, *Mech. Coh. Frict. Materials*, 1997, **2**, 121.
25. S. Torquato, *Random Heterogeneous Materials*, Springer, Berlin, 2001.
26. J. P. Bardet and J. Proubet, *Geotechnique*, 1991, **41**, 599.
27. J. Desrues, R. Chambon, M. Mokni and F. Mazerolle, *Geotechnique*, 1996, **46**, 529.
28. F. Radjai, M. Jean, J. J. Moreau and S. Roux, *Phys. Rev. Lett.*, 1996, **77**, 274.
29. D. M. Mueth, H. M. Jaeger and S. R. Nagel, *Phys. Rev. E*, 1998, **57**, 3164.
30. S. J. Antony, *Phys. Rev. E*, 2001, **63**, 011302.
31. L. Staron, J.-P. Vilotte and F. Radjai, *Phys. Rev. Lett.*, 2002, **89**, 204302.
32. F. Radjai and S. Roux in "Powders and Grains 2001", ed. Y. Kishino, Balkema, Tokyo, 2001, 21.

CHAPTER 7

Influence of Polymers on Particulate Dispersion Stability: Atomic Force Microscopy Investigations

SIMON BIGGS

Institute of Particle Science and Engineering, University of Leeds, Leeds LS2 9JT
Email: s.r.biggs@leeds.ac.uk

1 Introduction

Particulate dispersions are of great importance to an immense number of products and processes, both in the industrial and consumer sectors. Included here are fields as diverse as ceramics processing, foodstuffs, paints, household cleaning products, drug delivery systems, water treatment processes, and minerals processing. In each case, accurate control over the stability of the particulate systems is a desired, but not frequently achieved state.[1,2]

In almost every case of interest, the disperse phase is colloidal in its characteristic dimensions (1 nm < x < 1 μm). As a result, interfacial areas are large and surface properties are dominant.[3] At the simplest level of understanding, the creation of interfacial area costs energy (in direct proportion to the area created) and so, aggregation or coagulation of fine particulates (reduction of interfacial area) is thermodynamically favoured. The prevention or control of this aggregation (or coagulation) has therefore dominated particulate and colloid research throughout its history.

Developments in experimental capabilities have resulted in the direct measurement of interparticle surface forces becoming a well established approach for developing understanding of particulate stability.[4] Historically, the direct measurement of surface forces has a relatively long history starting with the pioneering work of Derjaguin.[5,6] However, the more routine approaches that are now used arose from the development of the so-called Surface Forces Apparatus (SFA) during the late 60's and early 70's.[7–10] The main features of standard colloid stability theory were confirmed through the use of the SFA. In addition, important non-standard interaction forces such as polymer steric layers and hydration layers were investigated for the first time.[4] The data achieved from such measurements has been central to our improved understanding in general of these interactions. In addition, surface force data has been used to enhance theoretical descriptions of particle-particle interaction forces, thereby improving our predictive capabilities.

In recent years, the development of scanning probe techniques has reinvigorated the measurement of surface forces.[11] Initially invented as a tool for imaging,[12] the Atomic Force Microscope (AFM) has now become an important device for the measurement of surface-surface interaction forces. In this review, the use of the AFM as a measurement tool to investigate particulate interactions, and in particular those in the presence of polymers, will be discussed.

2   Surface Forces and Particulate Dispersion Properties

The physical properties of any particulate dispersion are strongly influenced by the operative interaction energies. A colloidal dispersion may exist in a variety of states depending upon the exact nature of the interaction energy. For example, simple charge stabilised colloids can be described as stable (dispersed) or coagulated, depending upon whether repulsive or attractive interactions dominate. For coagulated systems, a range of possible states is possible depending upon the relative strength of the interaction, as shown schematically in Figure 1. For aqueous colloid dispersions, the states illustrated can be achieved through the simple addition of electrolyte which acts to suppress the repulsive barrier. For any dispersion at a volume fraction below the gel point, the establishment of a secondary energy minimum (energy 3–10 kT) occurs as the concentration of electrolyte is increased. This leads to a weak reversible coagulation of the particles. When the electrolyte concentration is sufficient to suppress the repulsive interaction, strong coagulation into the primary energy minimum can occur. The interaction energy profiles between the particles are important because they strongly influence the kinetics of aggregation/de-aggregation and also the final aggregate structures that form. In general, the stronger the attractive energy, the larger and more open are the aggregates that form.[13] The physical properties of aggregates will, in turn, influence the properties of other particulate networks such as sediment beds and filter cakes. Thus, an intimate understanding of the factors that influence aggregation and coagulation is critical in the processing of particulate dispersions.

A robust theory to explain the main features of colloid stability for an aqueous particle dispersion was first proposed by Derjaguin and Landau[14] and Verwey and Overbeek (DLVO).[15] Simple DLVO theory describes the total interaction energy (V_{TOT}) between particles as the sum of one repulsive (V_R) and one attractive component (V_A), where

$$V_{TOT} = V_R + V_A \tag{1}$$

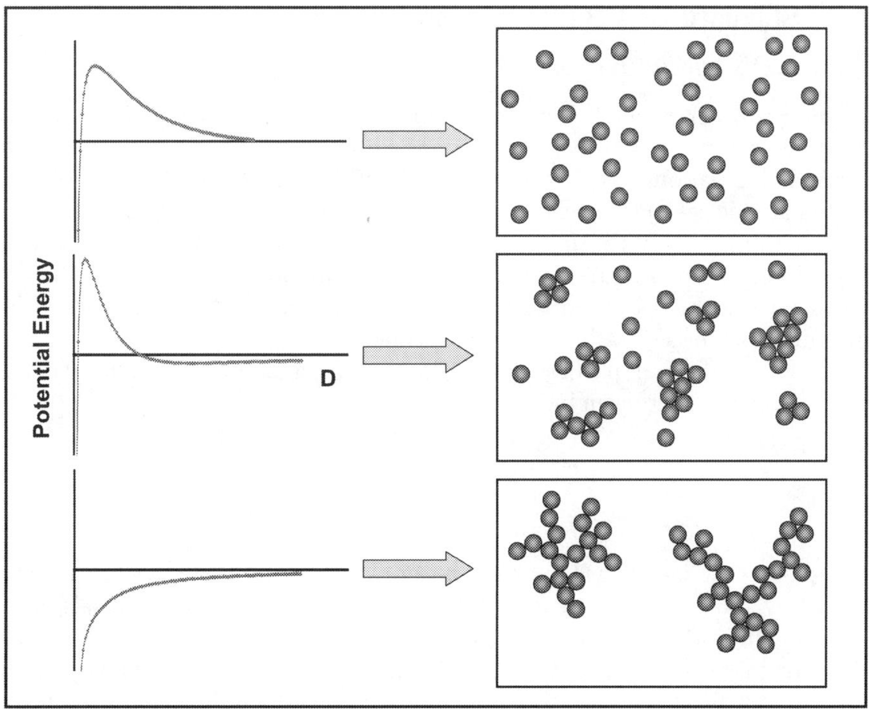

Figure 1 *Schematic illustration of the relationship between inter-particle potential energy and the resultant aggregate structures in a colloidal dispersion*

The repulsive component originates from the interaction between the overlapping electrical double layers as two particles collide. The attractive component is caused by long-range London-van der Waals dispersion forces. The origins and magnitude of each will be discussed briefly below. The essential strength of DLVO theory is its simplicity; in the theory it is assumed that all interactions are additive and can thus be separated. Subsequently, additional interaction forces have been measured or proposed.[4] Following the same approach, the total interaction energy is now expressed by another simple summation such that

$$V_{TOT} = V_{edl} + V_{steric} + V_{structural} + V_A \qquad (2)$$

where V_{edl} is the electrical double layer interaction, V_{steric} is a repulsive potential energy arising from the overlap of adsorbed polymer layers at the particle surfaces, and $V_{structural}$ is the potential energy caused by solution species that can develop long-range solid-like or liquid-like order which influences the particle-particle interactions. Although this simple summation approach is limited in its physical reality, it provides an excellent route to gain insight of what factors are important in the control of colloid stability.

2.1 Standard DLVO Interactions

2.1.1 Dispersion Forces

London-van der Waals forces are ubiquitous in colloid systems and originate from the spontaneous dipolar fluctuations within all atoms as a consequence of random fluctuations in electron clouds relative to the nucleus. In large objects (on an atomic scale) such as a colloidal particle, at any instant the sum total of all the atomic dipoles result in a net dipole for the particle. The dipoles of neighbouring particles are correlated since this minimises the energy. The net result is an attractive interaction between the particles which increases in its magnitude as the separation distance is reduced.

For two spherical particles of radius a separated by a distance D, the interaction energy is given by

$$V_{vdw} = -\frac{aA_H}{12D} \tag{3}$$

Where A_H is the so-called Hamaker constant. This equation is valid when a >> D.

Clearly, the magnitude of these attractive interactions becomes very large as the separation distance becomes small. The critical parameter in determining the magnitude of the interaction is the Hamaker constant, A_H. Various methods for determination of A_H exist, and the interested reader is directed to a number of excellent reviews.[4,16,17] It is sufficient to note here that typical values in water range from $0.3 \times 10^{-20} - 10 \times 10^{-20}$ J, with organic materials having the smallest values and conductive materials the largest.

Dispersion interactions are typically unaffected by the addition of salt or alterations in the temperature. They are, however, always present and must be off-set or overcome in some way if a colloid suspension is to have any stability at all.

2.1.2 Electrical Double Layer.

When a particle is immersed in water its surface will, in most cases, spontaneously acquire charge.[3] Mechanisms for this charging process include specific ion adsorption, differential ion dissolution and ionisation of surface sites. For oxide materials, the most common charging mechanism occurs through the protonation or de-protonation of surface hydroxyl groups according to

$$M-OH_2^+ \underset{-H^+}{\overset{+H^+}{\rightleftharpoons}} M-OH \underset{-H^+}{\overset{+H^+}{\rightleftharpoons}} M-O^- \tag{4}$$

The charge is therefore dependent upon the solution pH and the relative acidity of the surface groups. Hence, different materials can have opposite surface charges or different magnitudes of surface charge at any given pH.[3]

Regardless of the nature of the surface charge (or its origins), the net result of its presence is to perturb the surrounding solution by attracting counter-ions towards the surface and repelling co-ions away. The result is the development of an ionic cloud (the double layer) around the particle such that the net charge of the particle plus its ionic double layer is zero. The total ionic concentration in the double layer is greater than that of the surrounding medium. For a symmetric z:z electrolyte, the extension of this double layer into the bulk solution is defined by the Debye length, $1/\kappa$; where κ is given as

$$\kappa = \left(\frac{F^2 \sum_i N_i z_i^2}{\varepsilon_0 \varepsilon_r kT} \right) \tag{5}$$

Where z is the ionic valence, N_i is the number concentration of ions in the solution, e is the electronic charge, ε. is the dielectric constant of the bulk solution and ε_0 is the permittivity of free space.

When two particles approach one another, their double layers are forced to overlap. The result is a local increase in the ionic concentration above the equilibrium value that can be supported by the surface charge and hence a rise in the free energy. Solutions of the relevant theory are complex and no exact analytical expression is available. However, for two equal sized particles of the same material, and when there is small double layer overlap, an approximate solution is given by[18]

$$V_{edl} = 2\pi\varepsilon\varepsilon_0 a \psi_o^2 \exp\left[-\kappa D\right] \tag{6}$$

Where ψ_0 is the surface potential.

Examination of the above equations indicates that the stability of a charge stabilised colloidal dispersion can be altered by changes in electrolyte concentration and/or pH. Increases in the electrolyte concentration will decrease the stability by reducing both $1/\kappa$ and the ψ_0. Changes in solution pH affect the stability of oxides by altering the magnitude and sign of ψ_0. Further influence on the system stability can be obtained through the addition of ionic species that specifically adsorb at a particle surface altering ψ_0.

2.2 Steric Interactions

The adsorption of molecules at the solid-liquid interface can have a profound influence on the stability of any particulate dispersion.[19] If the physical size of the adsorbed molecules is large enough, then the attractive dispersion forces may be effectively overwhelmed. When two particles carrying adsorbed polymer collide, the overlap of the polymer layers leads to a loss of configurational entropy. This is unfavourable and so, a repulsive interaction ensues. The magnitude of this repulsion depends primarily upon the segment density profile of the polymer way from the interface. The segment density profile is influenced by many factors including the molecular weight of the polymer, the solvent quality, the concentration of the polymer and particles in solution, and the adsorption energy per segment of polymer at the interface.[20] Examples of typical segment density profiles are shown below in Figure 2.

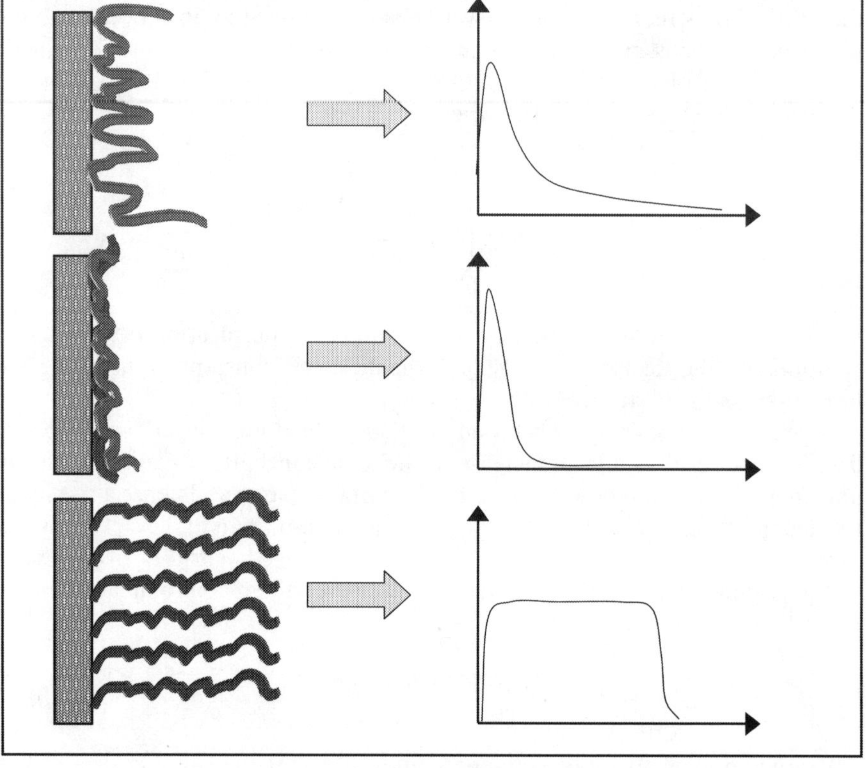

Figure 2   *Schematic illustration of some adsorbed polymer structures and the corresponding segment density profiles normal to the interface. Systems described include
freely adsorbed chain with "loops and tails", freely adsorbed chain lying flat to
the interface, and end attached chains*

2.2.1   Adsorption

2.2.1.1   Non-ionic Polymers.   Polymer adsorption at an interface is a complex
process. Predictions of conformations and adsorbed amounts rely upon the
attainment of full thermodynamic equilibrium for an adsorbed polymer coil.
However, the large size of polymer chains and the relatively large number of
different conformations available can result in a large variety of kinetically
stable conformations. Under full equilibrium conditions, the conformation of a
polymer chain at an interface will be determined by both entropic and enthalpic
considerations.[19] The enthalpy calculations rely upon knowledge of the relative
energies of segment-segment, segment-solvent, solvent-solvent, solvent-surface,
and segment surface interactions. Obviously, a favourable free energy for seg-
ment-surface interactions is a pre-requisite for adsorption. On the entropy side,
attachment of a polymer chain to an interface is always accompanied by a loss
of configurational and translational freedom and is hence unfavourable. Thus,
adsorption is largely controlled by the strength of the energy of attachment to

the interface. Silberberg[21] defined the net adsorption enthalpy using the parameter χ_s, which defines the net enthalpy of exchange for a segment at the surface with a solvent molecule in the bulk. If the polymer segment is preferred to the solvent molecule at the interface, χ_s is positive and adsorption can occur.

When the adsorption energy for the polymer segments at the interface is large, initial attachment of the chains to the interface is followed by a rapid relaxation of the coil segments to the surface and a relatively flat conformation with a high number of surface-segment attachment points result. As the interaction energy per segment decreases, there is a decreased driving force for relaxation to the interface since the enthalpy gain is insufficient to overcome completely the entropy loss that must result. Thus, the polymer coils will adopt an increasingly extended (normal to the interface) "loops and tails" conformation. A complication of the situations described above occurs when adsorption is initiated from a high bulk concentration onto a bare surface. Under these conditions, a high transport rate of the chains to the interface will result and lateral relaxation of coils when initially attached may be hindered (Figure 3) resulting in a more

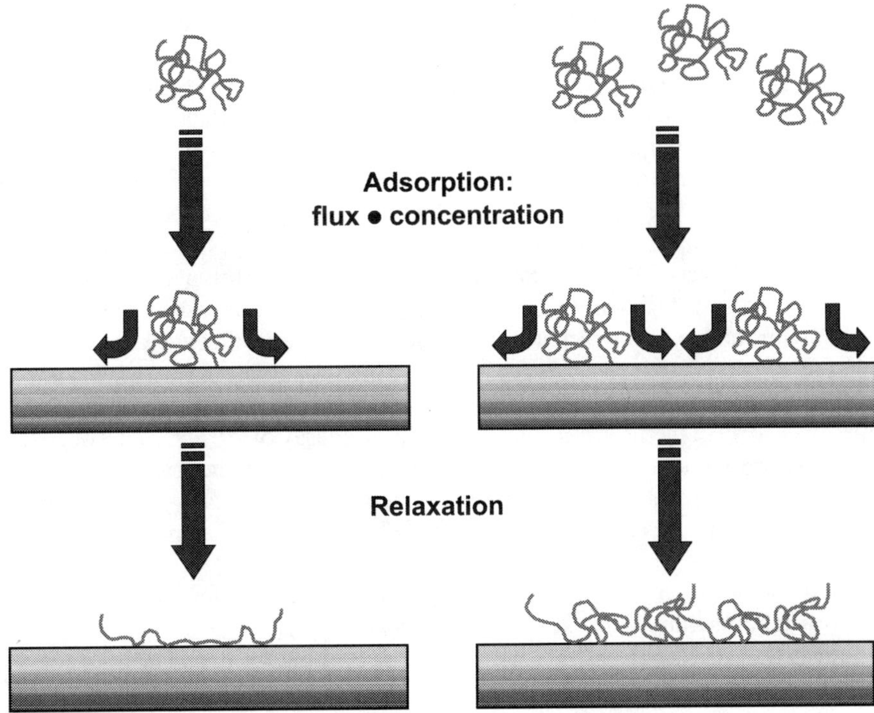

Figure 3 *Schematic illustration of the adsorption and relaxation process for polymer chains at the solid-liquid interface. The rate of flux of polymers to the interface will depend directly upon the bulk concentration. The opportunity for the polymer to relax to the interface will depend upon the ratio of this adsorption rate and the relaxation rate at the interface. The relaxation rate is dependent upon the adsorption energy per segment; if it is large then the relaxation rate is large*

extended conformation than predicted under full equilibrium. Obviously, if attachment is weak then a rapid exchange of polymer with the bulk solution can occur and establishment of equilibrium may be expected after a short time. However, if the initial adsorption energy is large, exchange with the bulk will be slow and a kinetically trapped conformation can result. A full description of the factors that control polymer adsorption at an interface is beyond the scope of this review; interested readers are directed elsewhere.[20] A number of theoretical treatments exist to describe the conformations of polymers at an interface based upon both mean-field[22,23] and scaling[24,25] approaches.

2.2.1.2 Ionic Polymers. The adsorption of ionic polymers introduces a number of complications when compared to their non-ionic analogues.[20] Obviously, the charge state, sign and magnitude, of both the surface and the polymer will play an important role. When both are highly charged and carry the same sign, adsorption is unlikely. In contrast, a high charge for both with an opposite sign is predicted to result in rapid and strong adsorption. The conformation adopted under these conditions depends upon the background electrolyte concentration; a low electrolyte concentration (<0.01M) results in a flat conformation and a high electrolyte concentration tends to give a loops and tails conformation. Adoption of a loops and tails conformation at the high electrolyte concentration is caused by competition between the ions and the polymer for surface adsorption sites that result in a decreased number of attachment points. A decrease in the charge density of either the polymer, the surface, or both can also promote a more extended conformation of the polymer away from the interface for similar reasons. Adsorption of a similarly charged polymer to a surface can also occur under conditions of high electrolyte or reduced charge (for either species). It has also been shown that adsorption of like charged polymers at a surface can be promoted by the use of multivalent ions such as calcium or magnesium which adsorb at the solid surface, effectively reversing its charge.

The adsorption of an oppositely charged polymer to a charged surface is obviously driven by the strong electrostatic free energy of attraction between the charged segments and the surface. In general, this results in a very favourable energy of adsorption and so the problems of kinetically trapped non-equilibrium conformations are generally increased for polyelectrolytes.

2.2.1.3 Adsorption Isotherms. The general equation for describing the adsorption of some species at an interface is given by

$$\Gamma = \frac{C_{eq}V}{mA_s} \tag{7}$$

where Γ is the adsorbed amount in moles per gram of adsorbent, C_{eq} is the equilibrium concentration of polymer in solution, V is the total volume of solution, m is the mass of adsorbent, and A_s is the specific surface area of the solid.

In general, polymer adsorption isotherms are of the high affinity type and are characterised by a very steep increase in the adsorbed amount in the initial part of the isotherm. At higher polymer concentrations the isotherm reaches a plateau. The transition from the rapid adsorption region to the plateau is sharp for monodisperse polymer samples but tends to be increasingly rounded as the polydispersity increases. The plateau amount of adsorption increases with increased molecular weight or decreased solvent quality for the polymer. The conformation of the polymer at the interface will typically be different in different regions of the isotherm for the reasons described above.

2.2.2   *Influence of Polymer Adsorption on Interaction Forces*

The presence of adsorbed polymers at the solid-liquid interface will have a profound influence on the interaction forces that are operative. The adsorbed polymer will affect both the double layer and dispersion force properties, as well as generating steric interaction forces. Clearly, adsorption of charged polymers at the interface can have a profound effect on any double layer at the solid liquid interface. Adsorption of a like charged polymer at an interface will increase the surface charge. In the case of an oppositely charged polymer and surface, adsorption can lead to either the reduction, neutralisation, or reversal of the charge. Regardless of the exact nature of the system, adsorption of a charged polymer introduces considerable uncertainty in how to represent the interface for analysis and modelling. Of particular difficulty is the location of the plane of charge. In simple approaches the plane of charge is assumed to be located at the outer edge of the adsorbed polymer layer. Clearly, this is a simplistic model of the true situation and it seems more likely that the actual plane of charge must be located somewhere between the underlying surface and the outer edge of the adsorbed layer.

Non-ionic polymers can also affect the properties of a charged surface electrical double layer. Adsorption of the polymer will perturb both the Stern layer, through the presence of trains (adsorbed segments), and the diffuse layer, through the loops and trains (Figure 4). Polymers adsorbed at an interface can also affect the dispersion forces experienced in a particle system. Typically, the presence of a polymer layer which is largely solvent results in a significant reduction of the Hamaker constant which in turn reduces the attractive interaction forces.

Theoretical descriptions of the steric interaction forces are extremely complex and require an accurate knowledge of the polymer segment density profile, both normal and parallel to the interface. It is also important to know whether the polymer is grafted to the surface or whether it is freely adsorbed and, if the latter, what the adsorption energies are. Finally, the solvency properties with respect to the polymer must be known. A wide variety of analytical techniques have been used to obtain information about segment density profiles for adsorbed polymers. Amongst these have been direct force measurements, light scattering, neutron scattering, electrokinetics, hydrodynamics, and ellipsometry.[20] The results have been extremely useful in theory development.

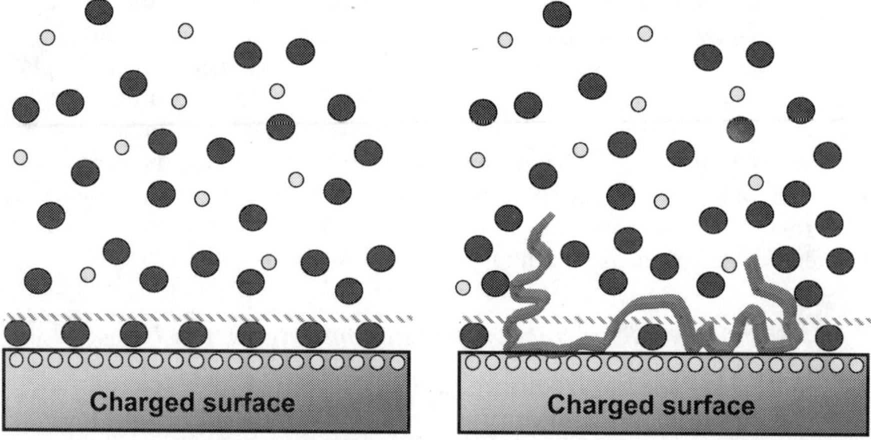

Figure 4 *Schematic illustration of the perturbation of ions within the Stern layer of a typical electrical double layer as a result of the adsorption of polymer chains*

Despite this, the problems alluded to above mean that most theories still require further development before they can become fully quantitative.

The simplest case remains that for the interaction between two surfaces coated by an end-grafted polymer brush layer. At low surface coverage, there is no lateral interaction between the chains and the attached polymers are best described as "mushrooms" (Figure 5). Under these conditions, scaling theory predicts the layer thickness to be[4]

$$\delta \approx N^{1/2} l \qquad (Theta\ solvent)$$

$$\delta \approx N^{3/5} l \qquad (Good\ solvent) \qquad (8)$$

Where N is the number of monomer segments in the chain and l is the size of an individual segment. Clearly the dimensions of the layer are solvent dependent

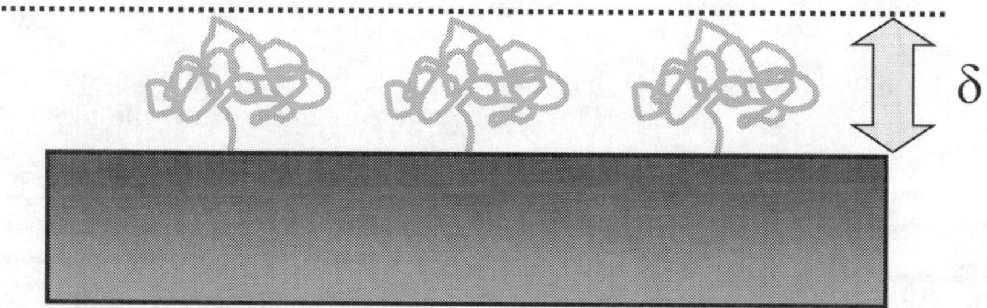

Figure 5 *Structure of adsorbed polymer chains terminally attached at the solid-liquid interface. At low adsorption densities, the chains adopt a "mushroom" appearance*

with the layer extension having a stronger dependence on polymer molecular
weight in a good solvent. Allowing for this simple description of the surface
polymer layer, the interaction forces can be determined by assuming that each
chain interacts with the second surface entirely independently of its nearest
neighbours. In a theta solvent, the resultant energy per unit area is given by

$$W(D) \approx 36\Gamma kT \exp^{-D/R_g} \qquad (9)$$

where Γ is the number of chains per unit area of surface, D is the surface-surface
separation distance and R_g is the polymer chains radius of gyration.

Increases in the surface coverage eventually lead to a lateral crowding of the
chains which forces them to extend normal to the interface. When the crowding
is sufficiently great, the extensions of the chains in the "brush" will approach the
contour length of the an individual polymer chain (i.e. $\delta = Nl$). The interaction
of two surfaces coated with polymer brush layer, and in a good solvent, are then
described by[26]

$$W(D) \approx \frac{kT}{s^3}\left(\left[\frac{2\delta^{9/4}}{1.25D^{5/4}} + \frac{D^{7/4}}{1.75(2\delta)^{3/4}}\right] - \left[\frac{8\delta}{5} + \frac{8\delta}{7}\right]\right) \qquad (10)$$

where s is the mean distance between polymer attachment points ($s^2 = 1/\Gamma$). This
equation can be simplified for $0.2 < D/2\delta < 0.9$ to

$$W(D) \approx \frac{100\delta}{\pi s^3} kTe^{-(\pi D/\delta)} \approx \frac{100\delta}{\pi s} \Gamma kTe^{-(\pi D/\delta)} \qquad (11)$$

The typical form of this interaction energy as a function of separation distance
D is given in Figure 6.

The situation for freely adsorbed polymers is considerably more complex. As
a function of time, the proportion of polymer segments present as loops, trains
and tails in any adsorbed chain can alter.[20] Furthermore, full equilibrium should
result in the desorption of any adsorbed chain as two surfaces approach such
that the interaction forces are always attractive. Indeed, this was predicted in the
initial theoretical treatments.[27] Subsequent theories have recognised the need to
impose kinetic restrictions on this desorption to obtain realistic predictions for
comparison to theory. Experimentally, direct force data have shown clearly that
an equilibrium force can take many hours to be attained.[28] Using a restricted
equilibrium model, scaling theory predicts that for two surfaces carrying
adsorbed homopolymers the interaction energy during initial overlap of the
diffuse steric region is described as[27]

$$W(D) \approx kT\left(\frac{1}{D}\right)^2 \qquad (12)$$

and in the region of significant polymer-polymer overlap

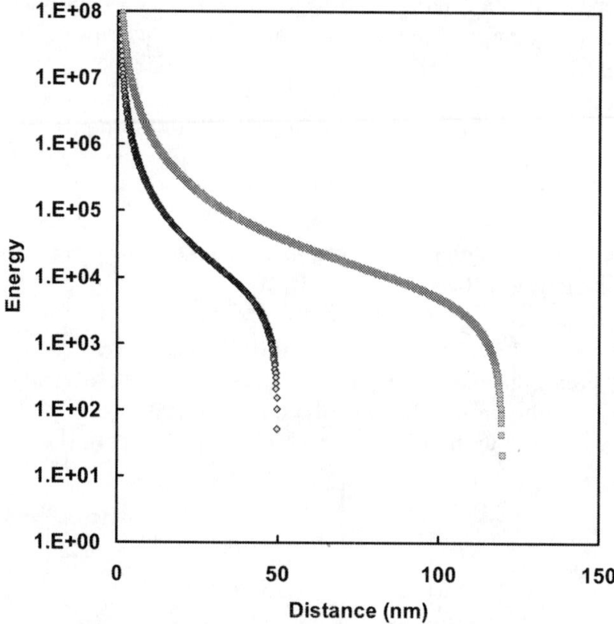

Figure 6   *The energy per unit area as a function of separation distance for the interaction between two surfaces coated in terminally attached polymer chains. The polymer samples are assumed to be in a good solvent environment. The data sets shown were calculated using the values; s = 9 nm, $\Gamma = 1.25 \times 10^{16}$ molecules/m². The two sets have the values of L = 25 nm and L = 60 nm, respectively*

$$W(D) \approx kT \left(\frac{1}{D}\right)^{5/4} \tag{13}$$

Subsequently, Luckham has suggested that the scaling theory predictions for terminally attached polymers can also be used for freely adsorbed homopolymers.[26] Comparison of the theory with data suggest a good agreement, the implication being that freely adsorbed polymer appears to behave as a brush layer. This is supported by the view that interactions are dominated by the presence of a few diffuse tails.[29]

2.3   Polymer Bridging Interactions

The interaction of polymer covered surfaces is strongly dependent upon both surface coverage and solvent quality.[29–31] Low surface coverage or poor solvency can give rise to attractive interaction forces. When two surfaces come together in a poor solvent, the outer segments are attracted to one another. The predicted interaction energy versus separation, as a function of polymer solvency, is given in Figure 7. It should be noted that, after the initial attraction, compression of the surfaces eventually results in steeply repulsive interaction regardless of the solvency.

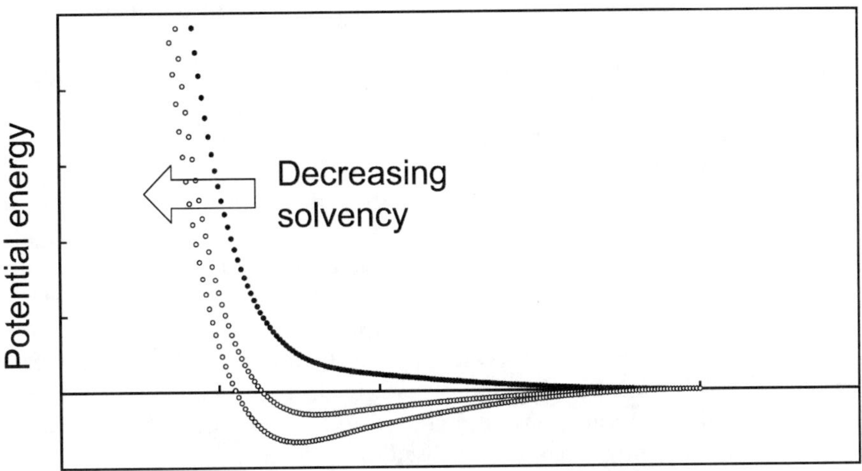

Figure 7 *Schematic illustration of the interaction between two surfaces carrying freely adsorbed polymers. The data curves shown illustrate the transition between a marginally good solvent through a theta condition to a marginally poor solvent*

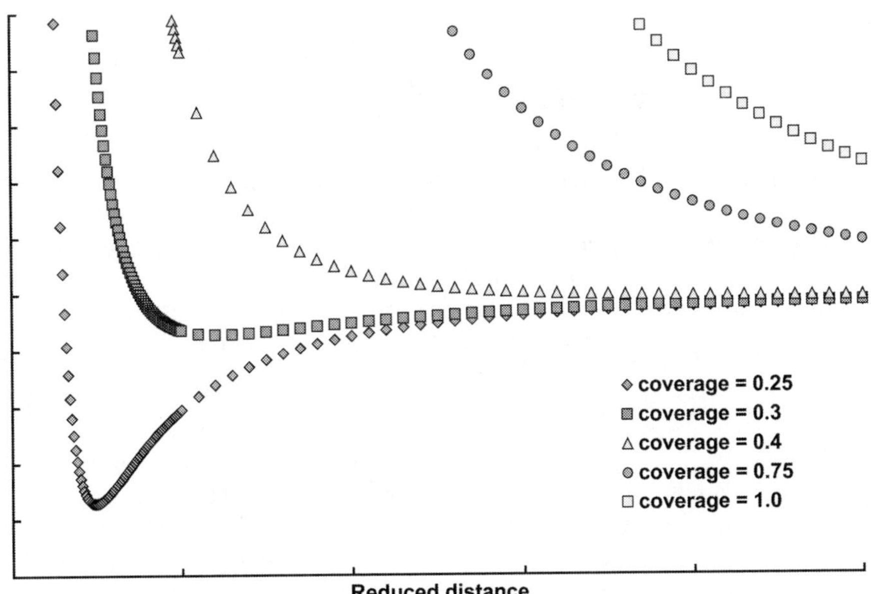

Figure 8 *The influence of surface coverage for the interaction of two surfaces carrying freely adsorbed polymer. The data curves illustrated were calculated according to the theoretical predictions of Klein and Rossi*[31]

The consequences of a change in the surface coverage of polymer have also recently been predicted by Rossi and Klein.[31] Figure 8 shows that as the coverage decreases, an increased attraction between the surfaces is predicted. The attraction arises from the adsorption of polymer segments from one surface at free sites on the second surface. Such attractive interactions can be very strong and can result in significant adhesion between the surfaces.

In a particulate dispersion, the kinetics of polymer-particle mixing and the molecular weight of the polymer play an important role in determining if bridging flocculation will occur. In general terms, it is important to have a high molecular weight polymer that is capable of sweeping out a large volume and collecting many particles. In addition, it is important that multiple particle collisions occur before the polymer has an opportunity to relax back to the interface, thereby more efficiently covering the surface. It has been known for some years that bridging flocculation is favoured at around 50% surface coverage of a dispersion.[32] More recently, it has been realised that this state may be transitory and it will be strongly linked to the mixing and shear conditions.[33,34] These features are generally true whether the flocculants is an ionic or non-ionic polymer.

For charged polymers, other "bridging" flocculation mechanisms have been proposed. In the first case, simple surface neutralisation may be sufficient to allow flocculation of a particulate sample. This is achieved when a flat adsorbed conformation for the polymer is favoured. Gregory[35] has also proposed an alternative flocculation mechanism that occurs when charged surfaces are only partially neutralised with an oppositely charged polymer. Under these conditions, the patchwise character of the surface charge can result in a strong attraction between the surfaces.

2.4 Polymer Depletion and Structural Forces

In the presence of a non-adsorbing polymer, close approach of two surfaces to within the characteristic diameter of the polymer in solution will result in the exclusion of any polymer from the gap between the surfaces. Thus, there is a solvent rich region between the surfaces surrounded by a polymer solution. The net result is that solvent moves out of the gap and there is an attractive osmotic pressure that acts to force the surfaces together. A classical description of the potential energy of interaction between two spheres is given by[20]

$$V_{dep} = - (2\pi/3)\Pi(2\Delta - D)^2(3a + 2\Delta + D/2) \tag{14}$$

where Π is the osmotic pressure, a is the sphere radius, D is the distance of separation and Δ is the depletion layer thickness (equal to the polymer radius of gyration). Typically these interactions are weak and can be difficult to measure or detect. The magnitude is dependent both on the polymer concentration (colligative property, so number concentration) and the size of the polymer coils (through D).[20]

Recently, it has been recognised that polymers which interact strongly in solution, even at low concentrations, such as polyelectrolytes can induce a bulk order which is sufficient to perturb the interactions between two surfaces.[36,37] The bulk order manifests itself as an oscillatory interaction potential with increasing amplitude as the surface separation decreases. The periodicity of the oscillation is directly related to the characteristic interaction lengths of the polymers in solution. As yet, the development of a theoretical interpretation for these forces is in its infancy.[38,39]

2.5 Relationship of Energy and Force: The Derjaguin Approximation

Standard theories of colloid and surface science rely upon determinations of the potential energy as a function of surface separation distance. Comparison of these theoretical predictions to direct force measurements requires some method for conversion of force to energy. The most widely accepted and flexible approach is the so-called Derjaguin approximation[40] which, for a sphere interacting with a flat surface, is given as

$$W(D) = \frac{F(D)}{2\pi R} \tag{15}$$

where R is the radius of the sphere. This equation is valid when the surface separation, D, is small compared to the radius of the probe. Such a restriction is achieved when using a colloid probe (see below) but may be problematic for simple AFM cantilever tips (radius 10–100 nm).

3 Atomic Force Microscopy

Initially developed as a high resolution imaging device,[41] it was quickly realised that atomic force microscopes (AFM) could be used to measure surface forces. Since then, there has been an enormous growth in both the variety of systems examined and the number of papers published. The AFM has proven itself to be amongst the most versatile of instruments ever developed.

3.1 Principles of Force Measurement

The main features of a typical AFM are shown in Figure 9. When performing a standard force measurement, the sample surface of interest is driven towards and away from a sensing cantilever probe using a piezoelectric drive. As the surface approaches the probe, it can respond to any operative surface forces by deflecting towards (attractive) or away (repulsive) from the oncoming surface. Deflections of the cantilever sensor are measured using an optical lever technique. A focussed light source (often a laser) is reflected from the end of the cantilever into a position sensitive photodetector. In this way, deviations from the stationary undeflected state are easily measured.

A typical force curve recorded using an AFM is shown in Figure 10. Deflections of the cantilever are recorded as a voltage (V) from the photodetector and

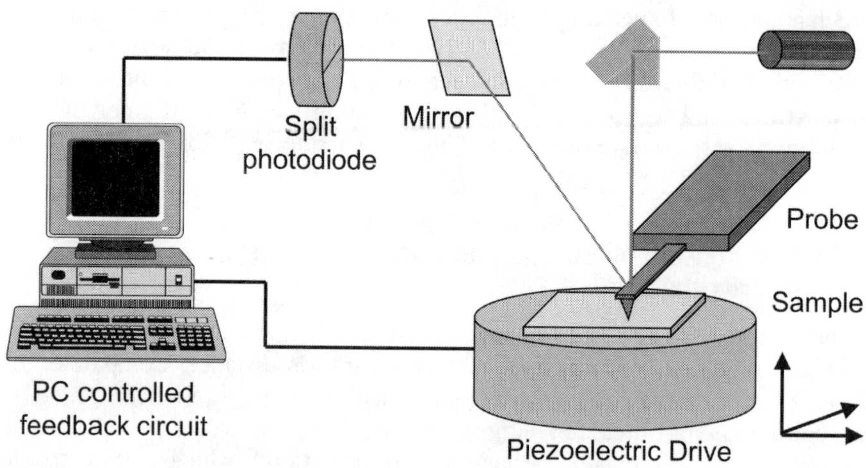

Figure 9 *Schematic illustration of the major features of a commercial AFM system*

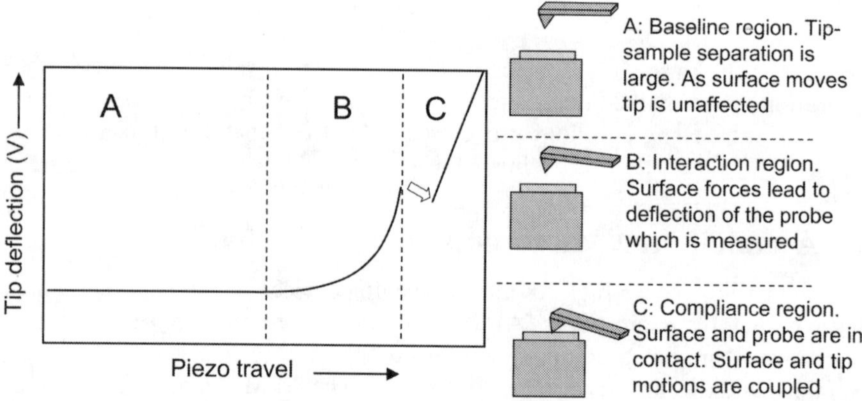

Figure 10 *A typical raw data curve for the interaction between a standard AFM cantilever tip and a flat surface. The data shown illustrates the expected interaction between the surfaces in an aqueous electrolyte solution when the two surfaces approach from an initially large separation. Data for the retraction of the surfaces is not shown*

are plotted as a function of the distance moved (z) by the piezoelectric translator normal to the surface. The conversion of this raw data to a force-distance data set requires both a baseline value for the deflection (zero force) and a region of constant compliance (zero distance).[42,43] The baseline value of the cantilever deflection is found when the surfaces are far apart and movements of the surface do not result in any deflections of the cantilever (region A). In this region of the data, no surface forces are acting on the cantilever. Zero distance is defined when the surfaces are in contact and a unitary movement of the surface results in a unitary deflection of the cantilever in the same direction (region C). All data in this region are defined as being at zero distance. Separation distances are then determined by calculating the cantilever deflection (in nm) and adding it to the piezoelectric position (relative to zero). Conversion of cantilever deflection to force simply requires the application of Hooke's Law, $F = k_s x$. (x = deflection distance) more complete descriptions of the data conversion process are given in a number of excellent reviews.

Correct conversion of the raw data requires accurate calibration of both the piezoelectric transducer and the cantilever spring constant. The piezoelectric transducer is usually calibrated through images of a sample of known dimensions. If the voltages applied to the piezoelectric are known, a conversion factor of volts to nm can be determined. Calibration of the spring constant is considerably more involved. However, a number of techniques now exist with the most popular being based upon determinations of resonance frequency data.[44] A more complete discussion is given below.

3.2   Accurate Force-Energy Relationships: Probe Radii

Use of the Derjaguin approximation (§2.5) for the conversion of force to energy requires an accurate knowledge of the probe geometry, and in particular the probe radius. This allows all force data to be normalised and direct comparison to be made both between different experiments and with theoretical predictions. In most cases, AFM data are recorded using a supposed flat surface and a spherical probe on the cantilever. The state of the flat surface, and the quantification of roughness present, is easily investigated using in-situ AFM images. Thus, the main issue in correct conversion of the data from force to energy is the determination of the probe radius. In the great majority of literature reports, data have been collected using standard AFM probes with an integral tip attached. The apex of these probes is usually assumed to be spherical, although this is rarely proven. Nonetheless, the popularity of measurements using integral tips has led to the development of methods for characterising the shape and an apparent radius of interaction.

Direct measurements of tip shape and radius involve so-called reverse imaging techniques.[45] These techniques utilise the images that are generated when the features on a surface are "sharper" than the imaging probe. Under these conditions, an image of the probe results and careful analysis can result in both an accurate shape and radius for the probe. Nano-fabricated standard calibration gratings are now available commercially and have been successfully utilised to obtain probe images. An example of such an array and the images that result is given in Figure 11. An alternative method for the determination of tip radii involves imaging an atomically sharp faceted surface such as $SrTiO_3$.[46] Although this technique does not give a full 3-dimensional representation of the tip, potentially it can give a good estimate of the tip radius along the scan direction.

In-situ measurement of the 'effective' radius for a probe used in measurements of long-range surface forces has also been proposed.[47] In this method, surface forces are measured between two surfaces in the presence of adsorbed layers of the surfactant cetyltrimethylammonium bromide (CTAB). At given concentrations, literature data for the surface potential of such CTAB layers is available. The measured data are then fit by adjusting the radius of the probe at a fixed surface potential using standard DLVO theory (rather than the opposite more normal approach).

Whilst a number of techniques do now exist for the determination of the size and shape of integral AFM probes, significant errors may result from their use in measurements of surface forces. For example, the Derjaguin approximation is valid only for large objects and sharp probes may not fulfil this criterion. Also, the in-situ approach described above may give a reasonable estimate of the effective probe radius acting in the double layer at finite separations but this may have no validity when the separation distances become small (< 10 nm). Examination of the probe image in Figure 11, clearly indicates that the complex shape will result in its effective size changing as a function of the separation distance. An alternative approach utilises the so-called "colloid probe technique" where the integral tip is replaced by a colloidal sphere (radius 1–10 μm).[42,43,48] This is now a well-established and accepted technique for the measurement of surface forces using a known geometry under conditions that are applicable to the Derjaguin approximation. An electron micrograph of a colloid probe is shown in Figure 12.

The colloid probe technique offers a number of significant advantages for the measurement of surface forces. The radius of the probe is easily determined to a high degree of accuracy using electron microscopy. A single probe can be re-used multiple times in a series of experiments and a wide range of surface chemistries are available. The main limitations are that the particle should be

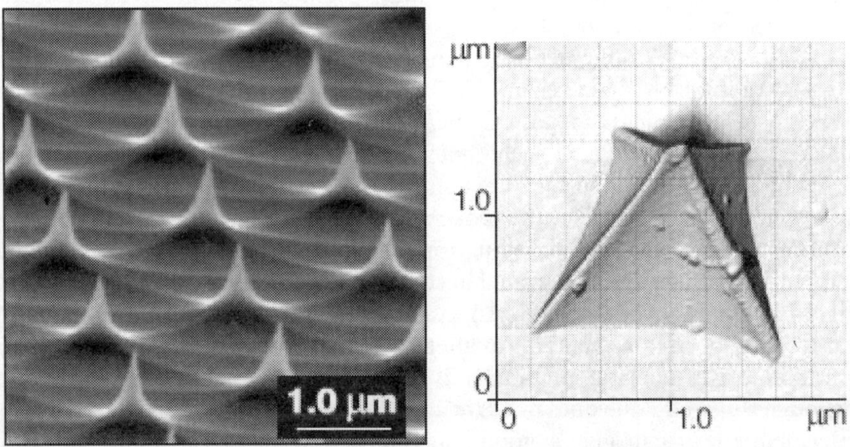

Figure 11   *Scanning electron micrograph of a nano-fabricated calibration grating used for the analysis of AFM cantilever tip shapes. A typical image of a standard tip resulting from scanning the grating is also shown*

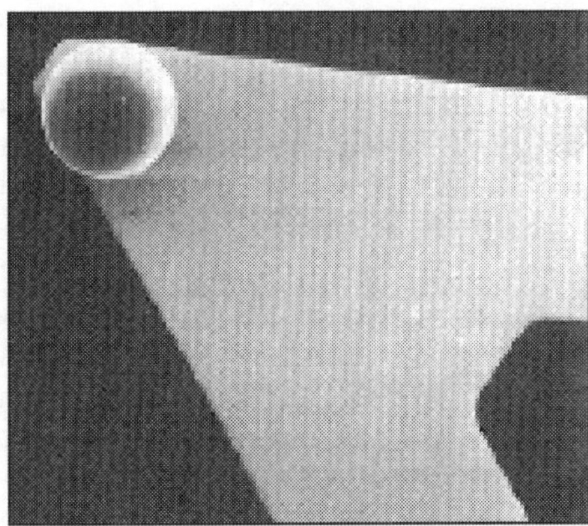

Figure 12 *Scanning electron micrograph of a colloid probe. The probe shown here is a 10 μm radius borosilicate glass sphere attached to a standard 100 μm triangular cantilever. The sphere was attached to the cantilever using a small amount of an epoxy glue*

smooth and spherical and that it should be of a size that can be easily manipulated in probe preparation (> 1 μm).

Recently, accurate analysis of the state of the interacting surface for colloid probes has been obtained from reverse images.[45] Such images are very useful since they allow characterisation of surface roughness on the probe surface. Moreover, the roughness is analysed exactly in the area of the probe which dominates the interaction with the surface. This is particularly important in developing an understanding of data collected at small separation distances where surface asperities will dominate the interaction.

3.3   Spring Constant Determination

Accurate determination of the cantilever spring constants remains one of the major challenges in the quantitative measurement of surface forces using AFM technology. The problems are mainly caused by the very small dimensions of the cantilevers (100–300 μm in length). Accurate knowledge of the cantilever dimensions and its material properties can be used, in principle, to determine the spring constant.[49,50] Such an approach, using a finite element analysis of the cantilevers has been reported. The main difficulty in this approach arises from uncertainties in the exact chemical composition of the cantilevers as well as uncertainties in features such as the density homogeneity. Thin coatings of a reflective metallic coating (such as gold) are also commonly used on cantilevers. This further complicates the analysis and produces significant uncertainty in the determined spring constants. As a result, more direct methods for the determination are usually favoured.

The simplest approach involves measuring the static deflection of the cantilever under the action of a known mass.[51] However, accurate measurements of such small deflections and the need for numerous repeats make this approach somewhat problematic. The most commonly accepted method utilises measurements of the resonant frequency as a function of added mass.[44] The resonant frequency is related to the end mass through a simple relationship with a constant of proportionality that is equal to the spring constant. The analysis is based upon the mass being attached to the end of the spring. This is the main difficulty in the experiment. In a typical measurement, a series of masses are used (typically more than four) which reduces errors caused by misplacement of the added mass.[49] It is also common practice to measure only a randomly selected sub set of cantilevers from a larger wafer. The assumption being that all cantilevers across the wafer have similar properties and dimensions and hence, the same spring constant. Recently, simpler analyses that require only a measurement of the unloaded resonant frequency of the cantilevers and their quality factors, Q, have been proposed.[52] Another approach that is frequently employed involves pressing a cantilever with unknown spring constant against another reference cantilever, with known spring constant.[50] Measurement of the total deflection allows determination of the spring constant.

It is well known that changes in materials properties as well as geometric alterations can affect the spring constant of a cantilever.[49] Thus, the preparation of a colloid probe (Figure 12) by attachment of a particle using glue can be expected to alter the spring constant of the cantilever. It has also been shown that the exact position of the sphere can affect the spring constant. When using a colloid probe to measure surface forces, therefore, it may be preferable to measure the spring constant of each individual colloid probe assembly. Craig and Neto have recently reported a method that uses an analysis of hydrodynamic data to calculate the spring constant of a colloid probe.[53] In this approach, accurate knowledge of the probe radius is required. A best fit to the data is then performed through adjustments of the spring constant. An alternative method that is also based upon hydrodynamic drainage has recently been proposed by Notley et al.[54] In this method, a small amplitude oscillatory excitation of the system is used as the surfaces are brought together. The response of the probe allows calculation of the viscous drainage between the surfaces. If the intervening fluid is Newtonian with a known viscosity, the spring constant can be determined.

3.4　Accurate Determination of Separation Distance

One of the two crucial measurements for obtaining correct force-distance data is the determination of zero distance from the constant compliance region. For hard incompressible surfaces, this measurement is straightforward and relatively reliable. The main concern arises from surface roughness and this can introduce small errors in the distance measurements although some idea of the uncertainty can be obtained from images of the interacting surfaces. A more significant problem occurs when one or both of the surfaces is compressible. This can arise if the bulk material itself is compressible, such as for a liquid or soft polymer, or if hard surfaces are coated by adsorbates such as polymer steric layers. In these cases, initial contact between the surfaces is followed by deformation of the soft surface(s) under the compressive load. For adsorbed polymer layers, a typical interaction force profile is shown in Figure 13. When compared to the standard force curve for charge stabilised surfaces a number of differences are apparent. As the polymer layers initially overlap, a repulsive interaction force results in deflection of the probe. Further movement of the surfaces towards one another causes an increasing compression of the polymer layer which is analagous to an increase in the effective spring constant of this layer. At some level of compression, the spring constant of the intervening polymer layers exceeds that of the cantilever probe. After this point, further compression results in all motion being absorbed by the cantilever and it appears as though the surfaces are in hard contact. However, although the constant compliance region is linear and allows accurate calibration of the photodiode, the presence of compressed polymer between the surfaces prevents accurate knowledge of the zero distance. For high molecular weight polymers, this could introduce a significant error in the reported data. If the adsorption is relatively weak, it is possible that desorption can occur under the action of a compressive force. In these cases, the data shows discrete jump(s) as the molecules desorb. In such cases, it is possible that hard surface contact could be established.

The problems of zero separation determination are particularly important when using AFM to measure steric interaction forces. In these cases, care has to be taken in the interpretation of the data, particularly when comparing experiment to theory.

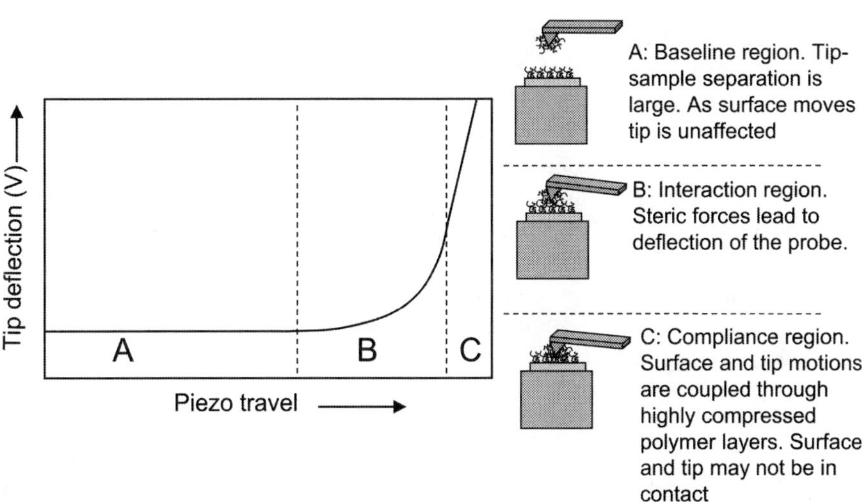

Figure 13 *A typical raw data curve for the interaction between a standard AFM cantilever tip and a flat surface. The data shown illustrates the expected interaction between polymer-coated surfaces in an aqueous medium, when the two surfaces approach from an initially large separation. Data for the retraction of the surfaces is not shown*

4   AFM Measurements of Polymer Induced Particle Interaction Forces

The growth in research publications that report images collected using AFM techniques has been phenomenal during the last decade. Although this is the main commercial driver for AFM sales, imaging without knowledge of the forces acting on a probe is a dangerous proposition. Especially since the image generated using AFM is a derivative of the forces that the probe experiences. In many cases, force data is only presented as an after-thought or as a minor component of the publication. The general growth in papers that deal exclusively with force data measurement has also been impressive. The rapid development is related to the broad multi-disciplinary application of the AFM technology. In this review, we will concentrate on research whose primary focus was the determination of interaction forces in the presence of polymer solutions.

4.1   Steric Interaction Forces: Adsorbed Polymer Layers

4.1.1   Non-ionic Polymers

4.1.1.1   Freely Adsorbed Polymer.   One of the earliest reports of surface force measurements using AFM technology involved the interaction between a standard AFM cantilever tip and a mica surface in the presence of poly(ethylene-oxide) [PEO]. Lea et al.[55] measured steric interaction forces in the presence of two different PEO samples with molecular weights of 200 and 900 KDa. No adhesion was seen between the surfaces in the presence of the polymer suggesting that, even at constant compliance, some polymer remained between the surfaces. In a second paper,[56] the same authors reported measurements between a bare tip and a mica surface with a grafted layer of monomethoxypolyethylene glycol [PEG] with a molecular weight of 2000 Da. In a 0.1 M KNO_3 solution, the interaction again exhibited a typical monotonic steric repulsion and no adhesion. Measurements in a poor solvent for the PEG, a Mg_2SO_4 solution, resulted in the absence of a steric interaction. It is worth noting that the data analysis in these early reports was unsophisticated. However, the importance of these papers is the simple observation of steric interactions, the effect of solvency variations, and the lack of an adhesive interaction between stabilised surfaces.

   Measurements of steric interactions between non-ionic polymer adsorbates using the colloid probe technique[57] were first reported in 1996. Braithwaite et al reported data for the interaction of a glass bead and a glass surface in the presence of PEO with a molecular weight of 56 kDa. The chosen measurement conditions ensured that the polymer was always in a good solvent environment. Data were collected as a function of adsorption time onto the initially bare surfaces. A fully equilibrated layer was seen to form only over a timescale of 24 h. The presence of adsorbed polymer on the two surfaces resulted in a measurable steric interaction of > 100 nm with little or no adhesion between the surfaces. However, during the initial adsorption (1–3 h) the polymer is seen to form a relatively flat layer and some evidence for a bridging interaction between the sparsely covered surfaces is presented. Comparison of the data with both theory and previous SFA work indicated that the extension of polymer from the surface was of the correct magnitude even allowing for the presence of compressed polymer at constant compliance. The difficulty of obtaining a full equilibrium measurement using the colloid probe approach was also highlighted in this paper. A more complete analysis of the effects of non-equilibrium interactions between adsorbed polymer layers was presented by Biggs.[58] Measurements between a zirconia sphere and flat were performed in the presence of freely adsorbed poly(vinyl pyrrolidone) [PVPy] having a molecular weight of 40 kDa. The effects of subtle variations in the collision rate of the two surfaces on the measured steric forces were probed. An increase in the repulsion was seen with increased collision rate. This was attributed both to the decreased time available for chain relaxation and to an increase in drainage forces. Importantly, these data were collected at collision rates which approximate those seen for colloids colliding under Brownian conditions or under the action of mild shear. Data for steric interactions using a similar sample of PVPy (Mw = 40 kDa) were also reported by Craig et al.[59] Although this paper was not primarily related to steric measurements, the length scale of the interactions between a silica sphere and silica surface were directly comparable to that reported by Biggs. Interestingly, Craig et al inferred that the presence of a constant compliance region was due to desorption of polymer during the collision of the surfaces. Numerous repeat runs showed no hysteresis leading the authors to the conclusion that re-adsorption was rapid. This is at odds with the report of Braithwaite et al.[60] and is unexpected for the adsorption of a large molecule. It seems more likely that the polymer was still present between the surfaces in the constant compliance regime in a highly compressed state. Recently, Sakai et al.[61] have also reported data for interaction forces between oxide surfaces in the presence of PVPy. Again, the polymer had a molecular weight of 41 kDa. In this case, however, the adsorption to the alumina surfaces used was weak and the polymer produced little or no steric barrier between the surfaces.

In a subsequent investigation, Braithwaite and Luckham[60] investigated the adsorption of poly(ethylene oxide) (PEO) of molecular weights 56, 205, and 685 kDa onto glass surfaces using a colloid probe technique. Data were presented for the polymers adsorbed onto one and two surfaces at low and high coverage of adsorbed polymer. In the case of polymers adsorbed onto one surface, or adsorbed onto both surfaces at low coverage, a strong adhesion was seen on separation of the surfaces. However, attraction was rarely noted as the surfaces were brought together. This was attributed to the rapid rate of approach not allowing sufficient time for polymer rearrangement and adsorption to occur. At high coverage of polymer repulsive steric interactions were seen. The magnitude of these increased with the molecular weight of the adsorbed polymer; the adsorbed layer thickness scaled with the molecular weight M through a power law with exponent 0.4.

Giesbers et al.[62] have also systematically investigated the effect of molecular weight for the interaction between silica surfaces in the presence of PEO. The molecular weights investigated were 23, 56, 105 and 246 kDa, and the samples were relatively monodisperse ($M_w/M_n < 1.10$). Adsorption to the surfaces was performed from a 100 ppm polymer solution; interaction forces were recorded after solvent exchange for polymer free electrolyte solutions. In no cases were significant steric interaction forces observed although some evidence for polymer bridging between the surfaces was seen. The results presented in this paper were significantly different to the earlier paper of Braithwaite et al.,[57] as well as even earlier SFA data, where a steric layer extending out to many times the polymer radius of gyration (R_g) was seen. A significant factor in this discrepancy may be due to the adsorption time; Giesbers et al only allowed the polymer to contact the substrate for 20 min whereas Braithwaite et al. did not see full equilibrium before 24 h. It is also worth noting that the work in Reference 57 was performed at 0.1M KNO_3 whereas that in Reference 62 was at 0.001M. The higher electrolyte concentration may lead to a more efficient competition of counter-ions with the polymer for surface adsorption sites and hence, a more extended polymer conformation.

Zauscher and Klingenberg[63,64] have examined the interaction between a cellulose bead and a cellulose surface in the presence of freely adsorbed polyacrylamide (PAM). The PAM sample was of a high molecular weight and resulted in long-range steric interactions extending out to over 600 nm. This is approximately $3R_g$ per surface and agrees well with theoretical prediction. Interestingly, these authors also reported problems in obtaining equilibrium results with hysteresis being seen as a function of collision rate.

4.1.1.2 Polymer Brush Layers. The end grafting of polymers or the adsorption of block coplymers with a short anchor block can result in the formation of a dense polymer brush layer at the solid-liquid interface. Kelley et al.[65] were amongst the first to investigate such brush layers using AFM techniques. In their initial study, these authors investigated the interaction of a bare AFM tip with a mica substrate coated with a poly(2-vinylpyridine)-b-polystyrene [P2VP-b-PS] sample. The force data were collected in a good solvent environment for the extended PS layer (P2VP is the anchor). Two types of force profile were consistently observed; in the first, little or no repulsion was seen whilst in the second a strong steric repulsion was recorded. In the second type of interaction, the repulsion extended to approximately 30 nm which is consistent with data collected using an SFA. Comparison to the size of the polymer in solution (Rg = 5 nm) implied that the layer was an extended polymer brush. In all cases, the magnitude of the steric repulsion was less than was seen using the SFA. The occasional lack of repulsion (type 1 interaction) was thought to be due to a "splaying" of the brush layer under the approaching tip. This agreed well with an earlier theoretical prediction by Murat and Grest.[66] Such a problem is not seen in the case of the SFA, where the macroscopic interaction area precludes such molecular detail. Further evidence for the local nature of interactions recorded when using AFM tips as the force probe was provided by Butt et al.[67] These authors examined two types of polymer brush: an end-grafted PS layer and a freely adsorbed polyethylene oxide-b-polymethacrylic acid [PEO-b-PMAA] copolymer, where the PMAA acted as an anchor block. The fundamental difference between the two systems is that the PS chains are totally immobile whereas the physisorbed PEO-b-PMAA can respond to an unfavourable environment by desorption or lateral mobility across the surface. Comparison of data between the two systems in a good solvent environment show important differences that support the concept of chain mobility for the block copolymer. In the case of the end-grafted PS, the data showed a strong steric repulsion and little or no hysteresis between the approach and retract sets of data. In contrast, the block copolymer data generally exhibited a weaker interaction that could not be fit with the same simple theory used for the PS brush. Furthermore, the approach and retract data showed considerable hysteresis. On the basis of the results presented, Butt et al. concluded that the PEO-b-PMAA copolymer chains migrated laterally as the tip approached the surface thereby reducing the repulsive potential as predicted by theory.[68–70] The reasons for favouring lateral mobility over desorption were not explained, and no discussion of the relevant energy for each process was given. The similarities with the earlier work of Kelley et al. are also strong although the paper was not referenced and no comparison between the "splaying" and "escape transition" theories was given.

These early AFM measurements of polymer brushes were directly comparable to previous SFA measurements. In general, the observed features were equivalent although the higher spatial resolution in the AFM does allow observation of features such as the "splaying" described above. This may be important in developing our understanding of how colloidal nanoparticles interact with surfaces, for example. Recently, Yamamoto and co-workers[71,72] have presented data for interaction measurements recorded against a "dense" polymer brush. This brush was prepared by synthetically growing polymers directly off a surface. The polymer used was a poly(methylmethacrylate) [PMMA] and the chain length of the samples could be adjusted by changing the synthetic conditions. The authors claim brush densities of up to 0.7 chains/nm^2 which is an order of magnitude greater than the typical systems previously reported. The increased chain density resulted in a number of important differences with earlier data for the lower density brushes. Scaling theory predicts that the thickness of the layer scales with graft density according to

$$L_e \propto L_c a^n \tag{16}$$

where L_e is the brush layer thickness, L_c is the chain contour length, and a is the graft density. For typical brushes (moderate graft densities), the exponent n has a value of 0.33; for the high graft density brushes the value of n was approximately 0.5. The higher exponent indicates a strongly extended brush. Also, as the chain contour length was increased, the force-distance interactions were seen to alter and the layer was seen to become more resistant to tip penetration with a thicker brush. For normal brush layers no such thickness effect is seen and the force relationship scales with distance. The work of Yamamoto and co-workers is notable for one other feature; by careful scoring of the surface they were able to remove part of the brush and images revealed the thickness of the layer. This thickness was then used to correct the force data and give true force-distance information even when the tip and surface did not achieve contact. Evidence for the thickness of brush layers from images is rare and the success of this work may well be due to the density of the brush film. However, the idea of directly measuring the thickness from images is appealing and is worthy of a more general examination for a variety of systems.

The effect of solvent quality on brush layers has also been investigated recently by a number of authors. Kidoaki et al.[73] investigated a temperature responsive brush layer formed by the chemical grafting of poly(N-isopropylacrylamide) [PNIPAm] at a surface. PNIPAm switches from a good solvent environment at low temperatures to a poor one at higher temperatures. The transition is usually fairly sharp and typically occurs at 30 to 35°C. A template surface structure, having both bare and polymer covered regions, was achieved through the use of photomasks during the synthesis of the brush layer. Force distance data were collected on the grafted regions of the surface at two temperatures, 25°C and 40°C. At the lower temperature the force distance data exhibited a typical steric interaction profile. The steric interaction was entirely absent at 40°C; interestingly, the force-distance data did not exhibit any attraction and also didn't show a short-range repulsion. The reasons for this are not clear and were not discussed in the paper. The presence of a polymer layer in a poor solvent might be expected to cause a weak attraction on approach of the tip and significant adhesion on its retraction. Images of the brush layer across an edge feature revealed a possible difficulty in using images to extract brush layer thicknesses. The recorded thickness of the brush was seen to be a function of the imaging force, both for the extended and the collapsed brush layers. The authors propose that this is due to penetration of the tip into the layer; the amount of the penetration is related to the strength of the imaging force as is shown schematically in Figure 14. This information is important, particularly if we are to use images to give information about polymer layer thicknesses, as discussed above.

Further information about the role of solution conditions on the extensions of polymer brush layers away from an interface has been presented by Sindel at al.[74] These authors investigated the effects of electrolyte on a PEO brush formed by adsorption of PEO-b-PMAA at the barium titanate/water interface. In this case, the data were collected using a colloid probe and in good solvent conditions (low electrolyte) the interaction forces were dominated by a long-range electrostatic component. The short-range steric interaction was only seen at highly compressive loads. At high electrolyte levels, > 0.1M, the steric interaction dominates the interaction forces. No evidence for an attractive interaction was seen even at electrolyte levels of > 1M. This is a surprising result since under these conditions the PEO chains are known to de-hydrate and in bulk solution precipitate. The continued presence of a steric barrier suggests that a brush layer may be more resistant to the de-hydration of the polymer.

Colloid probe measurements of interaction forces have also been used to investigate the presence and properties of inherent polymer brush layers at the surface of certain materials. Rutland and co-workers have examined the interactions of cellulose surfaces with sometimes conflicting results.[75,76] In an early

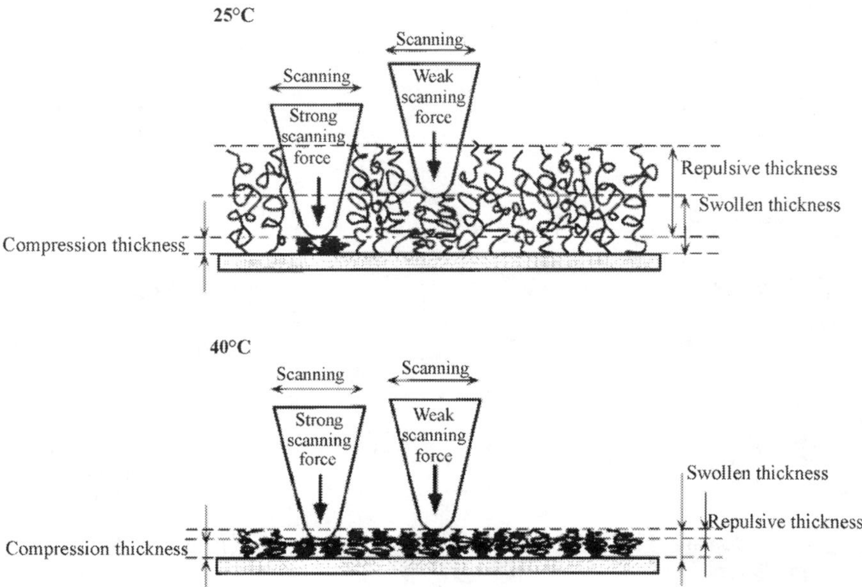

Figure 14 *Schematic representation of the influence of imaging force on the apparent recorded height of an adsorbed polymer layer. The two situations shown are for an adsorbed layer of thermo-responsive PNIPAm which is swollen at 25 and collapsed at 40°C*
(Illustration is taken from Reference 73)

investigation of cellulose surfaces interacting across electrolyte solutions using AFM, no evidence for a steric layer caused by a 'hairy' surface layer was seen. This was despite previous data from the SFA technique showing clearly the presence of such a hairy layer. The discrepancy was discussed in a subsequent publication where the importance of hydrodynamic drainage forces between the approaching surfaces was highlighted. For weak interactions, such as those seen in the presence of these hairy layers, poor control of the approach conditions can make data analysis problematic. These reports highlight a continuing problem where AFM data, which is essentially collected in a dynamic environment, is compared to static (equilibrium) interaction theory. Very few authors have considered the linked problems of hydrodynamic drainage and polymer relaxation between the approaching surfaces; both can be expected to cause discrepancy with static theories.[64,76] Prescott and co-workers have investigated similar 'hairy' layers at the surface of latex particles using the colloid probe technique.[77] Such layers form as a result of the synthesis conditions and the use of polymer stabilisers in the initial emulsion. Currently, very little information is available about how these layers might influence the interaction of latices with a surface. The importance of such interaction in the application of latex particles either in paints or as carrier materials in pharmaceutic applications should not be underestimated. It is precisely for this system specific type of information that the AFM and colloid probe approach are ideally suited.

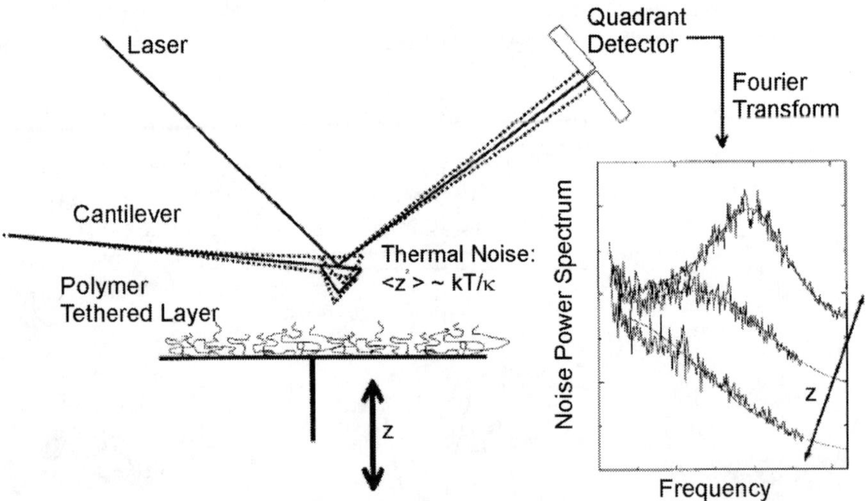

Figure 15   *Schematic description of the experimental setup used by Roters et al.[78] for the noise analysis of tip motions in the presence of an adsorbed polymer layer. The illustration also shows examples of typical data obtained when using this approach*

Roters et al.[78] have investigated the influence of solvent quality on an end-anchored PS brush using a novel approach to interaction force measurement that relies upon changes in the thermal noise spectrum of a cantilever as two surfaces approach. A schematic of the basic features of this measurement approach is given in Figure 15. Best fits to the data enable extraction of an elastic and viscous component and enable an interpretation in terms of viscoelastic parameters. The presence of a brush layer is easily detected through the elastic component and the authors claim this allows an extremely sensitive measure of brush thickness. Evidence for this sensitivity is presented from measurements of the brush in good and poor solvent conditions; significant collapse being seen in the poor solvent. The authors do not attempt to extract more detailed information about the brush layer from their data although comparison to viscoelastic and hydrodynamic theories for swollen brushes will be interesting as a future development of this approach.

4.1.2   Ionic Polymers.

4.1.2.1   Freely Adsorbed Polymer. Whilst non-ionic polymers are of undoubted interest, the range of phenomena and applications where they are utilised is small when compared to that for polyelectrolytes. This is in no small part due to the incredible variety of polyelectrolytes found in nature.[79] However, the complexity of polyelectrolytes, particularly their solution behaviour, means that both theory and experiment have tended to lag behind when compared to the non-ionic analogues. Despite this, the AFM, and in particular the colloid probe technique, has been employed by a number of researchers to gather further information about the interaction of colloidal solids in the presence of polyelectrolyte layers.

A major focus of the AFM research reported to date has been driven by a continued requirement to control the stability of aqueous colloidal slurries. Particulate slurries are important in a range of industries including minerals processing, paper manufacture and ceramics production. In particular, measurements of the stability of ceramic particles have received significant attention. The control of particle stability is critical in the formation of a green body; this is frequently achieved through the use of a polyelectrolyte adsorbate. The colloid probe approach has proven particularly successful due to the wide variety of substrates that can be examined, allowing direct comparison with other techniques such as rheology. Biggs and Healy were the first to report direct measurements on a relevant ceramic system when they investigated the influence of a low molecular weight ($M_w = 2000$ Da) polyacrylic acid (PAA) sample on zirconia stability.[80] This report is noteworthy for a number of reasons as well as being the first report of this type. The authors compared the force-distance data, collected as a function of pH at a fixed concentration, with yield stress results collected using equivalent materials. As a result, it was possible to relate the strength of the yield stress observed directly to the magnitude of the electro-steric barrier to aggregation. In the case of PAA, the degree of charge on the polymer increases as the pH increases. This results in an expansion of the adsorbed polymer layer and an increase in the electro-steric barrier to aggregation. To accurately characterise a charged polymer layer, it is important to know both the segment density profile of the adsorbed polymer layer and the effective plane of charge. Theoretical understanding of such layers has been hindered by uncertainties in this information and in particular in the relationship of one with the other. In the simplest cases, theories assume that the plane of charge is either located at the solid surface or at the outer region of the polymer film. Biggs and Healy utilised data taken from zeta potential measurements, on equivalent zirconia colloids with an adsorbed PAA layer, to analyse the force-distance data from the AFM and gain insight about the plane of charge relative to the steric interaction distance. At low pH, where the polymer carries little charge, the steric layer and the plane of charge were coincident at around 1 nm above the surface. As the pH was increased, the steric thickness was observed to increase dramatically out to almost 10 nm. The position of the plane of charge did not move as far; across the same pH range it simply doubled out to about 2 nm. The importance of these data are that they indicate clearly that simplistic theories that position the plane of charge at either extremity of the polymer layer are likely flawed. The influence of a high molecular weight PAA sample (Mw = 750 kDa) on the same zirconia surfaces was subsequently reported by Biggs.[81] Incubation of the two surfaces at a large separation distance over 24 h was seen to result in long-range steric interactions between the surfaces as they were brought together. Little or no hysteresis was seen when the surfaces were retracted. In a second series of experiments, the surfaces were incubated with a polymer solution for a shorter time and whilst they were much closer together. Under these conditions, the force-distance data showed considerable adhesion when the surfaces were retracted after contact. In some cases, the surfaces showed evidence of being 'tethered' together and could not be separated even at large separation distances. The shorter incubation time and limited mobility of chains at small separation distances was believed to have prevented the polymer from forming a dense steric layer. Instead, bridging of polymer between the surfaces was seen to occur.

Since these initial articles, other workers have examined the structures of common dispersants used in ceramics processing. Kamiya et al.[82] have investigated the stabilisation of alumina with copolymers of ammonium polyacrylate [NH$_4$PAA] and poly(methylacrylate) using AFM force data and rheological measurements. A series of copolymers were examined having varying degrees of ionic character. Highly charged polymers were observed to adsorb and form relatively flat layers; the force data showed no steric barrier. As the degree of 'hydrophobic' character was increased, the polymer was induced to adsorb in a more 'loops and tails' conformation resulting in the presence of a larger steric layer. Bergstrom and co-workers[83–86] have recently examined the influence of low molecular weight PAA (Mw = 10 kDa) at both zirconia and silicon nitride surfaces. On zirconia, their findings are in broad agreement with the earlier report of Biggs and Healy. However, in a significant extension of the earlier work, these authors examined how initial adsorption conditions and subsequent solution changes affected the observed steric layer. For example, adsorption of the PAA at low pH (pH < pH$_{iep}$) tended to result in a high adsorbed amount but a relatively collapsed layer whereas adsorption at high pH (pH > pH$_{iep}$) gave an extended layer but at a low adsorption density. Alteration of the pH to a higher value in the first case gives a highly extended electro-steric layer but still at a relatively high adsorbed amount. In the second case, lowering the pH gives rise to a flat layer that exhibits no steric barrier. This occurs since the low density of chains allows a complete relaxation to the interface as the charge on the polymer is reduced. So, the initial adsorption conditions and subsequent treatment can be used to generate a wide variety of adsorbed structures and hence system stability. The adsorption of the PAA onto silicon nitride surfaces was seen to give a relatively flat adsorbed layer under all conditions, even those where the zirconia gave an extended steric layer. Laarz et al.[85,86] postulate that this is caused by a relatively weak physisorption process on the silicon nitride which favours a flat conformation. In contrast, they argue that PAA adsorbs through a chemisorption process on zirconia which locks in a more extended conformation. No evidence for this chemisorption is provided and no indication of what the chemical reaction might be is given. Whilst the analysis fits the data it is not clear that this proposed reason for the differences is the correct one.

A number of researchers have investigated the adsorption of a strong anionic polyelectrolyte, sodium poly(styrenesulfonate) [NaPSS], at the alumina/water interface[61,87,88] in the presence of a co-adsorbing surfactant. Once again the exact conditions under which the polymer adsorbs is seen to influence strongly the surface conformation. Again, the conformation is inferred directly form AFM force-distance measurements. Adsorption of the polymer results in a flat conformation as a result of the strong electrostatic attraction between the cationic alumina and the anionic PSS. As a result, the force data show significant changes in the electrostatic interactions but no steric barrier. Addition of a surfactant to this pre-adsorbed layer has little effect on the conformation, confirming the strength of the interaction. However, adsorption of mixtures of surfactant and PSS, whether sodium dodecylsulfate (SDS)[61] or cetyltrimethylammonium bromide (CTAB),[87,88] results in a more extended conformation. In the case of SDS this is due to competition with the polymer for surface adsorption sites. In contrast the CTAB is strongly attracted to the polymer and this fundamentally changes the nature of the polymer and adjusts its adsorption at the interface.

The interaction forces that arise from the adsorption of a weakly basic polymer, P2VP, at the silica water interface have been investigated by Biggs and Proud.[89] The polymer sample had a low molecular weight (Mw = 7 kDa) and is soluble only in acidic solutions, precipitating at any pH >6. The effects of alterations in pH on layers adsorbed at pH 3 were investigated using the colloid probe technique. At pH 3 the polymer is strongly charged and will adsorb easily on the anionic silica surfaces. The interfacial conformation of the polymer layer was seen to be concentration dependent. A low concentration (5 ppm) resulted in a flat polymer layer whilst higher concentrations ranging from 10 to 50 ppm gave an ever increasing extended layer. Of particular note in this paper were measurements of the magnitude of steric interactions as a function of surface collision rate. The data supported earlier findings that a faster collision rate results in a larger steric force. Again this is related to fluid drainage and polymer relaxation effects. No attempts were made to separate these different effects or to accurately quantify them.

Force interactions measured in the presence of highly charged cationic polymers have been the subject of a number of reports. Such polymers are routinely used as flocculants in a number of industries, including water treatment, and this has acted as a significant driving force for the work. Holmberg et al.[90] have investigated the interaction forces between either silica and cellulose in the presence of a high molecular weight sample of poly[(2-(propionyloxy)-ethyl)trimethylammonium chloride] [PCMA]. Such a system is relevant to the paper manufacturing industry. Only one polymer concentration (20 ppm) was investigated and no evidence of a significant steric interaction was seen. Analysis of the resultant interaction forces indicated the presence of polymer on both surfaces although the cellulose surface was at a much lower resultant charge compared to the coated silica sphere. The main feature of the data analysis was that the adhesion between these surfaces was weak. This is important, since PCMA is used in the paper industry as a retention aid to increase the binding of paper and pigment. With respect to this review, an important feature is the ability of AFM to measure interactions in non-symmetric systems, and in particular the use of 'real' system analogues. The use of direct force measurements to illuminate the mode of action for an analogous polymer flocculant, poly[2-(methacryloyloxy)ethyltrimethylammonium chloride] [PMCMA], was reported by Bremmell et al.[91,92] The force data on silica surfaces were compared with equivalent data for particulate dispersion (silica) stability, adsorption isotherms, electrophoretic mobilities and aggregate sizing. The adsorbed amount of polymer and the adsorbed layer conformation correlated strongly with the initial solution concentration of polymer. Under conditions where the force data indicated a relatively low coverage and a flat adsorbed layer, rapid flocculation and an unstable silica dispersion were seen. At high polymer concentrations, the force data indicated a significant electro-steric barrier to flocculation. This correlated with extremely high dispersion stability. The differences in polymer conformation at the interface are related to the adsorbed amount and the available space for relaxation of a chain to the interface (Figure 3).

Liu et al.[93] have reported force-distance data for the interaction of a silica AFM tip and a silica surface in the presence of either poly(diallyldimethylammonium chloride) [PDADMAC], poly-L-lysine hydrobromide, or poly-(vinylbenzyltrimethylammonium chloride) [PVBTAC]. Adsorption of any of these polymers from low electrolyte solutions onto the oppositely charged silica surfaces resulted in the formation of a flat polymer layer and the slight overcompensation of the surface charge. In contrast, adsorption from 10 mM electrolyte resulted in an extended polymer layer and the observation of an electro-steric layer in the force data. Again, the surface charge was overcompensated by the adsorbed polymer layers. Rinsing of the adsorbed polymer layers was seen to lead to no desorption in the first case and an almost complete desorption in the second. These differences were simply attributed to the number of attachment points to the surface per polymer chain and the relative increase in desorption time expected for a flat adsorbed chain. The differences in the adsorbed layer conformations were attributed to the increased competition for surface adsorption sites in the presence of the electrolyte.

Information about the spatial coverage of a silica surface in the presence of adsorbed PDADMAC have been presented by Kramer et al.[94] Adsorption from low electrolyte conditions result in a flat polymer conformation, as described above. Force data for the interaction of a silicon nitride tip and the silica surface at pH 5.6 showed a standard repulsive double layer interaction and little or no adhesion. In contrast, in the presence of the adsorbed polymer there was no repulsive interaction and the force data showed a long-range electrostatic attraction and a significant adhesion. The attraction originates from the different charge character of the tip (negative) and the polymer coated surface (positive). Importantly, the interaction force differences were then used to map the polymer coverage of the surface in a so-called "force-volume" image. Variations in surface coverage were then mapped as a function of polymer concentration with a resolution of a few nm over an image of 2 x 2 µm. Such high resolution imaging of polymer covered surfaces on the basis of differences between the forces in coated and uncoated areas represents an important advance in the analysis of polymer adsorption.

The interaction forces between silica surfaces in the presence of an adsorbed polyampholyte, gelatin, have been reported by Braithwaite et al.[95] The gelatin has an isoelectricpoint at pH 4.9 and is expected to have an increasing negative charge as the pH increases. The force-distance data indicated that the adsorbed layer expanded with increasing pH. This was attributed to the decreasing interaction between the negative surface and the polymer. At low pH, the opposite charge of the surface and the polymer results in a relatively flat layer. When force data were measured between a pre-coated surface and an uncoated silica sphere the results were seen to be the same as for two coated surfaces. The implication was that the gelatin transfers rapidly from the coated to the uncoated surface.

Recently, modifications to the standard AFM force balance to allow measurement of dynamic interactions and viscoelastic parameters have been reported.[54,96] The approach, developed independently by two groups, involves excitation of the system through a high frequency low amplitude excitation of the surface. Detection of the response for the cantilever or colloid probe allows determination of both the phase lag and the amplitude attenuation between the drive and detected signals. This approach offers important opportunities for the study of dynamic interactions, hydrodynamic drainage, and viscoelastic interactions.

4.1.2.2 Ionic Polymer Brush Layers. Surprisingly, there have been very few investigations of fundamental force-distance data for adsorbed ionic polymer brush layers. Kelley et al.[65] have investigated the interactions between an AFM tip and a surface coated with a brush layer of poly(4-tert-butylstyrene)-b-sodium poly(styrene-4-sulfonate) [PtBS-b-NaPSS] as a function of the electrolyte concentration. The brush height was shown to depend upon the electrolyte concentration according to a power law relationship with an exponent of 0.4. This was consistent with theory predictions for medium density polyelectrolyte brushes. Hartley et al.[97] have recently reported observations of electro-steric interactions between an uncoated silica sphere and a polymer surface coated with carboxymethyldextrans. The extension of the layer away from the surface was increased through the use of an aminated pre-coating of the surface which increased the adsorbed amount of the polysaccharide and hence lateral repulsions within the brush layer. Likewise, increases in the charge on the polysaccharide layer were seen to increase its extension away from the surface.

Woodward et al.[98] have reported the interactions between a silica colloid probe and a microgel particle of poly(N-isopropylacrylamide-acrylic acid) [PNIPAm-AA]. The microgel has a surface brush layer of charged polymers that arise from the synthesis procedure. These charged hairy layers are polyelectrolytes and respond as expected to changes in ionic strength and pH.

Gelbert and co-workers[99] have recently reported the noise spectrum analysis for a grafted polyelectrolyte brush. As for the earlier report of the non-ionic system[78] discussed above, the noise analysis measurements offer a sensitive measure of the brush height. Again, shrinkage of the brush in the presence of added electrolyte was observed. The value of the noise analysis approach is that it offers viscoelastic information about the adsorbed layer. Close to the collapse point of the brush it was observed that the brush became highly compressible with an increase in the viscous dissipation. The implication of this for the rheology of colloidal dispersions was discussed.

4.2   Natural Polymer Systems

Razatos et al.[100] have investigated the forces of interaction between a surface coated with a PEG brush layer and an AFM probe that was prepared by the attachment of an *E.coli* D21 bacterium to an AFM cantilever. The data indicated that in the presence of a PEG brush layer there was a long-range steric interaction and no adhesion was seen. PEG layers have previously been proposed as surface coatings to prevent bacteria adsorption; the data in this publication confirmed that these layers do indeed act in this way. Further information about polymers that are used to reduce bio-adhesion have been reported by Morra and Cassinelli.[101] Measurements of force-distance data were reported between an AFM tip and two types of surface; one coated with an alginic acid and the other which is essentially the same surface except it has been treated with a glow discharge. Previous research showed that the glow discharge treated surface was not resistant to bacterial adhesion whereas the simple alginic acid surface was. Interestingly, the force-distance data showed no differences between the two samples; both surfaces showed a steric interaction with the tip. The conclusions were that the bio-adhesion resistance was not caused by steric interactions but instead is caused by changes to the local hydration of the adsorbed polymer layers.

Considine and co-workers have investigated the interactions of a biocolloid, *Cryptosporidium parvum* with silica.[102,103] Force-distance data indicated that the surface of *C.parvum* was coated in a polyelectrolyte layer that gives rise to an electro-steric layer. Adhesion data indicated that this surface polyelectrolyte layer can bridge to the silica surface after contact between the two. Velegol and Logan[104] have also investigated the interactions at the surface of a bacterium, *E.Coli* K 12. The force-distance information were collected between an uncoated AFM tip and surface immobilised bacteria. The surface of the bacterium is coated with lipopolysaccharides and the force data indicated the presence of a steric surface layer.

Interaction forces between polymer surfaces in the presence of a natural protein lactoferrin have been reported recently by Meagher and Griesser.[105] Lactoferrin is a protein found in tear fluid and is one of the molecules responsible for the biofouling of contact lenses. The force-distance data indicated electro-steric interactions between the surfaces. The data were interpreted using a model in which a significant proportion of the proteins were adsorbed in an end-on conformation.

Specific interactions between ligands and receptors have also received considerable attention in the literature. The research deals primarily with adhesion measurements and there is little information about long-range interaction forces.

4.3 Polymer Structural and Depletion Forces

The interactions between colloidal particles in the presence of non-adsorbing polymers are very important in a number of industrial and natural processes. When the adsorption energy is unfavourable polymer molecules are excluded from a region close to the surface due to conformational restrictions. As a result, close approach of two surfaces results in the exclusion of the polymer chains from between the surfaces. This results in an attractive interaction potential between the surfaces. In addition, if the polymer molecules themselves exhibit a strong interaction in the bulk then long range structural oscillations in the interactions can be observed.

The first report of an attractive depletion interaction was given by Milling and Biggs.[106] The interaction forces between stearylated silica surfaces in the presence of a solution of poly(dimethylsiloxane) [PDMS] in a good solvent (cyclohexane) showed a simple attractive depletion interaction whose range was comparable to twice the radius of the polymer (Figure 16). Milling and co-workers have subsequently reported a series of investigations between non-adsorbing polyelectrolyte samples and silica surfaces using the colloid probe technique.[37,107,108] The most striking feature of these reports is the detection of significant oscillatory forces. An example of these data are shown in Figure 17. The presence of oscillations is not predicted by the simple depletion theories

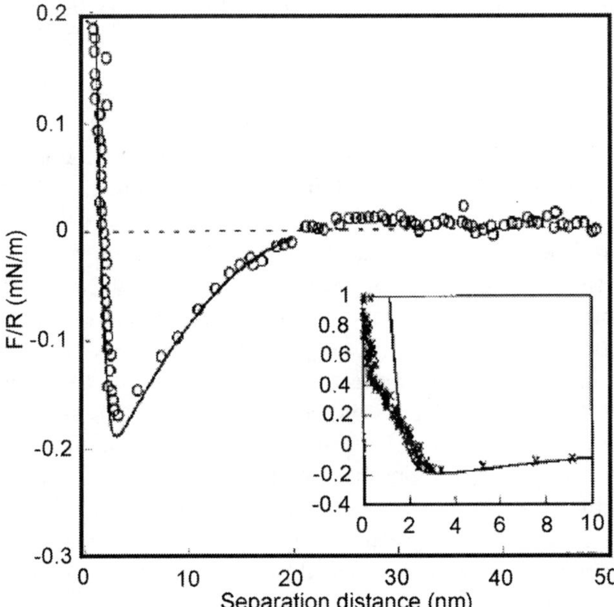

Figure 16 *The force as a function of separation between a stearylated silica glass sphere (radius = 3.8 μm) and a stearylated silica surface in a solution of PDMS (M_w = 120 kDa) in cyclohexane. The inset shows the full data set and the solid line is a best-fit to a standard depletion interaction model.*

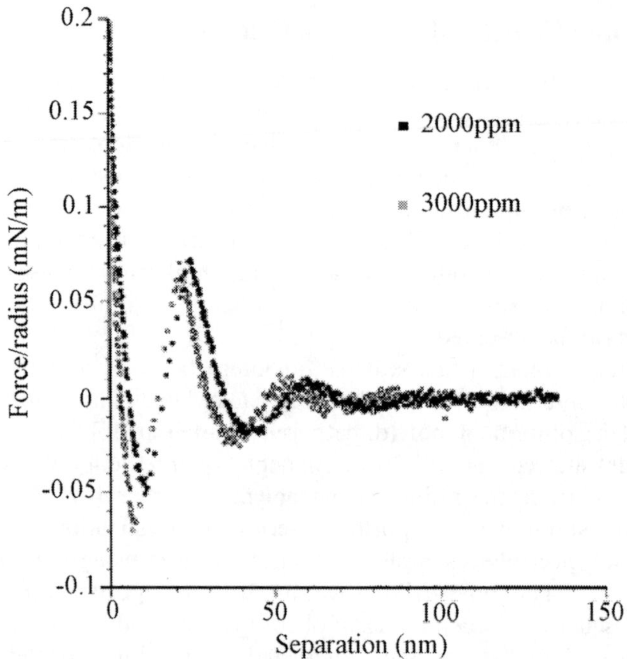

Figure 17   *Reduced force versus surface separation curves for a silica sphere (of radius 4.7 μm) interacting with a flat silica plate at two NaPSS (molecular weight 200 kDa) concentrations (as indicated) in the absence of added electrolyte. The pH was maintained in the range 5.5–6*

available, although these were primarily developed for non-ionic polymers. Current physical understanding of these interactions is that they arise from bulk order of the polymer in solution. The presence of a depletion interaction disrupts this order close to the surface; importantly, the surfaces do not induce the order. Milling[37] reported that the periodicity of the oscillations for poly-(styrenesulfonate) [PSS] was dependent upon factors such as molecular weight, electrolyte concentration, pH, and polymer concentration. In general, changes that result in a reduction in the effective polymer coil size in solution were seen to result in a decrease in the periodic length of the oscillations. Increases in the PSS concentration caused a reduction in the periodic length that was coupled with an increase in both the number of oscillations seen and the amplitude of these oscillations. Milling proposed that these oscillatory interactions were caused by a surface induced electrostatic ordering mechanism. This is now thought to be incorrect and the oscillations are simply a manifestation of the bulk order present in a polyelectrolyte solution. Milling and Kendall[108] have since confirmed the main features of these data in a subsequent paper. Biggs et al. have also investigated the depletion interactions between silica surfaces in the presence of PSS.[109] A systematic investigation of the influence of molecular weight on the extent and magnitude of the depletion interactions was reported.

The depth of the secondary minimum was seen to increase as a function of both molecular weight and concentration.

Milling and Vincent[107] have reported the measurement of depletion interactions between silica surfaces in the presence of a poly(acrylic acid) [PAA] sample. The novel feature of this system was that the polymer is known to adsorb at the solid-liquid interface and so the depletion interactions occur beyond these steric layers. Evidence for the presence of oscillatory interactions was poor. The AFM data were compared to bulk stability data for the flocculation of a silica dispersion in the presence of PAA; correlation between the two were excellent. In the AFM data, depletion forces initially increased in magnitude up to about 1500 ppm. Between 1500 and 2500 ppm the depletion interaction decreases and eventually disappears altogether leaving only a long-range steric interaction. These changes were used to fully understand the phase diagram of the equivalent dispersion system.

The effect of polydispersity in the depletant has recently been reported by Piech and Walz.[110] Measurements of controlled variations in the polydispersity at both a constant number density and a constant volume fraction of the depletant molecules was reported. At a constant number density, the magnitude of the depletion interaction increased with an increasing proportion of the higher molecular weight component. The opposite was seen for a constant volume fraction. The data were compared to an earlier theory from the same group and showed semi-quantitative agreement.

Burns and co-workers[111,112] in a series of publications have recently reported correlations between measured depletion interaction forces and the features of bulk dispersions, such as yield stresses and aggregate structures when under similar solution conditions. In general, the correlations between the main features recorded using each technique were excellent. In particular, a plateau in the yield stress at higher polymer concentrations was predicted directly from the AFM force data where a similar maximum in the magnitude of the attractive depletion force was seen.

5   Conclusions and Future Developments

The work reviewed here was restricted to recent developments in the measurement of polymer induced colloid interaction forces using AFM based technologies. Of necessity, this is a limited sub-set of the current and rapid expansion of research using AFM equipment. Within this body of work, substantial progress has been made during the last decade. It still remains for the problem of zero separation to be solved and until this is done, fully quantitative research into steric interaction layers will not be possible. However, numerous researchers have shown that expected scaling relationships of force with distance can be observed from the data available without full knowledge of the absolute distance.

A major benefit of the AFM based approach lies in the possibility of probing local variations in interaction forces. This has only been tested by a few authors so far, although it offers great hope for the future, particularly in the characterisation of segment density profiles parallel to the interface. Another oft quoted benefit of the AFM is the ability to examine specific surface-polymer systems. This has yet to be totally exploited although some initial progress has been made. In particular, reports of biological systems where cells are used as the "colloid probes" demonstrate the great potential.

The main benefit of AFM force measurements in particle science arises from the opportunity to match exactly the measurement system to comparable bulk measurements such as rheology. This too has not been extensively investigated, as yet, and there is a need to develop relationships between single particle and bulk particle systems.

References

1. R. S. Farinato and P. L. Dubin, Eds., *Colloid-Polymer Interactions: From Fundamentals to Practice*, Wiley-Interscience, New York, 1999, pp 417.
2. T. F. Tadros, *Advances in Colloid and Interface Science*, 1993, **46**, 1.
3. R. J. Hunter, *Introduction to Modern Colloid Science*, Oxford University Press, Oxford, 1993.
4. J. N. Israelachvili, *Intermolecular and Surface Forces*, Academic Press, London, 1992.
5. B. V. Derjaguin and I. I. Abrikossova, *Discussions of the Faraday Society*, 1954, **18**, 24.
6. B. V. Derjaguin, I. I. Abrikossova and E. M. Lifshitz, *Quarterly Review of the Chemical Society*, 1956, **10**, 295.
7. J. N. Israelachvili and D. Tabor, *Proceedings of the Royal Society London, A*, 1972, **331**, 19.
8. J. N. Israelachvili and D. Tabor, *Progress in Surface and Membrane Science*, 1973, **7**, 1.
9. J. N. Israelachvili and G. E. Adams, *Journal of the Chemical Society-Faraday Transactions*, 1978, **74**, 975.
10. D. Tabor and R. H. S. Winterton, *Proceedings of the Royal Society London, A*, 1969, **312**, 435.
11. H. J. Butt, M. Jaschke and W. Ducker, *Bioelectrochemistry and Bioenergetics*, 1995, **38**, 191.
12. M. Murat and G. S. Grest, *Macromolecules* 1996, **29**, 8282–8284.
13. M. Elimelech, J. Gregory, X. Jia and R. A. Williams, *Particle Deposition and Aggregation: Measurement, Modelling and Simulation*; Butterworth-Heinemann, Woburn, MA, 1995.
14. B. V. Derjaguin and L. Landau, *Acta Physicochim. URSS*, 1941, **14**, 633.
15. E. J. W. Verwey and J. T. G. Overbeek, *Theory of the Stability of Lyophobic Colloids*, Elsevier, Amsterdam, 1948.
16. R. J. Hunter, *Foundations of Colloid Science*, 2 ed., Oxford University Press, Oxford, 2001.
17. R. H. French, *Journal of the American Ceramic Society*, 2000, **83**, 2117.
18. R. Hogg, T. W. Healy and D. W. Fuerstenau, *Transactions of the Faraday Society*, 1966, **62**, 1638.
19. D. H. Napper, *Polymeric Stabilization of Colloidal Dispersions*; Academic Press, London, 1983.
20. G. J. Fleer, J. M. H. M. Scheutjens, M. A. Cohen-Stuart, T. Cosgrove and B. Vincent, *Polymers at Interfaces*, Chapman and Hall, London, 1993.
21. Silberberg, *Journal of Chemical Physics*, 1968, **48**, 2835.

22. J. M. H. M. Scheutjens and G. J. Fleer, *Journal of Physical Chemistry*, 1979, **83**, 1619.
23. J. M. H. M. Scheutjens and G. J. Fleer, *Journal of Physical Chemistry*, 1980, **84**, 178.
24. P. G. de Gennes, *Scaling Concepts in Polymer Physics*, Cornell University Press, Ithaca, NY, 1979.
25. P. G. de Gennes, *Macromolecules*, 1981, **14**, 1637.
26. P. F. Luckham, *Advances in Colloid and Interface Science*, 1991, **34**, 191.
27. P. G. de Gennes, *Macromolecules*, 1982, **15**, 492.
28. J. Klein and P. F. Luckham, *Macromolecules*, 1984, **17**, 1041.
29. J. M. H. M. Scheutjens and G. J. Fleer, *Macromolecules*, 1985, **18**, 1882.
30. G. Rossi and P. A. Pincus, *Europhysics Letters*, 1988, **5**, 641.
31. J. Klein and G. Rossi, *Macromolecules*, 1998, **31**, 1979.
32. T. W. Healy and V. K. LaMer, *Journal of Colloid Science*, 1964, **19**, 323.
33. S. Biggs, M. Habgood, G. J. Jameson and Y. D. Yan, *Chemical Engineering Journal*, 2000, **80**, 13.
34. S. M. Glover, Y. D. Yan, G. J. Jameson and S. Biggs, *Chemical Engineering Journal*, 2000, **80**, 3.
35. J. Gregory, *Journal of Colloid and Interface Science*, 1973, **42**, 448.
36. C. Bechinger, D. Rudhardt, P. Leiderer, R. Roth and S. Dietrich, *Physical Review Letters*, 1999, **83**, 3960.
37. A. J. Milling, *Journal of Physical Chemistry*, 1996, **100**, 8986.
38. M. Piech and J. Y. Walz, *Langmuir*, 2000, **16**, 7895.
39. A. Yethiraj, *Journal of Chemical Physics*, 1999, **111**, 1797.
40. B. V. Derjaguin, *Kolloid Zeitschrift*, 1934, **69**, 155.
41. G. Binnig, C. F. Quate and C. Gerber, *Physical Review Letters*, 1986, **56**, 930.
42. W. A. Ducker, T. J. Senden and R. M. Pashley, *Langmuir*, 1992, **8**, 1831.
43. H. J. Butt, *Biophysical Journal*, 1991, **60**, 1438.
44. J. P. Cleveland, S. Manne, D. Bocek and P. K. Hansma, *Review of Scientific Instruments*, 1993, **64**, 403.
45. C. Neto and V. S. J. Craig, *Langmuir*, 2001, **17**, 2097.
46. R. G. Cain, M. G. Reitsma, S. Biggs and N. W. Page, *Review of Scientific Instruments*, 2001, **72**, 3304.
47. C. J. Drummond and T. J. Senden, *Colloids and Surfaces a-Physicochemical and Engineering Aspects*, 1994, **87**, 217.
48. W. A. Ducker, T. J. Senden and R. M. Pashley, *Nature*, 1991, **353**, 239.
49. J. E. Sader, I. Larson, P. Mulvaney and L. R. White, *Review of Scientific Instruments*, 1995, **66**, 3789.
50. Y. I. Rabinovich and R. H. Yoon, *Colloids and Surfaces a-Physicochemical and Engineering Aspects*, 1994, **93**, 263.
51. T. J. Senden and W. A. Ducker, *Langmuir*, 1994, **10**, 1003.
52. J. E. Sader, J. W. M. Chon and P. Mulvaney, *Review of Scientific Instruments*, 1999, **70**, 3967.
53. V. S. J. Craig and C. Neto, *Langmuir*, 2001, **17**, 6018.
54. S. M. Notley, V. S. J. Craig and S. Biggs, *Microscopy and Microanalysis*, 2000, **6**, 121.
55. A. S. Lea, J. D. Andrade and V. Hlady, *Acs Symposium Series*, 1993, **532**, 266.
56. A. S. Lea, J. D. Andrade and V. Hlady, *Colloids and Surfaces a-Physicochemical and Engineering Aspects*, 1994, **93**, 349.
57. G. J. C. Braithwaite, A. Howe and P. F. Luckham, *Langmuir*, 1996, **12**, 4224.
58. S. Biggs, *Journal of the Chemical Society-Faraday Transactions*, 1996.
59. V. S. J. Craig, A. M. Hyde and R. M. Pashley, *Langmuir*, 1996, **12**, 3557.

60. G. J. C. Braithwaite and P. F. Luckham, *Journal of the Chemical Society-Faraday Transactions*, 1997, **93**, 1409.
61. K. Sakai, T. Yoshimura and K. Esumi, *Langmuir*, 2002, **18**, 3993.
62. M. Giesbers, J. M. Kleijn, G. J. Fleer and M. A. C. Stuart, *Colloids and Surfaces a-Physicochemical and Engineering Aspects*, 1998, **142**, 343.
63. S. Zauscher and D. J. Klingenberg, *Nordic Pulp & Paper Research Journal*, 2000, **15**, 459.
64. S. Zauscher and D. J. Klingenberg, *Journal of Colloid and Interface Science*, 2000, **229**, 497.
65. T. W. Kelley, P. A. Schorr, K. D. Johnson, M. Tirrell and C. D. Frisbie, *Macromolecules*, 1998, **31**, 4297.
66. G. Binnig, C. F. Quate and C. Gerber, *Phys. Rev. Lett.*, 1986, **56 (9)**, 930–933.
67. H. J. Butt, M. Kappl, H. Mueller, R. Raiteri, W. Meyer and J. Ruhe, *Langmuir*, 1999, **15**, 2559.
68. E. M. Sevick and D. R. M. Williams, *Macromolecules*, 1999, **32**, 6841.
69. G. Subramanian, D. R. M. Williams and P. A. Pincus, *Europhysics Letters*, 1995, **29**, 285.
70. M. C. Guffond, D. R. M. Williams and E. M. Sevick, *Langmuir*, 1997, **13**, 5691.
71. S. Yamamoto, M. Ejaz, Y. Tsujii, M. Matsumoto and T. Fukuda, *Macromolecules*, 2000, **33**, 5602.
72. S. Yamamoto, M. Ejaz, Y. Tsujii and T. Fukuda, *Macromolecules*, 2000, **33**, 5608.
73. S. Kidoaki, S. Ohya, Y. Nakayama and T. Matsuda, *Langmuir*, 2001, **17**, 2402.
74. J. Sindel, N. S. Bell and W. M. Sigmund, *Journal of the American Ceramic Society*, 1999, **82**, 2953.
75. M. W. Rutland, A. Carambassis, G. A. Willing and R. D. Neuman, *Colloids and Surfaces a-Physicochemical and Engineering Aspects*, 1997, **123**, 369.
76. A. Carambassis and M. W. Rutland, *Langmuir*, 1999, **15**, 5584.
77. S. W. Prescott, C. M. Fellows, R. F. Considine, C. J. Drummond and R. G. Gilbert, *Polymer*, 2002, **43**, 3191.
78. A. Roters, M. Schimmel, J. Ruhe and D. Johannsmann, *Langmuir*, 1998, **14**, 3999.
79. H. Dautzenberg, W. Jaeger, J. Kötz, B. Philipp, C. Seidel and D. Stscherbina, *Polyelectrolytes*, Hanser, Munich, Germany, 1994.
80. S. Biggs and T. W. Healy, *Journal of the Chemical Society-Faraday Transactions*, 1994, **90**, 3415.
81. S. Biggs, *Langmuir*, 1995.
82. H. Kamiya, Y. Fukuda, Y. Suzuki, M. Tsukada, T. Kakui and H. Naito, *Journal of the American Ceramic Society*, 1999, **82**, 3407.
83. H. Guldberg-Pedersen and L. Bergstrom, *Acta Materialia*, 2000, **48**, 4563.
84. H. G. Pedersen and L. Bergstrom, *Journal of the American Ceramic Society*, 1999, **82**, 1137.
85. E. Laarz, A. Meurk, J. A. Yanez and L. Bergstrom, *Journal of the American Ceramic Society*, 2001, **84**, 1675.
86. E. Laarz and L. Bergstrom, *Journal of the European Ceramic Society*, 2000, **20**, 431.
87. I. Muir, L. Meagher and M. Gee, *Langmuir*, 2001, **17**, 4932.
88. L. Meagher, G. Maurdev and M. L. Gee, *Langmuir*, 2002, **18**, 2649.
89. S. Biggs and A. D. Proud, *Langmuir*, 1997, **13**, 7202.
90. M. Holmberg, R. Wigren, R. Erlandsson and P. M. Claesson, *Colloids and Surfaces a-Physicochemical and Engineering Aspects*, 1997, **130**, 175.
91. K. E. Bremmell, G. J. Jameson and S. Biggs, *Colloids and Surfaces a-Physicochemical and Engineering Aspects*, 1998, **139**, 199.

92. K. E. Bremmell, G. J. Jameson and S. Biggs, *Colloids and Surfaces a-Physicochemical and Engineering Aspects*, 1999, **155**, 1.

93. J. F. Liu, G. Min and W. A. Ducker, *Langmuir*, 2001, **17**, 4895.

94. G. Kramer, K. Estel, F. J. Schmitt and H. J. Jacobasch, *Journal of Colloid and Interface Science*, 1998, **208**, 302.

95. G. J. C. Braithwaite, P. F. Luckham and A. M. Howe, *Journal of Colloid and Interface Science*, 1999, **213**, 525.

96. G. J. C. Braithwaite and P. F. Luckham, *Journal of Colloid and Interface Science*, 1999, **218**, 97.

97. P. G. Hartley, S. L. McArthur, K. M. McLean and H. J. Griesser, *Langmuir*, 2002, **18**, 2483.

98. N. C. Woodward, M. J. Snowden, B. Z. Chowdhry, P. Jenkins and I. Larson, *Langmuir*, 2002, **18**, 2089.

99. M. Gelbert, M. Biesalski, J. Ruhe and D. Johannsmann, *Langmuir*, 2000, **16**, 5774.

100. A. Razatos, Y. L. Ong, F. Boulay, D. L. Elbert, J. A. Hubbell, M. M. Sharma and G. Georgiou, *Langmuir*, 2000, **16**, 9155.

101. M. Morra and C. Cassinelli, *Colloids and Surfaces B-Biointerfaces*, 2000, **18**, 249.

102. R. F. Considine, D. R. Dixon and C. J. Drummond, *Langmuir*, 2000, **16**, 1323.

103. R. F. Considine, D. R. Dixon and C. J. Drummond, *Water Research*, 2002, **36**, 3421.

104. S. B. Velegol and B. E. Logan, *Langmuir*, 2002, **18**, 5256.

105. L. Meagher and H. J. Griesser, *Colloids and Surfaces B-Biointerfaces*, 2002, **23**, 125.

106. A. Milling and S. Biggs, *Journal of Colloid and Interface Science*, 1995, **170**, 604.

107. A. J. Milling and B. Vincent, *Journal of the Chemical Society-Faraday Transactions*, 1997, **93**, 3179.

108. A. J. Milling and K. Kendall, *Langmuir*, 2000, **16**, 5106.

109. S. Biggs, J. L. Burns, Y. D. Yan, G. J. Jameson and P. Jenkins, *Langmuir*, 2000, **16**, 9242.

110. M. Piech and J. Y. Walz, *Journal of Colloid and Interface Science*, 2002, **253**, 117.

111. J. L. Burns, Y. D. Yan, G. J. Jameson and S. Biggs, *Colloids and Surfaces A-Physicochemical and Engineering Aspects*, 2000, **162**, 265.

112. Y. D. Yan, J. L. Burns, G. J. Jameson and S. Biggs, *Chemical Engineering Journal*, 2000, **80**, 23.

Applications

Applications of Atomic Force Microscopy to Granular Materials: Inter-Particle Forces in Air

ROBERT JONES[1] and CHRISTOPHER S. HODGES[2]

[1]Department of Physics, University of Lancaster, Lancaster LA1 4YB, UK
E-mail: robert.jones@lancaster.ac.uk
[2]Department of Chemical Engineering, University of Leeds, Leeds LS2 9JT, UK
E-mail: c.s.hodges@leeds.ac.uk

1 Introduction

Particulate, or granular, matter has many fascinating properties that are apparent at molecular dimensions (macromolecules, colloids, life processes) and at astronomical dimensions (agglomeration processes involved in the formation of planetary systems). In the length scales between, we can study sand castles, garden soil, avalanches and moon dust[1] in the natural environment, and a wide range of important wet and dry commercial processes. Typical processes are hopper flow, pneumatic conveying, fluidized bed reactions, wet granulation, electrostatic precipitation, and agglomerate dispersal for controlled drug delivery. All are controlled by inter-particle forces, and this topic was well summarized about a decade ago in a series of books.[2-4] As a simple illustration, Kendall showed that for an assembly of close packed uniform 1µm silica spheres in dry adhesive contact, the application of contact mechanics theory[5, 6] gives a contact spot diameter of just 2% of the particle diameter and a separation force of 47nN, and this contact is responsible for all the powder behaviour.

Traditionally, bulk processes have been modeled and understood using a continuum mechanics approach, with experimental input from shear testers and related devices. Although the importance of inter-particle contacts is recognized, the basic assumption is made that there is a scale above which fluctuations due to the particulate nature can be ignored. This assumption has proved in many cases to be reasonable and has allowed many useful applications in process design over many decades. However, the approach is essentially empirical and for the most part does not use fundamental parameters describing contacts at the single particle level, such as surface energy and elastic modulus. For this reason it lacks the predictive capability for new materials that we might expect from a comprehensive theory of contact mechanics linking the single particle with the bulk. The need for such links is well recognized.[7]

For this new approach two key methods, both in their infancy at the time of the earlier reviews,[2–4] offer much promise. One is the experimental technique of Atomic Force Microscopy (AFM) invented in 1986.[8] Initially used mainly as an imaging technique, it was rapidly developed to study surface adhesive forces, friction forces and elastic properties over an unprecedented range of force sensitivity (approximately 10pN to 1mN) and spatial resolution (approximately 1nm–100μm). The earliest studies involved mostly small contacts, of the order 20nm, between AFM tips and planar surfaces, but now studies with particles up to 100μm in size are routine. The other method is numerical modeling by the Distinct Element Method (DEM).[9] This is used to simulate bulk properties such as granular shear and agglomerate strength using contact mechanics theory and input parameters for single particle contacts (surface energy γ, friction coefficient μ, elastic modulus E and Poisson's ratio ν). The DEM method is now used intensively and is capable of making predictions of shear strength, size segregation and grain fracture of considerable interest to engineers.[10–12] Neither tool has yet had significant impact on the design of everyday commercial processes, but we believe that this is only a matter of time, and we return to this topic at the end of the chapter.

In this chapter we focus on the measurement by AFM of forces in dry systems between individual particles or grains in the approximate size range of 1–100μm, or sometimes between small agglomerates. Here "dry" is taken to include humid air and conditions under which the particle surface may be modified by liquid films or coatings, since such conditions are of great technological importance. Particles in the liquid environment are reviewed elsewhere, since both the theories used to describe the interaction forces and the applications in large scale commercial processes are distinctly different. Coverage of the theory of inter-particle forces will be minimal because of the availability of many reviews. However, we will discuss some applications of AFM to "model" contacts (AFM tips, planar surfaces) because the lessons learned on adhesion, hardness, tribology and wear are relevant to our main topic of real granular materials.

2   The Forces Acting Between Granular Materials In Air

Many reviews are available dealing with inter-particle forces in both dry and wet systems, as studied by AFM, the surface force apparatus[13] and other techniques, for example those of Pollock,[7] Israelachvili[13] and Hodges.[14] For granular materials in normal air, strongly attractive contact forces are not usually in evidence, and we see mainly the effects of the weaker van der Waals forces, liquid

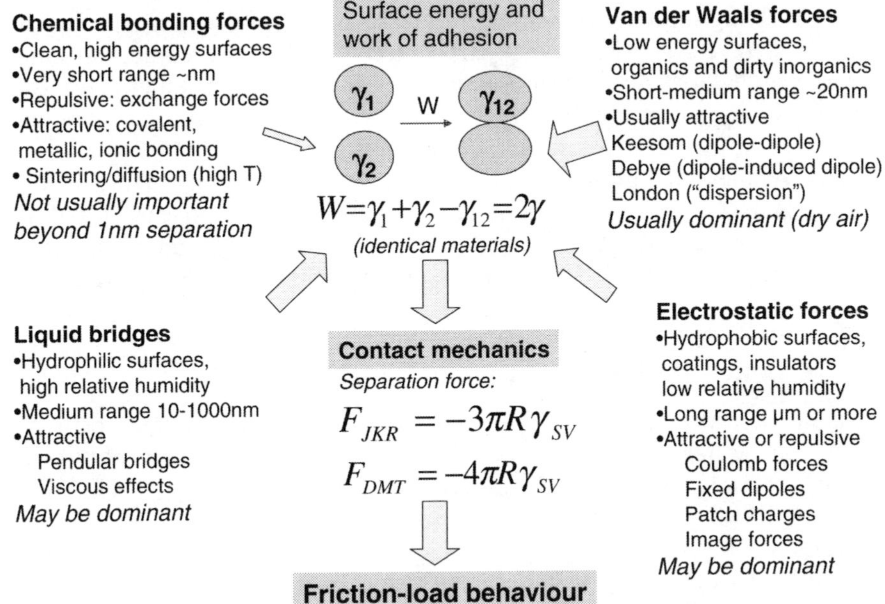

Figure 1 *The main types of force acting between granular materials in air. The forces determine the surface energy and the work of adhesion, which in turn, via contact mechanics theory and the elasto-plasticity and roughness of grains, determine the experimentally measurable properties of pull-off force and friction-load behaviour*

bridges and electrostatic forces (Figure 1). All the forces are greatly modified by the effects of roughness and elasto-plastic properties.[15,16]

2.1   Chemical Bonding Forces

These are rarely apparent in everyday contacts, but the possibility that they may contribute to adhesion or cause fusion of grains should always be considered, particularly for atomically flat surfaces, strongly ionic surfaces, metals, high temperature situations, or in the presence of liquid films that might facilitate chemical reactions. Softer, polymeric materials may, of course, fuse under the action of much weaker forces.

2.2   Van der Waals Forces

In dry air the van der Waals contribution to adhesion appears to be dominant not only in polymeric materials, as expected, but also in a wide range of inorganic commercial powders with rather similar adhesion and tribological properties, consistent with organic contamination.[17] The force-separation or energy-separation dependence is described by the Hamaker coefficient (H) for the energy of interaction of two materials via a third medium, together with a factor allowing for the geometry dependence of the interaction (sphere-flat, sphere-sphere, etc.).[7,13,18] For most particle technologists, the related, but more experimentally accessible quantities, the work of adhesion (W) and interfacial energy (γ) per unit area (both in Jm^{-2} or Nm^{-1}) and the pull-off force (in N) are more likely to be of interest (Figure 1). Pull-off force is one of the commonest parameters measured by AFM and is related to γ by contact mechanics theory (Figure 1). However, the simple relationships given are only applicable for idealized contact geometries in the absence of plastic deformation, and other factors than surface energy (liquid bridges, etc.) often contribute to pull-off force.

2.3 Contact Mechanics Theory

In its simplest form (Hertzian theory) contact mechanics theory treats perfectly elastic spheres in non-adhesive contact, but a wide range of subsequent theories [Johnson, Kendall, Roberts (JKR),[5] Derjaguin, Müller, Toporov (DMT),[19] and Maugis[20]] have taken account of adhesion, elasto-plastic behaviour and fracture mechanics in a variety of ways. The "JKR approximation" (Figure 1) is applicable to highly deformable sphere-on-flat contacts of large surface energy γ_{SV}, large radius R and low elastic modulus E, whereas the "DMT approximation" is applicable for the other extreme of small deformations. The essential point to note is that for the perfect elasticity assumed, the force is independent of maximum applied load and the elastic modulus, although these parameters do appear in the expressions for total contact area under load, which is important for friction-load data. Thus a load-dependent pull-off force would suggest plastic effects, and a time-dependent pull-off force would suggest visco-elastic effects.

2.4 Friction Forces

Friction arises from a variety of interactions loosely described as (1) adhesive processes (2) "Coulombic" processes (interlocking of asperities), and (3) wear and fracturing processes. All are energy dissipative to a greater or lesser extent, and related to processes that are not fully reversible[7,13] (storage of strain energy, bond breaking and formation) and hence to hysteresis in loading and unloading curves in adhesion studies. This explains why high adhesion surfaces are not necessarily high friction surfaces. The adhesion model of friction[21] has been combined with Hertzian theory for the elastic deformation of smooth spherical asperities,[22] and later with JKR theory, to give relationships[23] between friction force (F) and applied load (W) of the type:

$$F = \pi\tau_0\left(\frac{3WR}{4E^*}\right)^{2/3} + aW \tag{1}$$

where τ_0 is the interfacial shear strength, R is the contact radius, E^* is the composite elastic modulus and α is the pressure coefficient.[23] This has led to the successful prediction of experimental friction results in systems of well-defined geometry, using more fundamental parameters as input. The exponent in the power law load dependence is known as the load index, n. For "smooth" contacts, n is predicted to be between ⅔ and 1, depending on which version of contact mechanics theory is applicable, and for "rough" (multi-asperity) contacts n is predicted to be 1. This is the linear behaviour of common experience where the concept of a friction coefficient has some validity.[24] On all models, n = 1 at sufficiently high loads, unless some new friction mechanism, e.g. wear, becomes operative.

2.5 Capillary Bridges

A good overview is provided by several articles[7,13,25,26] and the references therein. The simplest models assume the usual sphere-flat or sphere-sphere geometry ("smooth" contacts) and a small solid-liquid contact angle θ, i.e. a wettable surface (Figure 2). As the relative humidity (RH) increases, vapour will spontaneously condense on surfaces, and the meniscus curvature is related to the relative vapour pressure p/p_{sat} (or RH for water) by the Kelvin equation:

$$\left(\frac{1}{r}+\frac{1}{x}\right)^{-1} = r_k = \frac{\gamma V}{R_g T \log(p/p_{sat})} \tag{2}$$

where r and x are the smaller (negative) and larger (positive) radii of curvature of the annular meniscus, respectively, r_k is the Kelvin radius, γ is the surface tension of water, V is the molar volume and R_g is the gas constant. The Laplace pressure in the liquid is given by:

$$P = \gamma\left(\frac{1}{r}+\frac{1}{x}\right) \approx \frac{\gamma}{r} \tag{3}$$

and acts on an area πx^2, pulling the surfaces together. For smooth sphere-flat geometry with a contact radius x small compared with the sphere radius R (Figure 2a), a simple expression for the Laplace pressure contribution to the adhesion force may be derived:

$$F = 4\pi R\gamma \cos\theta \tag{4}$$

Equation 4 has been experimentally verified in smooth contacts, but for the practical contacts likely to be encountered by particle technologists it falls well short of reality in many ways. Since r disappears in the derivation, so does any RH dependence. In addition, it ignores the effect of surface tension acting around the circumference of the bridge, the effects of asperity contacts (Figure 2b), condensed liquid films present before contact, which may stabilize liquid

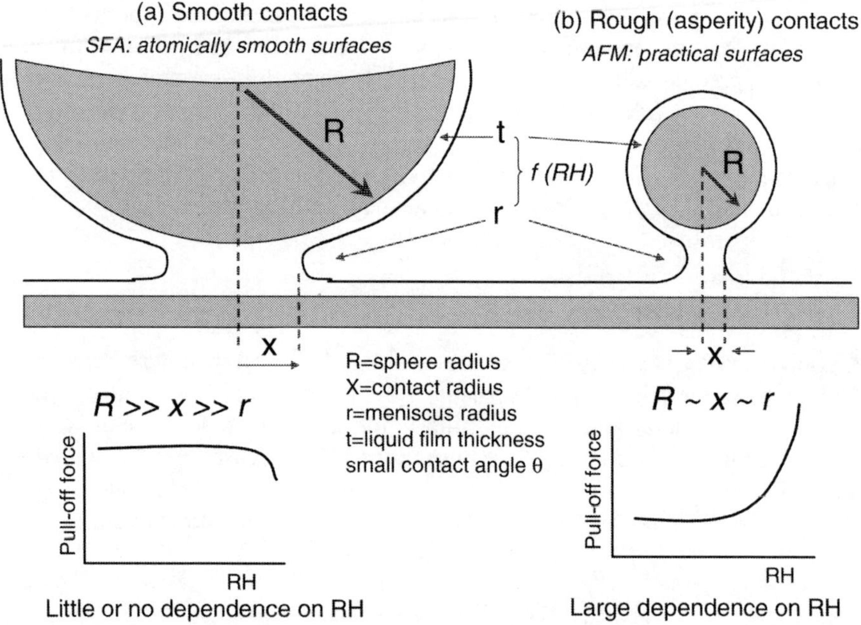

Figure 2 *The humidity sensitivity of granular materials is strongly dependent on contact geometry as well as surface chemistry. For smooth hydrophilic contacts (a) both experiment and theory indicate very small RH dependence, but for rough contacts (b) experiment and theory indicate that the adhesion should increase with RH*

bridges, and dynamic (e.g. viscous) effects. Dynamic pendular bridge models, where the time and separation dependence of the attractive bridge forces are considered, offer a closer approach to reality, but the complex effects of contact geometry remain a considerable barrier to a full understanding of RH-dependent adhesion. It is often supposed that adhesion forces between particles in water are much smaller than in air due to the absence of liquid bridges. However there is symmetry in the behaviour of hydrophilic surfaces in air and hydrophobic surfaces in liquid. Young's equation[13] gives:

$$\gamma_{LV} \cos \theta = \gamma_{SV} - \gamma_{SL} \tag{5}$$

where the subscripts S, L, and V refer to the solid, liquid and vapour phases respectively. If θ is close to 90 degrees (as for many polymers), adhesive forces in air and liquid may not be very different.[13,14]

2.6 Electrostatic Forces

These may arise in a variety of ways (Figure 1) and are again difficult to model, even if the separation distance is large compared with the particle size. Many different force-distance relationships are possible[7,18,27] for different electrical effects and particle geometries. Simple Coulomb charging effects, which are readily seen on a macroscopic scale, are thought to be of most significance in powder behaviour. Fixed dipolar layers may also be important for reactive or moisture-coated surfaces. Much smaller electrical fields are produced by patch charges, which arise from work function inhomogeneity over grains of overall electrical neutrality. They were originally proposed to account for behaviour in AFM force curves[28] that could not be explained by the much weaker and shorter range van der Waals forces.

3 Basic AFM: Imaging

AFM is unique in providing within the same instrumental set-up both images and surface force measurements. Although the emphasis here is on the microme-chanics of surfaces, it may be noted that variants on the proximal probe method can now also provide a wide range of localized electrical, optical, thermal and spectroscopic information of considerable use for the chemical analysis of sur-faces. AFM operates equally well in vacuum, the gaseous environment and the liquid environment.

3.1 Principle of the Atomic Force Microscope

The basic principle of the atomic force microscope is that a cantilever with a particular stiffness or spring constant k (Nm^{-1}) is used to detect normal forces as two surfaces are brought together. The detection system in commercial instru-ments is now almost universally based on a laser beam reflected off the canti-lever, and a detector that either monitors the static deflection of the cantilever or the damping of an oscillating cantilever as it experiences different forces. In principle, AFM images can be obtained using almost any parameter that can be detected by the proximal probe method, such as normal force, lateral (friction) force, or thermal effects. For topographic imaging, contact mode uses a feed-back system to move the tip in the normal (z) direction to maintain a constant lever deflection and normal force while the sample is scanned raster fashion in x–y, using piezoelectric tubes. In "tapping" mode the lever is oscillated, usually at a resonant frequency, and feedback involves maintaining a constant damping level of the oscillations during scanning. It is preferred for soft materials since the intermittent contact minimizes lateral forces, but consequently cannot be used for friction imaging.

Traditionally, AFM experiments have been conducted using highly planar sample surfaces and silicon tips (nominal radius of the order 20nm) attached to silicon nitride or silicon cantilevers which are commonly V-shaped or rectangular, giving k in the approximate range 0.01–100Nm^{-1}. For the particle technologist a much wider range of contacts is of interest, and may now be investigated:

1) Tip against particle attached to substrate (for imaging and elastic modulus);
2) Particle attached to cantilever against a flat surface (for particle-wall adhesion and friction);
3) Particle attached to cantilever against particle attached to substrate (for particle-particle adhesion and friction).

3.2   Imaging of Granular Materials

AFM imaging is easiest for very flat surfaces and is now routine for an enormous range of materials, e.g. semiconductors, metals, crystals, ceramics, polymers, and thin molecular films. The typical maximum range of the AFM is about 150μm laterally and 10μm vertically, limiting what samples may be examined. The imaging of granular material is usually straightforward, but complications may arise because of roughness, agglomeration, and the need to fix the granule to a substrate. Two-part epoxy adhesives have proved very satisfactory and provide the necessary non-compliant base. Granules above about 10μm in size will usually be studied individually, with the AFM scan covering only part of the surface. Below about 5–10μm, it is difficult to isolate single grains, and the surface of an agglomerate will usually be studied. Very rough surfaces are best scanned in non-contact mode using stiff cantilevers with high aspect ratio tips, but limits are imposed by the feedback mechanism and the overall limit of about 10μm in the z-piezo movement, and scanning electron microscopy may then be the only solution. However the biggest advantage of AFM imaging for the particle technologist, the detailed quantitative analysis of topography and roughness, would then be lost.

4   Normal Forces and Adhesion

4.1   AFM Force Curves

The theory and practice of AFM force-distance studies (force curves or force spectroscopy) for the period up to 1998 have been exhaustively reviewed.[27] The most important considerations in any force curve study are choice of cantilever, choice of acquisition parameters, and calibration and correction of the force and distance scales. Normal forces in the pN to mN range can be studied using an appropriate choice of cantilever but the most useful range for contacts in air is 1nN to 10μN, using cantilevers with k ~ 0.1–30Nm^{-1}. The most commonly varied acquisition parameters are maximum separation, maximum applied load, approach and retraction rates and time in contact. These are important if long-range electrical forces or extended liquid bridges are present (since complete separation of the surfaces and a flat baseline must be achieved between successive force curves), and if dynamic pull-off effects or plastic effects in the contact region are expected.

Figure 3 illustrates the main features of force curves likely to be seen, with examples taken from real granular materials. The experimental force curve, in which cantilever deflection is plotted against the z-piezo movement, is not a faithful representation of the force-separation behaviour. Discontinuities occur where the force gradient due to short-range attractive forces exceeds the spring constant of the cantilever ("jump to contact" and "pull-off"). This loss of information in the near-contact region can be minimized by using stiff cantilevers, but there is then a loss of force sensitivity. Also, the z-piezo movement frequently shows hysteresis on the loading and unloading curves. These artifacts

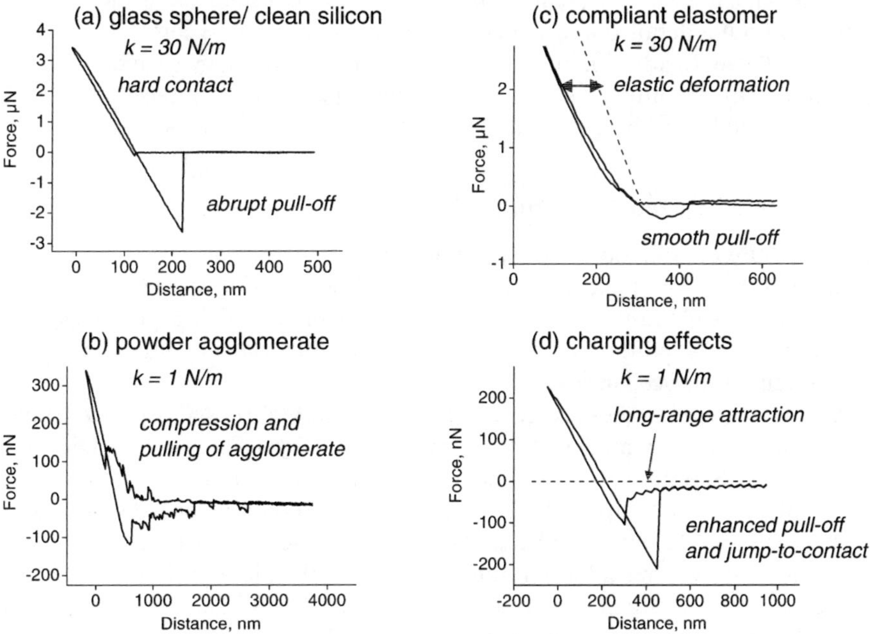

Figure 3 *Typical AFM force curves for a variety of materials in air. In the curves shown, the force scale has been calibrated but the distance scale is uncorrected for the cantilever deflection, i.e. it shows the z-piezo movement and not the true separation of the contacts. Note the differences in the force and distance scales in the four curves. In curve (a) the slope of the hard contact is equal to the spring constant, k. After correction, this region would become vertical and the pull-off would show a jump in the separation indicating a loss of information in this region. The other curves show a variety of effects and indicate the importance of establishing the baselines (dashed lines) corresponding to zero deformation (c) and zero force (d). Curve (b) was obtained for contact between two agglomerates of fine titania particles, curve (c) was for a soft rubber indented by an AFM tip, and curve (d) shows charging effects between hydrophobic (coated) glass surfaces*

may obscure weak short range forces of interest (particularly for soft contacts or colloidal probes in liquid) and elasto-plastic behaviour in the contact region. In addition, a correction must be applied to convert the piezo movement to a tip-sample separation, based on the sensor response (signal/nm) of the lever, measured from the slope of the force curve against a hard sample. For force calibration (nN/signal), both the sensor response and the spring constant (N/m) are required. Calibration procedures for normal forces are well understood and have been extensively reviewed.[14,27] Spring constants from the atomic force microscope manufacturer are rarely accurate to better than 10%, but may be determined by direct calculation, using fundamental properties of the lever,[29] from the resonant frequency,[30] from thermal fluctuations,[31] or by static loading using a two probe method.[32]

In the minority of cases where a smooth force-separation curve is obtained, it may be possible to integrate under the curve to obtain a work of adhesion for the contact. However, this cannot be translated into values of surface energy (Jm^{-2}) without a detailed knowledge of the contact area and geometry, which is rarely available. In addition, it may be a characteristic of many such curves that they represent dynamic processes,[27,33,34] involving very soft contacts or viscous films, and will be sensitive to load, contact time and separation velocity.

4.2   Model Systems and Small Contacts

The literature on adhesion of dusts, grains and powders in air is now huge[15,16,35] because of the technological importance where the adhesion is an essential part of a process, e.g. transfer of toner particles in photocopying processes, wet granulation and binding processes, or where the adhesion is undesirable, e.g. in semiconductor device fabrication.

A wide range of studies involving AFM tips, microfabricated asperity arrays, planar surfaces, and model particles is described by Cappella and Dietler,[27] Kendall and Stainton[35] and others. The commonest materials chosen are Si, Si_3N_4 and metal AFM tips, gold, graphite, glass, silica and silicon substrates, and glass, silica and polymeric spheres. Weisenhorn et al[36] performed many early force curve studies on adhesion, and Ducker et al[37] pioneered the "colloidal probe" concept by attaching silica spheres to cantilevers. The initial work and the bulk of subsequent work have been in the liquid environment. Work on contacts in air has focused on the relation between adhesive forces, elastic and plastic deformation, particle size, and surface roughness, using contact mechanics theory. Early work by Fuller and Tabor[38] indicated that such contacts would need to be almost atomically smooth, at least for hard particles in the micron size range and above, to give experimental adhesion in agreement with theory. Many subsequent AFM and other studies[39–50] have confirmed this generalization in more detail. Schaefer, Rimai and co-workers[15,39] studied adhesion mechanisms and roughness effects in glass and polystyrene spheres, using both AFM and the direct measurement of deformation.

In dry air, typical adhesive (pull-off) forces for model contacts that are comparable with the AFM tip size (R~20nm) are of the order 10–100nN. Wide variations[27] reflect uncertainties in contact size, surface energy, elastic modulus and hardness. Force curves for model contacts and rigid glass spheres are usually simple (Figure 3a). Systematic studies with asperity contacts, AFM tips, and well-characterised particles of different sizes, including the load dependence, provide useful lessons for the contact mechanical behaviour of more common granular materials.

4.3 Granular Materials: Inorganics

For a chemical engineer or particle technologist the range of materials above is only a small fraction of the diversity that may be encountered in industrial processes; however, very few attempts have been made to study inter-particle contacts in cohesive powders and grains of practical importance using the atomic force microscope. This reluctance can probably be explained by the large uncertainties in surface condition, morphology and roughness in practical materials. Pollock et al[51] and Jones et al[26,52] describe force curve and adhesion studies in both model systems (AFM tips and glass spheres) and a variety of cohesive powders. The powders (silica aerogel, hydrated alumina, limestone, titania and a zeolite) were selected because of their importance in commercial applications and because data from parallel studies in shear testers was available.[53] For small maximum applied loads (~20nN) pull-off forces were remarkably similar over the whole size range from AFM tips to 100μm grains, suggesting single asperity contacts, elastic deformation, and possibly that any large variations in the intrinsic surface energy of materials had been masked by contamination. For larger loads (200nN) pull-off forces increased, particularly for the larger particles, suggesting multi-asperity contacts and plastic deformation. Fine details in force curves of different materials appeared to be associated with drawing out particle chains, elasticity of agglomerates (Figure 3b), effects of coatings, liquid bridges, and electrostatic charging, but were in general difficult to analyze.

While many inorganic surfaces and a smaller number of grain surfaces have been characterized in air by AFM imaging, very few AFM studies other than those above have been published in which a study of inter-grain adhesion was the main object. Stark et al have studied adhesion and elastic modulus in silica aerogel particles,[54] and Weth et al[55] studied force curves from silica aerogel particles and attempted to relate the adhesion effects to powder flow properties. Finot et al[56] have studied adhesive forces between gypsum micro-crystals with particular reference to relative humidity.

The possibility that adhesion (and tribology) of typical inorganic materials in air could be dominated by adsorbates, contaminants and van der Waals forces should not be surprising since even in the ultra high vacuum of the lunar environment, surfaces are not ultra-clean, and adsorbates play an important role.[1]

4.4   Organics, Polymers and Biomaterials

Most of these studies have been performed under liquid, but even in air some of
the characteristic features of polymeric interactions, specific bond interactions
and viscous effects in force curves may be detected. The whole area of AFM tips
functionalized with biomaterials and affinity imaging microscopy has recently
become very active, but is outside the scope of the subject of granular materials.
Single chain events ("molecular pulling") can be detected with pull-off effects in
the region 10pN. Less specific interactions of polymer surfaces especially if
liquid-like in structure,[27] or even hard inorganic surfaces in the presence of
suitable liquid films or surfactants,[57] are seen as non-abrupt pull-off effects
(Figure 3c) but may easily be lost with weak cantilevers because of instability
(jump off).

The main effort in this area in granular materials has been with drugs and
pharmaceuticals. Adhesion interactions on lactose, commonly used as a carrier
for drugs, have been widely studied.[58–61] Dey et al[58] studied moisture adsorption
on lactose particles by AFM. A different approach involves functionalizing the
AFM tip with micronized drug particles and studying the interaction with different surfaces, including lactose. Eve et al[59] studied micronised salbutamol, and
Bérard et al[60] studied the adhesion of micronized zanamivir to lactose, and its
RH dependence. Adhesion studies of this kind, although mainly comparative
and difficult to interpret rigorously, are vital to understanding the controlled
release of drugs from dry powder inhalers

4.5   Coatings, Films and Humidity Effects

Much of the cohesive behaviour of granular materials in air, beneficial or detrimental, is dominated by surface films of one type or another and by relative
humidity, rather than by the intrinsic properties of the clean surface. Capillary
bridge formation in high RH may cause caking and simply be a costly nuisance
in many operations, or it may be an essential precursor in an agglomeration
process using a binder. The first comprehensive review of the adhesion of dust
and powder and its RH dependence was carried out by Zimon,[62] and the subject
was later reviewed again by Harnby et al.[63] A large range of studies have
addressed the effects of RH on adhesion, using both AFM and earlier techniques such as the surface force apparatus, but the majority have been confined
to model systems, i.e. AFM tips, planar surfaces and spheres. The essential
features are described by Cappella and Dietler,[27] Tyrrell,[25] and Jones et al.[26] The
key factors appear to be contact geometry and surface chemistry.

On hydrophilic surfaces (typically silica or mica) the most common behaviour
seen is for the adhesion (pull-off force) to increase monotonically with RH,
mostly above 40%RH.[25,26,62,64–66] Experiment and modeling suggest that the RH
dependence is more marked with asperity contacts, cones, or very small spherical contacts, than with smooth large area contacts.[25,26,67,68] A somewhat different
approach using micro-fabricated asperity arrays[69] indicated that pull-off force
was RH-independent for single asperities but RH-dependent for multi-asperity

contacts. Thus the influence of contact geometry and particle size on the RH sensitivity appears to be very complex. Although there is some scatter in reported values, typical pull-off forces for single-asperity contacts might be ~10nN in dry air and ~40nN at high RH.[26,70] On hydrophobic surfaces adhesion is usually small with little RH dependence, but anomalous behaviour has been observed in a number of studies. It is often possible to distinguish hydrophilic and hydrophobic surfaces, or granular materials with or without a coating (frequently an organosilane) by their RH sensitivity, or susceptibility to long range electrostatic forces.

Little work has been done on screening the moisture sensitivity of adhesion in different commercial powders by AFM, but by using a small glove box this can be done more rapidly than in parallel studies in bulk shear testers, where long conditioning is necessary. Jones et al[51,52] found that many common materials were rather insensitive to RH changes, the main exceptions being uncoated glass and silica particles and zeolites, where the moisture sensitivity was probably related to the pore structure and the highly polar surface.

Other workers[33,71,72] have addressed the problems of "pendular bridge" geometry, that is one contact between each pair of particles, with the bridge volume comparable with the particle volume. They studied the precise force-separation behaviour, and how this depends on changing bridge volume, fed by liquid films, or dynamic effects due to viscosity. Pepin et al[72] related the surface tension of a liquid binder, its viscosity, and the inter-particle friction, for a system of non-spherical particles, to the hardness of the bulk agglomerate, and discuss the relevance for wet granulation processes.

There is much interest in controlling the cohesive behaviour of powders by coating procedures, and ideally this could be switched on and off by changing the surface chemistry and nothing else.[6] In practice, even uncoated glass or polymeric particles may exhibit gel-like or plasticized surface layers. Surface treatments will generally alter the roughness and hardness as well as the surface energy, and all these properties will affect the adhesion forces.

4.6 Charging Effects

These are of immense commercial importance and many studies have been done by workers in the photographic, xerographic and semiconductor industries. Qualitative comparisons, e.g. between coated and uncoated particles, are easily made by AFM, but its use in quantitative studies is restricted unless provision is also made for the controlled charging of particles, the application of electric fields and the study of forces beyond the normal separation range of about 10μm. In some of the earliest AFM work on charged particles, Mizes[73] studied the effect of electric fields on pull-on and pull-off force, and the relative contributions of coulombic charging and surface energy to the adhesion, since commercial processes involving particle adhesion or removal can only be made more efficient if the mechanisms are understood. In such processes it is often important to understand both particle-substrate and particle-particle adhesion. In a later series of papers by Rimai, Gady and co-workers,[15,16,74] not exclusively using

AFM, further studies were done on contact electrification, roughness effects and the relative significance of Coulomb and van der Waals forces. Yamamoto and Matsuyama[75] attempted a combined approach in which impact electrification of polymer particles on a metal target was studied, together with analysis of AFM force curves to determine the charge on the particle.

Electrostatic forces can often dominate the near-contact, pull-on and pull-off, behaviour as well as extending to long range (Figure 3d), and the resulting adhesion forces in rough particles can be larger than predicted for smooth spherical particles. The likely mechanism was explained by the work of Hays[76] and involves the concentration of charge on asperities. This is different from the process described by Burnham et al[18] where patch charges arise from work function differences, but both mechanisms could in principle be involved in long range forces and be responsible for some of the complex dependence of the force-separation behaviour on particle geometry.

5 Elastic Modulus and Hardness

5.1 Indentation Measurements

In the same way as a conventional nano-indenter, the loading/unloading force curves in the contact region obtained by an AFM (e.g. Figure 3c) contain information on elastic modulus and hardness (plasticity). For most AFM studies the most important—and usually unknown—factor is the precise tip shape and size and this is also true when a tip is required to indént a sample. Figure 4 illustrates four possible configurations in which an AFM might be used to study these parameters in a granular material, together with the load-depth dependence computed for each geometry using Hertzian contact mechanics theory, assuming that the indenter is much harder than the sample.

The geometry of a cone or pyramid (a) is particularly convenient experimentally because the profile shape does not change with depth, and it also corresponds to the geometry of most silicon AFM tips or diamond nano-indenter tips. The sphere-on-flat geometry (b) has been the one most used historically for theoretical calculations. The arrangement (d), in which a grain is squashed between two hard flat surfaces while the load and deformation are recorded, can potentially provide very useful data for particle technologists, but has been studied least.

The load dependences in Figure 4 assume elastic behaviour and non-adhesive contacts. Plastic effects are invariably present, and as with conventional nano-indentation, the loading and unloading curves must both be studied in order to separate the elastic and plastic components of the deformation, i.e. the modulus $(E/1-v^2)$ and the hardness (H). The unloading curve must be analysed to study elasticity, but it may be complicated by high adhesion (pull-off), and hysteresis effects in the z-piezo movement. These often cause the un-physical apparent behaviour that unloading curves show a higher force at a particular separation than the loading curves—the "reverse path effect".

Conventional AFM can typically be used to study elastic moduli in the range

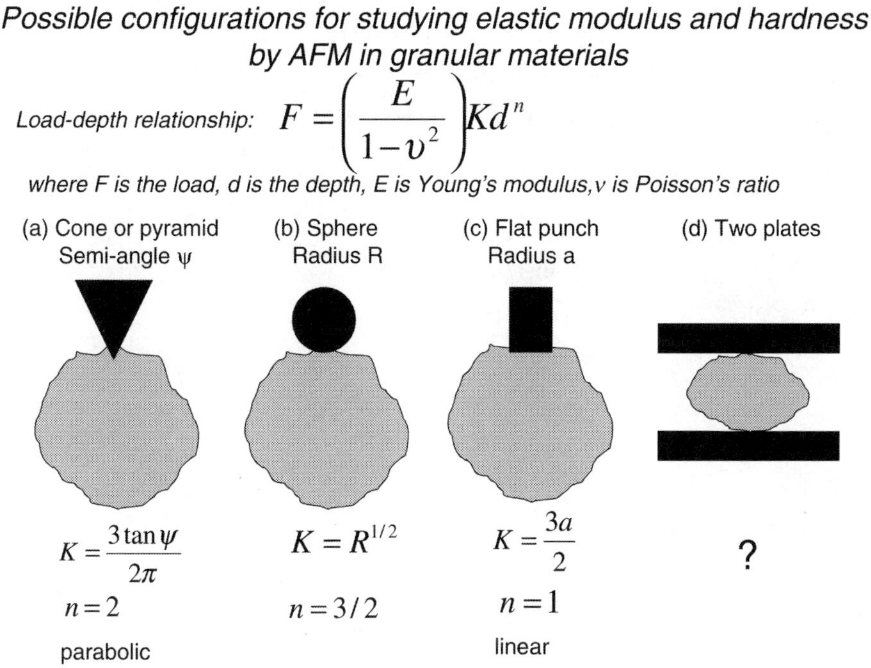

Possible configurations for studying elastic modulus and hardness by AFM in granular materials

Load-depth relationship: $F = \left(\dfrac{E}{1-v^2}\right)Kd^{n}$

where F is the load, d is the depth, E is Young's modulus, v is Poisson's ratio

(a): $K = \dfrac{3\tan\psi}{2\pi}$, $n = 2$, parabolic

(b): $K = R^{1/2}$, $n = 3/2$

(c): $K = \dfrac{3a}{2}$, $n = 1$, linear

(d): ?

Figure 4 *Different contact geometries for studying elasto-plastic properties of grains, and the corresponding load-depth dependence for the case of purely Hertzian contact where adhesion and plasticity are ignored. Geometry (a) is by far the most commonly used. Near-surface properties are of particular interest but may be difficult to study because of adhesion forces and roughness*

1 MPa to 10 GPa, corresponding to indentation depths of the order 1μm and 10nm respectively for stiff cantilevers (30Nm^{-1}) giving a typical maximum load of 20μN. This covers a wide range of polymers, but many harder inorganic materials of interest such as glasses, crystals, and ceramics have smaller indentation depths. For the harder materials, normal AFM cantilevers cannot provide the high loads and good depth sensitivity required, and force curves suffer from artifacts due to cantilever instability and piezo hysteresis. Conversely, conventional nano-indentation has poor lateral resolution and cannot easily be used to study very small samples or particles, or to provide spatial maps of surface mechanical properties, for example stiffness or hardness. Hence, a new hybrid nano-indenter has recently been developed that combines the useful features of both techniques.[77,78] Its key features are:

1) A capacitive transducer for accurate load control (up to 30mN) and displacement measurement;
2) Force modulation to provide improved (lock-in) detection, enabling near-surface deformations to be studied in detail, and from the dynamic response, allowing the isolation of the reversible and the lossy components of the deformation (Figure 5).

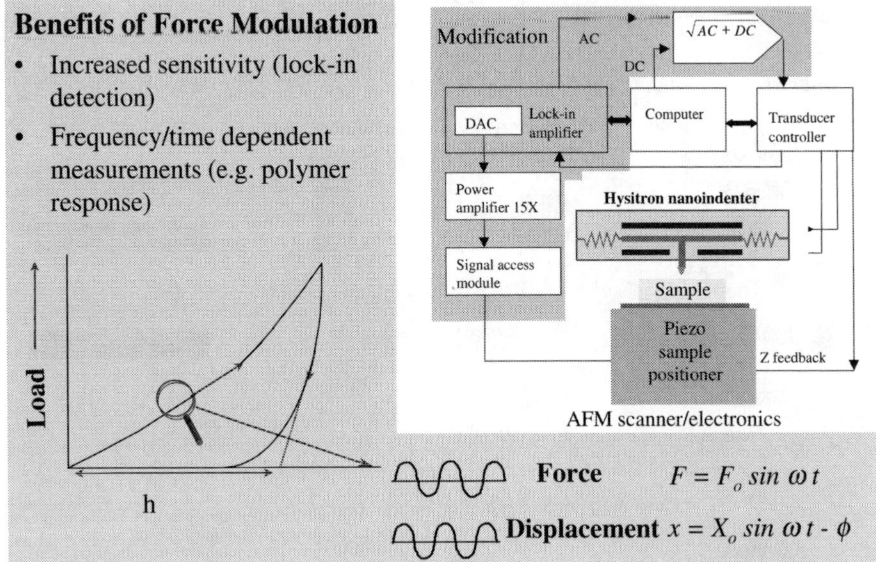

S.A. Syed Asif, K.J. Wahl & R.J. Colton, *Rev. Sci. Instrum.* **70**, 2408 (1999)

Figure 5 *Principle of operation of the hybrid nanoindenter, developed to combine the main advantages of conventional nano-indentation (true depth sensing and high loads) with those of AFM (high spatial resolution and imaging).*
    (Reprinted with permission from Review of Scientific Instruments. Copyright 1999, American Institute of Physics)

Developments of this kind will be of particular use for the study of granular materials where surface plasticization effects occur or coating procedures are used.

5.2   Model Systems and Related Studies

The literature on the closely related topics of hardness, elastic modulus, wear and tribology for flat surfaces and thin film coatings, using traditional techniques such as nano-indentation, and also AFM, is very large,[27,79,80] because of the commercial importance of polymer coatings, diamond-like carbon films, and the optimization of semiconductor device fabrication and the operation of hard drives. AFM is the ideal tool for studying the results of wear and nano-indentation from other procedures. Direct AFM nano-indentation studies have mostly focused on elastomers, graphite, gold and mica.[27] Significant early work was carried out by Burnham and Colton[81] and by Weisenhorn et al.[82] Several of these studies have shown that an AFM tip may behave like a sphere for small indentations and like a cone for large indentations, and that the deformation is entirely elastic at small loads and elasto-plastic at higher loads.

Using the hybrid AFM/nano-indenter, Syed Asif et al[77,78] showed that it was possible to study the different properties of the load-displacement curves and storage and loss moduli, i.e. elastic and plastic components, for localized regions of a carbon fibre-epoxy composite. Although this work was on extended surfaces, it demonstrated the principle of localized data collection with much better than 1μm resolution that could easily be extended to individual grains.

5.3 Measurements on Granular Materials

Work in this area is so far almost non-existent although potentially very informative and now easier with the availability of the hybrid nano-indenter. Stark et al[54] determined elastic moduli of silica aerogel powder particles by nano-indentation in an unmodified commercial AFM set up. The results were complicated by the fact that the 100μm "particles" studied were highly porous structures built from strings of 5nm particles. Thus, to a large extent, the behaviour studied involved collapse of the chains and the elasticity of the assembly as a whole rather than that of the basic particles. However for the particle technologist, the elastic behaviour of both individual particles and agglomerates may be of equal interest. More studies of this type would be very desirable, because it cannot be assumed that elastic and plastic properties measured for bulk materials can be applied to the same materials in granular form. These properties of grains are, if anything, more important than adhesion and friction in determining bulk properties such as shear strength.

5.4 Near-surface Measurements and Coatings

It is well known that elasto-plastic behaviour changes near to a grain surface even when a coating is not deliberately introduced. For example, some polymers plasticize near to the surface, particularly under conditions of high RH, and this could be manifested as hysteresis of force curves in the contact region and an increased, load-dependent, pull-off force. A very important commercial surface is that of Si with thermally grown SiO_2, and Syed Asif el al[83] have studied the increases of modulus and hardness with depth down to 40nm for 30nm oxide films, using the hybrid nano-indenter.

Huang et al[84] have studied the depth dependence of elastic properties for a range of rubbers using an unmodified commercial atomic force microscope, and discussed the limits on modulus measurement imposed by cantilever stiffness, and the uncertainties in near-surface measurements. Essentially, surface adhesive forces affect the load-depth behaviour in nano-indentation. So, if we are interested in near-surface properties, we cannot adopt the simple solution of calculating the modulus only for high load data where the Hertzian approximation is valid, and variations of modulus and hardness with depth are always subject to large uncertainties. In a similar study but using the hybrid nano-indenter Meyers et al[85] were able, using an abrasion-resistant Vitrinite coating, to selectively abolish the surface plasticity part of the loading-unloading curve of polycarbonate and leave just the purely elastic bulk response of the polymer.

Such clear-cut effects could easily be extended in principle to coatings on granular materials and would be extremely valuable in studying the relative importance of elastic and plastic effects in granular shear. DEM simulations have shown the importance of elastic modulus in shear[12] but have difficulty with accommodating plastic effects that may be at least as important in many materials.

6   Friction Forces

6.1   Friction-load Measurements by AFM

The fundamental principles of the method, and all work until 1997 has been thoroughly reviewed.[86] Lateral force images are acquired simultaneously with topographic images during scanning by monitoring the torsional motion of the cantilever due to frictional drag, detected as a displacement of the laser signal at right angles to the displacement produced by normal forces. Weak cantilevers ($k \sim 0.1$ Nm^{-1}) are commonly used, giving normal loads of the order 50nN, but larger forces can also be studied. Sloping surfaces give a lateral signal that mimics friction, due to the component of the normal reaction force on the surface, hence it is essential to monitor signals from both forward and reverse scans and obtain the difference, the so-called "friction loop", which is twice the true friction signal. The difference may be obtained from line scans or by averaging lateral force data over a complete image.

The calibration of the friction signal is important for quantitative studies, but this is much less straightforward than for normal forces. The torsional stiffness of cantilevers can be determined from first principles, using finite element analysis, or in terms of the normal spring constant,[29] but in both cases the lateral sensor response must be found independently. A very convenient practical solution to this is to perform an *in situ* calibration that determines the lateral force conversion (nN/signal) directly in terms of the normal force conversion by analyzing the friction-load data on two known slopes.[87] Ideally, calibrations should be done for every cantilever, but problems arise when macroscopic particles are attached, altering the mechanical properties and making conventional calibration very difficult. This is a new area of work and a variety of solutions to the calibration problem have been proposed.[51,87–90] Very useful comparative friction studies can be done using AFM, but it is doubtful whether the absolute values of friction forces are known to much better than 30% accuracy in most cases.

The friction of inter-particle contacts, as revealed by AFM, shows a much richer range of behaviour (Figure 6) than the simple numerical parameter, the friction coefficient μ, that we tend to associate with macroscopic contacts, and the details may be highly relevant to bulk processes in granular media. Thus we can study among other things:

1) Line profiles, and stick-slip phenomena;
2) Lateral force histograms from a scanned area, providing the force distribution and information relating to the statistics of friction;

3) Non-linear load effects;
4) Adhesive effects, the relation to the pull-off from force curves, and friction at negative load;
5) Effects of wear, e.g. by performing successive experiments at different loads and different scan areas.

6.2   Model Systems and Small Contacts

The extensive literature on AFM studies of tribology and wear of extended surfaces reviewed by Bhushan,[79] Bhushan and Sundararajan[91] and Carpick and Salmeron[86] forms a very useful background to the rather small number of studies using actual particles or grains as one or both of the contacting surfaces. In general the studies have tended towards either hard engineering surfaces[79] with large contact forces, or towards atomic scale processes[86] and effects of tip shape and contact area. For example, Putman and Kaneko,[92] using AFM tips, confirmed the ⅔ power law load dependence predicted by Hertzian contact mechanics theory at the atomic scale. In a series of studies using microfabricated

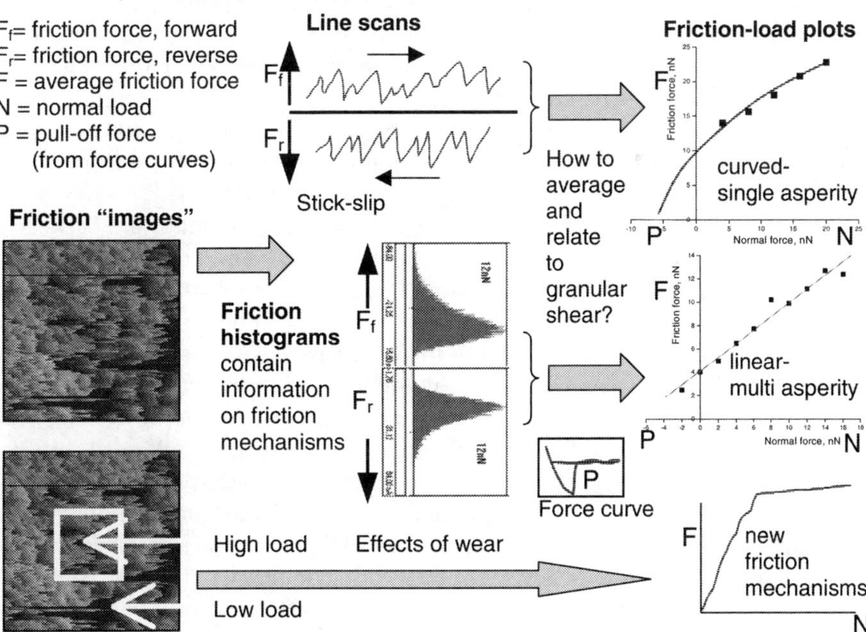

Figure 6   *The diversity of friction studies possible by AFM. On the scale of single asperity contacts (of the order 10nm in size) the statistics of the random fluctuations in friction can be studied, rather than macroscopic averages or simple friction coefficients often measured in large scale experiments. This information, normally averaged out, may be useful in understanding some bulk processes. Friction-load plots are essential in all studies to understand friction mechanisms. Wear processes are also apparent on the AFM force scale and can readily be studied*

asperity arrays, Ando[44,45,69] investigated pull-off force, friction-load, and their RH dependence. Bhushan and Sundararajan[91] studied effects of contact size, and Hu et al[93] showed that the friction on mica was non-linear at low load and linear at high load, with some evidence for wear at contact forces as low as 80nN.

6.3 Friction-load Measurements on Granular Materials

When two grains, or a grain and a flat surface, are in sliding contact the contact points are continuously changing. Neither the number of contacts nor the effective contact area can be determined explicitly, but many useful clues about the type of contact and friction mechanisms are provided by the various measurements possible (Figure 6). To a large extent the behaviour depends on whether the contact is "smooth" (which could be either a very small or large single contact) or "rough", involving multiple contacts, which in turn depends on particle size, roughness, elasticity and plasticity.

In an early study, Mullier et al[94] developed a single particle friction cell, but its use was limited to much larger grains and contact forces than can be routinely studied by AFM. Biggs et al[95] and Cain et al[57] studied contacts between glass and silica spheres and silica plates. Hydrophilic surfaces gave non-linear friction-load behaviour that fitted well to JKR theory and thus appeared to be molecularly smooth, possibly due to a gel-like surface layer. However this behaviour was only established after some wear had taken place, removing multi-asperity contacts. Stiff cantilevers were used, resulting in relatively high loads of up to $2\mu N$, corresponding to normal stresses of about 120MPa which are much larger than those normally encountered in shear testers (~100KPa).

Meurk et al[96] studied friction-load between two Si_3N_4 granules, and between a granule and a metal substrate, with particular interest in the interactions of a binder and the RH. Pollock et al[51] and Jones et al[90] studied the friction-load behaviour for particle-particle contacts between glass spheres, and for the same range of cohesive powders as used in force curve studies[51,52] and parallel shear tests.[51,53] Weak cantilevers were used, giving low contact forces of about 30nN that would appear to be of the same order as average contact forces in shear testers for particles in the 1–10µm size range.[97] At these low loads, glass spheres showed non-linear (single asperity) behaviour and the cohesive powders showed linear (multi-asperity) behaviour, probably associated with the different roughness and elasto-plastic properties. Particle-wall friction was also studied because of its significance in testing devices and bulk flow processes. It showed different load dependence from the inter-particle friction, except in some cases where a bound powder layer might be present. In addition to studying the average friction forces at a particular load, the fluctuations (stick slip behaviour and friction force histograms) may be studied (Figure 6), but it is not clear at present how to relate any of these measurements to bulk shear in a truly quantitative way.

6.4   Coatings, Films and Humidity Effects

Several studies have looked at the interactions between adhesion, friction, contact size, applied load, liquid films and RH for relatively simple contacts between AFM tips, asperities and flat surfaces. In general adhesion increased with increasing RH for hydrophilic contacts, the magnitude of effects depending on contact size. Friction was found to decrease with increasing RH by Hu et al[93] for mica, by Xu et al[98] for NaCl, and by Meurk et al[96] for Si_3N_4 granules in the presence of a binder. This is generally expected if the water film has a lubricant effect, but Ando[69] found that friction increased with RH for microfabricated asperity arrays in silicon at low sliding speeds, and thus the humidity dependence of friction is not always simple. As with the humidity effects discussed earlier, the type of behaviour observed may be determined by the relative size of the liquid menisci and the asperity contacts, i.e. whether the liquid film can bridge many asperities.

Many studies have looked at the effects of coatings and lubricants on the bulk flow and shear of granular materials, but relatively few studies have attempted to make parallel measurements at the single particle scale. Podczeck and Miah[99] studied the effect of grain size and shape and of lubrication by Mg stearate films on the shear properties of a range of powders. Pepin et al[72] carried out experimental and modeling studies on the effects of binder liquid bridges, viscosity and inter-particle friction on the bulk agglomerate properties. Under compression, the agglomerate deforms plastically until it breaks through crack propagation, i.e. like a wet snowball. It was shown that the agglomerate plasticity was controlled by the inter-grain friction for grains of irregular morphology, rather than by the elasto-plastic properties of individual grains.

In their AFM study on single silicon nitride granules, Meurk et al[96] proposed that the lowering of the friction coefficient and increased adhesion observed with increasing RH was due to softening of the poly(ethylene glycol) binder. In their study on lubricated and unlubricated shear using single glass beads and silica surfaces, Cain et al[57] reacted the hydroxylated surfaces with stearic acid and observed the results in AFM friction studies and shear tests. A similar lubricant film had previously been used to study the bulk shear of metal and metal oxide grains.

Studies of this type are important in understanding, at the single particle level, the large range of technological processes depending on lubricated powder flow and wet granulation. This will not be easy because of the complex interplay of the liquid film, its surface tension and viscosity, particle morphology, adhesion, friction and elasto-plastic properties of the agglomerate. A significant practical problem is wear and attrition, which has been shown to occur even at low contact forces. Unless some mechanism exists for the constant replenishment of thin lubricant films, their effects are rapidly lost. Thicker low friction coatings provide a partial solution, but these may also change other important surface properties such as roughness and hardness.

7 The Outlook: Linking AFM To Bulk Powder Mechanics

Most real granular materials are very different from our illustration in the introduction[6], important though it is as a starting point. Modeling studies[9-12] have shown the extent of the non-uniformity and anisotropy of contact forces even for this simple geometry. For granular materials of complex morphology, particularly in dense (highly consolidated) systems, grain interlocking, plastic deformation, attrition and fracture are all important. Single particle AFM measurements, where contact loads may be small and not much larger than adhesive forces, seem far removed from the bulk mechanics of such systems. The geometrical effects appear to be overriding, compared with the differences in surface energy, friction, and other parameters that can be studied by AFM. Similarly the DEM method, despite its interesting predictions, is unrealistic since until recently only spherical particles were modeled, plastic effects have not been included, and simple numerical inputs are used (γ, μ, E, ν). We saw that the concept of a simple friction coefficient μ is inappropriate on the scale of single contacts. So what hope do we have of relating AFM measurements to bulk powder mechanics?

Bulk properties that we would like to predict using AFM measurements and modeling[97] might include:

1) Granular shear and cohesion (loosely defined as the shear strength at zero normal load);
2) Agglomerate or compact tensile strength, tensile modulus and compressive modulus (elastic and plastic);
3) Powder-wall adhesion and friction.

We might begin by assuming, naively, that shear strength is related to inter-particle sliding friction, because AFM friction-load plots look very similar to shear stress-normal stress plots or yield loci, that agglomerate plasticity is related to grain plasticity, and tensile strength is related to pull-off force, etc. In fact, experimental and modeling studies indicate that shear strength may largely be controlled by elastic modulus,[12] at least in dense flows, agglomerate plasticity may be controlled by inter-grain friction,[72] and failure in shear and tensile tests are essentially fracture processes[100] with all the associated problems of stress concentrations and force chains. Thus scaling effects are far from simple, and the way forward is by using improved models and trusting their predictions, since these are often counter-intuitive.

However, in parallel with the development of the numerical simulations to include some of the effects described above, the atomic force microscope could be usefully employed in studying the size, morphology, and roughness of all kinds of granular materials in either air or liquid. Indeed, the atomic force microscope excels in obtaining quantitative data on these properties over other techniques and it is this flexibility that makes it so useful. Comparative studies of a semi-quantitative nature on the effects of coating procedures, lubricant films, specific drug interactions and relative humidity can be straightforward to perform, and we have seen several examples of this type. They may be very informative and be faster and less wasteful of valuable materials than the equivalent trials on bulk samples.

In other situations, the geometrical problems may prove to be less of a barrier to understanding bulk processes than was feared, for a variety of reasons. The average contact forces in many bulk processes and single grain AFM studies appear to be similar, at least for loose (unconsolidated) systems and small to medium grain sizes $(1–10\mu m)$[97] and we can have some confidence that the friction and wear processes will then be similar. Wall friction is important in many processes, and if plug flow occurs, the bulk process more closely resembles a direct scale-up of single grain-wall friction behaviour than in the case of the internal failure plane of a powder with its complex geometry (interlocking, sliding and rolling in a zone several particles in width). Systematic AFM studies of contacts between different grain sizes and wall materials of different roughness could be revealing for bulk behaviour. Most significantly of all, the future of granular materials lies increasingly with "designed" particles of controlled geometry, e.g. smooth mono-dispersed spheres. The specific adhesion and other surface interactions can ideally be studied by AFM and modeling, particularly in loose agglomerates and in liquid, with less need to worry about geometry, plasticity and damage.

In the long term, when the models have advanced to include some of the above effects, such as a non-uniform particle geometry, long-range forces etc., it should be possible to include the full range of single particle AFM data in these models rather than simple constant numerical values, as used at present, to give more realistic predictions of particulate behaviour. The AFM data, in the form of a complete force-separation curve, already includes the effects of elastic deformation, contact roughness, liquid bridges and charging effects, and the AFM friction-load dependence includes the effects of roughness and elasto-plasticity. Thus what is required is a way to incorporate these data directly into a computer simulation (e.g. by DEM), since this would allow the behaviour of many particles to be accurately predicted using "real world" information. Traditional bulk testing will still be needed since some behaviour may be collective and not a direct property of individual particles.

References

1. L-H Lee in 'Fundamentals of Adhesion and Interfaces', D. S. Rimai, L. P. DeMejo, and K. L. Mittal, (Ed.), VSP, Utrecht, 1995, 73.
2. B. J. Briscoe and M. J. Adams, (Ed.), *Tribology in Particulate Technology*, Adam Hilger, Bristol, 1987.
3. I. L. Singer and H. M. Pollock, (Ed.), *Fundamentals of Friction: Macroscopic and Microscopic Processes*, Kluwer Academic Publishers, London, 1992.
4. C. Thornton, (Ed.), *Powders and Grains 93*, Balkema, Rotterdam, 1993.
5. K. L. Johnson, K. Kendall, and A. D. Roberts, *Proc. R. Soc. Lond. A*, 1971, **324**, 301.
6. K. Kendall in 'Powders and Grains 93', C. Thornton, (Ed.), Balkema, Rotterdam, 1993, 25.

7. H. M. Pollock, *The Forces Acting Between Dry Powder Particles*, IFPRI Report SAR 12–09, 1994.
8. G. Binnig, C. F. Quate, and C. Gerber, *Phys. Rev. Lett.*, 1986, **56**, 930.
9. P. A. Cundall and O. D. L. Strack, *Geotechnique*, 1979, **29**, 47.
10. C. Thornton and S. J. Antony, *Phil. Trans. R. Soc. Lond. A*, 1998, **356**, 2763.
11. S. J. Antony and M. Ghadiri, *J. Appl. Mech. Trans. ASME*, 2001, **68**, 772.
12. S. J. Antony, *Phys. Rev. E*, 2000, **63**, 011302.
13. J. N. Israelachvili, *Intermolecular and Surface Forces*, 2nd Edn., Academic Press, London, 1992.
14. C. S. Hodges, *Adv. Colloid Interface Sci.* 2002, **99** 13–75
15. D. S. Rimai, L. P. DeMejo, and K. L. Mittal, (Ed.), *Fundamentals of Adhesion and Interfaces*, VSP, Utrecht, 1995.
16. L. P. DeMejo, D. S. Rimai, and L. H. Sharpe, (Ed.), *Fundamentals of Adhesion and Interfaces*, Gordon and Breach, The Netherlands, 1999.
17. M. J. Adams, in 'Fundamentals of Friction: Macroscopic and Microscopic Processes', I. L. Singer and H. M. Pollock, (Ed.), Kluwer Academic Publishers, London, 1992, 183.
18. N. A. Burnham, R. J. Colton, and H. M. Pollock, *Nanotechnology*, 1993, **4**, 64.
19. B. V. Derjaguin, V. M. Müller, and Y. P. Toporov, *J. Colloid Interface Sci.*, 1975, **53**, 314.
20. D. Maugis, *J. Colloid Interface Sci.*, 1992, **150**, 243.
21. F. P. Bowden and D. Tabor, *The Friction and Lubrication of solids*, Oxford Univ. Press, London, 1964.
22. J. F. Archard, *Proc. R. Soc. Lond. A*, 1957, **243**, 190.
23. M. J. Adams, B. J. Briscoe, and L. Pope, in 'Tribology in Particulate Technology', B. J. Briscoe and M. J. Adams, (Ed.), Adam Hilger, Bristol, 1987, 8.
24. P. J. Blau, *Tribology International*, 2001, **34**, 585.
25. J. W. G. Tyrrell, PhD Thesis, University of Surrey, Guildford, UK, 1999.
26. R. Jones, H. M. Pollock, J. A. S. Cleaver, and C. S. Hodges, *Langmuir*, 2002, **18**, 8045.
27. B. Cappella and G. Dietler, *Surf. Sci. Reports*, 1999, **34**, 1.
28. N. A. Burnham, R. J. Colton, and H. M. Pollock, *Phys. Rev. Lett.*, 1992, **69**, 144.
29. J. L. Hazel and V. V. Tsukruk, *Thin Solid Films*, 1999, **339**, 249.
30. J. P. Cleveland, S. Manne, D. Bocek, and P. K. Hansma, *Rev. Sci. Instrum.*, 1993, **64**, 403.
31. J. L. Hutter and J. Bechhoefer, *Rev. Sci. Instrum.*, 1993, **64**, 1868.
32. C. T. Gibson, G. S. Watson, and S. Myhra, *Nanotechnology*, 1996, **7**, 259.
33. M. J. Adams and V. Perchard, *I. Chem. E. Symp. Series*, 1985, **91**, 147.
34. G. Lian, C. Thornton, and M. J. Adams, *J. Colloid Interface Sci.*, 1993, **161**, 138.
35. K. Kendall and C. Stainton, *Powder Technol.*, 2001, **121**, 223.
36. A. L. Weisenhorn, P. Maiwald, H.-J. Butt, and P. K. Hansma, *Phys. Rev. B*, 1992, **45**, 11226.
37. W. A. Ducker, T. J. Senden, and R. M. Pashley, *Langmuir*, 1992, **8**, 1831.
38. K. N. G. Fuller and D. Tabor, *Proc. R. Soc. Lond. A*, 1975, **345**, 327.
39. D. M. Schaefer, M. Carpenter, B. Gady, R. Reifenberger, L. P. DeMejo, and D. S. Rimai, *J. Adhesion Sci. Technol.*, 1995, **9**, 1049.
40. P. Attard, *J. Phys. Chem. B*, 2000, **104**, 10635.
41. E. Barthel, *Colloids and Surfaces A*, 1999, **149**, 99.
42. J. Q. Feng, *J. Colloid Interface Sci.*, 2001, **238**, 318.
43. J. A. Greenwood, *Proc. R. Soc. Lond. A*, 1997, **453**, 1277.
44. Y. Ando, *J. Tribology Trans. ASME*, 2000, **122**, 639.

45. Y. Ando and J. Ino, *Wear*, 1998, **216**, 115.

46. S. Biggs and G. Spinks, *J. Adhesion Sci. Technol.*, 1998, **12**, 461.

47. B. J. Briscoe, K. K. Liu, and D. R. Williams, *J. Colloid Interface Sci.*, 1998, **200**, 256.

48. L. O. Heim, J. Blum, M. Preuss, and H.-J. Butt, *Phys. Rev. Lett.*, 1999, **83**, 3328.

49. G. Toikka, G. M. Spinks, and H. R. Brown, *J. Adhesion*, 2000, **74**, 317.

50. T. Stifter, E. Weilandt, O. Marti, and S. Hild, *Appl. Phys. A*, 1998, **66**, S597.

51. H. M. Pollock, R. Jones, D. Geldart, and A. Verlinden, *Inter-particle Forces in Powder Flow*, IFPRI Final Report FRR 32–04, 2001.

52. R. Jones, H. M. Pollock, D. Geldart, and A. Verlinden, *Powder Technology*, 2003, **132**, 196.

53. A. Verlinden, PhD Thesis, University of Bradford, UK, 2000.

54. R. W. Stark, T. Drobek, M. Weth, J. Fricke, and W. Heckl, *Ultramicroscopy*, 1998, **75**, 161.

55. M. Weth, M. Hofmann, J. Kuhn, and J. Fricke, *J. Non Cryst. Solids*, 2001, **285**, 236.

56. E. Finot, E. Lesniewska, J. C. Mutin, and J. P. Goudonnet, *Langmuir*, 2000, **16**, 4237.

57. R. G. Cain, N. W. Page, and S. Biggs, *Phys. Rev. E*, 2000, **62**, 8369.

58. F. K. Dey, J. A. S. Cleaver, and P. A. Zhdan, *Advanced Powder Technol.*, 2000, **11**, 401.

59. J. K. Eve, N. Patel, S. Y. Luk, S. J. Ebbens, and C. J. Roberts, *Int. J. Pharm.*, 2002, **238**, 17.

60. V. Bérard, E. Lesniewska, C. Andrès, D. Pertuy, C. Laroche, and Y. Pourcelot, *Int. J. Pharm.*, 2002, **232**, 213.

61. U. Sindel and I. Zimmermann, *Powder Technol.*, 2001, **117**, 247.

62. A. D. Zimon, *Adhesion of Dust and Powder*, Consultants Bureau, Plenum Publishing Company, New York, 1982.

63. N. Harnby, A. E. Hawkins, and I. Opalinski, *Trans. I. Chem. E.*, 1996, **74**, 605.

64. H. K. Christenson, *J. Colloid Interface Sci.*, 1988, **121**, 170.

65. T. Thundat, X. Y. Zheng, G. Y. Chen, S. L. Sharp, R. J. Warmack, and L. J. Schowalter, *Appl. Phys. Lett.*, 1993, **63**, 2150.

66. M. Binggeli and C. M. Mate, *Appl. Phys. Lett.*, 1994, **65**, 415.

67. R. W. Coughlin, B. Elbirli, and L. Vergara-Edwards, *J. Colloid Interface Sci.*, 1982, **87**, 18.

68. K. Iinoya and H. Muramoto, *J. Soc. Mater. Japan*, 1967, **16**, 352.

69. Y. Ando, *Wear*, 2000, **238**, 12.

70. M. Fujihira, D. Aoki, Y. Okabe, H. Takano, H. Hokari, J. Frommer, Y. Nagatani, and F. Sakai, *Chem. Lett.*, 1996, **7**, 499.

71. S. M. Iveson, J. D. Litster, K. Hapgood, and B. J. Ennis, *Powder Technol.*, 2001, **117**, 3.

72. X. Pepin, S. J. R. Simons, S. J. Blanchon, D. Rossetti, and G. Couarraze, *Powder Technol.*, 2001, **117**, 123.

73. H. A. Mizes, *J. Adhesion Sci. Technol.*, 1994, **8**, 937.

74. B. Gady, R. Reifenberger, and D. S. Rimai, *J. Appl. Phys.*, 1998, **84**, 319.

75. H. Yamamoto and T. Matsuyama, *Contact/Impact Charging of a Single Particle*, IFPRI Report ARR SOKA-01.

76. D. A. Hays, in 'Fundamentals of Adhesion and Interfaces', D. S. Rimai, L. P. DeMejo, and K. L. Mittal, (Ed.), VSP, Utrecht, 1995, 61.

77. S. A. Syed Asif, K. J. Wahl, R. J. Colton, and O. L. Warren, *J. Appl. Phys.*, 2001, **90**, 1192.

78. S. A. Syed Asif, K. J. Wahl, and R. J. Colton, *Rev. Sci. Instrum.*, 1999, **70**, 2408.

79. B. Bhushan, *Wear*, 2001, **251**, 1105.

80. J. Malzbender, J. M. J. den Toonder, A. R. Balkenende, and G. de With, *Materials Sci. Eng. R: Reports*, 2002, **36**, 47.

81. N. A. Burnham and R. J. Colton, *J. Vac. Sci. Technol. A*, 1989, **7**, 2906.

82. A. L. Weisenhorn, M. Khorsandi, S. Kasas, V. Gotzos, and H.-J. Butt, *Nanotechnology*, 1993, **4**, 106.

83. S. A. Syed Asif, K. J. Wahl, and R. J. Colton, *J. Mater. Res.*, 2000, **15**, 546.

84. Z. Huang, S. A. Chizhik, V. V. Gorbunov, N. K. Myshkin, and V. V. Tsukruk, in 'Microstructure and Microtribology of Polymer Surfaces', V. V. Tsukruk and K. J. Wahl, (Ed.), American Chemical Society, Washington DC, 1999, 177.

85. G. F. Meyers, B. M. DeKoven, M. T. Dineen, A. Strandjord, P. J. O'Connor, T. Hu, Y.-H. Chiao, H. M. Pollock, and A. Hammiche, in 'Microstructure and Microtribology of Polymer Surfaces', V. V. Tsukruk and K. J. Wahl, (Ed.), American Chemical Society, Washington DC, 1999, 190.

86. R. W. Carpick and M. Salmeron, *Chem. Rev.*, 1997, **97**, 1163.

87. D. F. Ogletree, R. W. Carpick, and M. Salmeron, *Rev. Sci. Instrum.*, 1996, **67**, 3298.

88. C. T. Gibson, G. S. Watson, and S. Myhra, *Wear*, 1997, **213**, 72.

89. R. G. Cain, S. Biggs, and N. W. Page, *J. Colloid Interface Sci.*, 2000, **227**, 55.

90. R. Jones, H. M. Pollock, D. Geldart, and A. Verlinden, 2003, submitted for publication.

91. B. Bhushan and S. Sundararajan, *Acta Materialia*, 1998, **46**, 3793.

92. C. Putman and R. Kaneko, *Thin Solid Films*, 1996, **273**, 317.

93. J. Hu, M. Salmeron, D. F. Ogletree, and X.-D. Xiao, *Surf. Sci.*, 1995, **327**, 358.

94. M. Mullier, U. Tuzun, and O. R. Walton, *Powder Technol.*, 1991, **65**, 61.

95. S. Biggs, R. Cain, and N. W. Page, *J. Colloid Interface Sci.*, 2000, **232**, 133.

96. A. Meurk, J. Yanez, and L. Bergström, *Powder Technol.*, 2001, **119**, 241.

97. R. Jones, *Granular Matter*, 2003, **4**, 191.

98. L. Xu, H. Bluhm, and M. Salmeron, *Surf. Sci.*, 1998, **407**, 251.

99. F. Podczeck and Y. Miah, *Int. J. Pharm.*, 1996, **144**, 187.

100. K. Kendall, in 'Tribology in Particulate Technology', B. J. Briscoe and M. J. Adams, (Ed.), Adam Hilger, Bristol, 1987, 110.

CHAPTER 9

In-Process Measurement of Particulate Systems

CORDELIA SELOMULYA and RICHARD A. WILLIAMS

Institute of Particle Science and Engineering,
School of Process, Environmental, and Materials Engineering,
University of Leeds, Leeds LS2 9JT, United Kingdom
E-mail: r.a.williams@leeds.ac.uk

1. Introduction

A weakness in the ability to develop sophisticated granular and dense phase models is the lack of suitable experimental validation methods. Ideally, these methods should be *in-situ* and not perturb the process being modeled in any way. This chapter considers several examples of such methods that can be applied to dry and wet particulate systems.

For model validation, detailed information on the characteristics of the system is often required on a process (macro-) and local (micro-) scale. Just as a theoretician must consider modeling across the length scales, an experimentalist too must adopt multi-scale measurements. It is a fact that few techniques can measure at micro- and macro-scales simultaneously. Consequently, a range of experimental approaches is required to cover both spatial and temporal ranges. Some examples of these basic needs have been discussed elsewhere.[1-4]

For powders and granular systems, measurements of average and local specific solid concentration are normally required. In addition, the trajectory of a single particle or a population of specified particles is desirable for verification of Lagrangian type models. Other cases may require thermal properties, coupled with details on the flow of the phases in a dispersed system. The purpose of this brief review is to highlight methods that are sufficiently advanced and reliable for providing tools to support any modeling work. The examples cited consider in-process measurements of particle concentration and velocity (Sections 2.1–2.3) and particle size and form (Sections 2.4–2.5).

2   Process Measurements

2.1   Dielectric Imaging of Granular Materials

The difference in dielectric constant between most powders and the dispersed fluid can be sensed through a capacitance measurement, provided that the fluid is electrically non-conducting. If a multiplicity of sensing electrodes is placed around the process pipelines in which the granular mixture is flowing, sufficient information can be obtained by rapid sequential determination of the capacitance between pairs of electrode plates – enabling an image of the dielectric distribution to be captured. Such an image can be computed by reconstructing the pair-wise measurements by inversion of an ill-conditioned data set. The results can be spectacular. Figure 1 shows the first real-time visualization of powder slugs in a pneumatic conveyor,[5] from which details on the geometric form and velocity of the slugs and moving beds can be obtained.

More detailed studies have sought to elucidate the formation and transport properties of different slugging characteristics in horizontal and vertical pipes, as well as providing comprehensive information on the shapes of the slugs' noses and tails to interpret granular flow models.[6] Integration of solid concentration, at a point, can also be used to estimate the mass flow. Such techniques can be used to monitor flow dynamics in fluidized beds, powder chutes, mixers, and conveying systems.[7–8] Figure 2 shows a flow regime characterisation via an on-line statistical analysis of fluctuations in the concentration – a more sophisticated version of this method can therefore be used to extract specific flow phenomena (e.g. swirling, etc). At the micro-scale of inspection, the passage of discrete, micron-sized particles can also be sensed using miniaturized wall-mounted dielectric sensors. The application of these sensors to determine particle velocities and the shape of individual particles has been reported.[9]

Figure 1   *Real time visualisation of granular flows in pneumatic conveying using electrical capacitance tomography*[5]

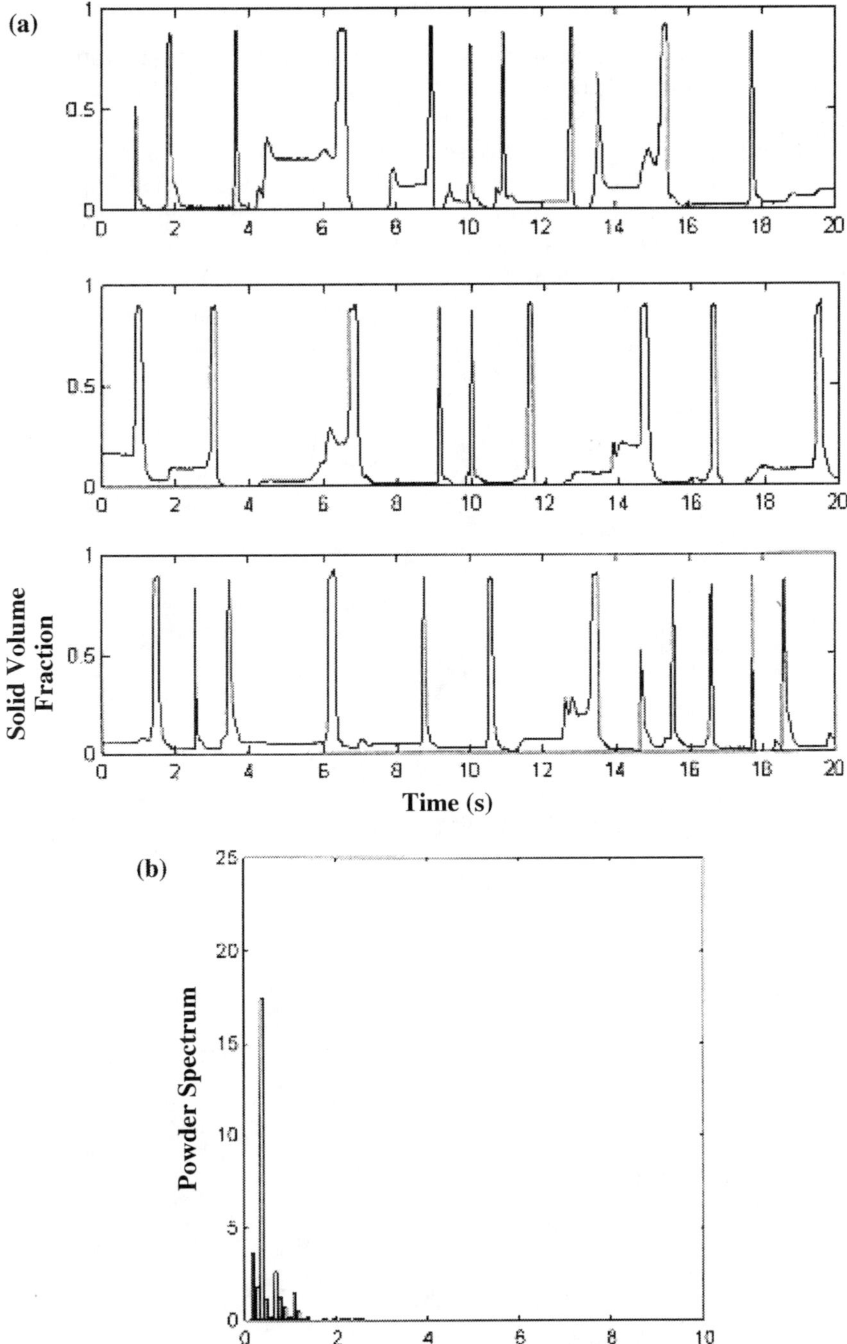

Figure 2 *Examples on measurements of (a) fluctuations of the relative volume fraction of solids across at a single plane during dense phase conveying (slugging) and (b) extraction of the characteristic power function through autocorrelation analysis*[5]

2.2 Positron Emission Particle Tracking

The above tomographic method is well suited for mapping concentration variations, but cannot easily be applied to track the motion of an individual particle or an assembly of particles of a given type. The ability to map the trajectory and velocity of individual particles is a requirement of verification for many particle flow codes such as the Distinct Element Method (DEM). By labeling specific particles with a positron emitting species, the β-decay yields back-to-back gamma rays that can be used to pinpoint the location of the source (i.e. particle) along each line of sight within the field of two planar detectors. A thorough discussion of this tracking procedure has been given by Parker et al.[10]

Figure 3 shows a photograph of a small attritor used in bead milling,

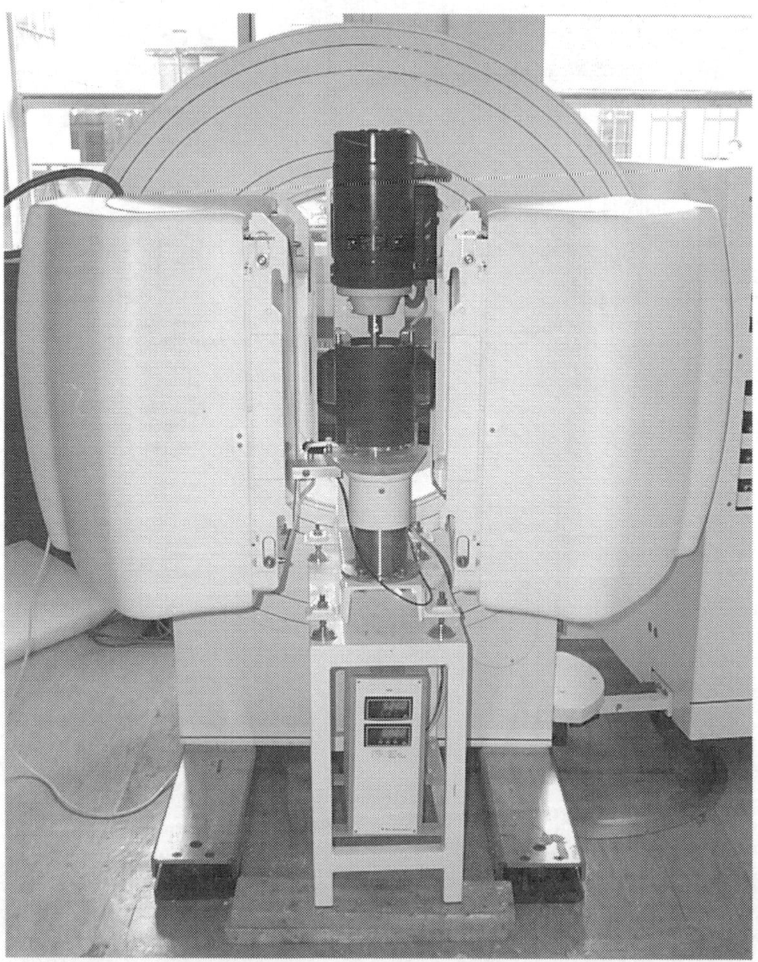

Figure 3   *Positron emission particle tracking applied to the measurement of beads in a 2dm³ laboratory scale stirred ball mill*

positioned between two planar detectors in a positron emission camera. By using a labeled bead, it is possible to follow the trajectory of the media, thus producing detailed velocity maps; and to compute the occupancy of the bead at different locations in the mill. Figure 4 shows the occupancy information as a function of the mill stirring rate.[12] The effect of fill on stirring rate and the regions of highest occupancy are clearly evident. Such information could be useful to estimate the energy dissipation occurring in the mill, to be used in conjunction with estimates for particle-particle collisions from DEM.

Similar methods have likewise been applied to observe size segregation effects in suspension mixing. Here, small populations of sieved, rather than single, particles are labeled – enabling the homogeneity of the mixing process to be visualized (Figure 5). Future developments are expected to allow the use of different emitters that can facilitate multiple-species tracing, and to follow the phenomena of agglomeration and breakage.

2.3 Acoustic Emission Spectroscopy

Flow of granular and wet suspensions often provides a convenient means for monitoring since the particles themselves produce noise. Such 'passive acoustic emissions' can arise from a variety of sources, such as the turbulent characteristic of the fluid, particle-particle collisions, and particle-wall collisions. The resulting acoustic spectrum may also reflect properties such as particle size (large particles having high momentum collisions) and the presence of ultrafine particles, or viscosifying chemical reagents (which may dampen the acoustic response as the continuous phase viscosity or density increases). An alternative to the passive methods is to introduce an acoustic source, and then listen to how the particulate flow modifies this source by analyzing the spectrum. This is known as active acoustic emission.

Passive methods are used industrially to monitor powder flow and for contamination monitoring, e.g. sanding in oil processing.[13] It is essential to audit the measurement environment since external sources of process noise (pipe vibration, other nearby operations) can interfere with the signals. Nevertheless, through a judicious choice of signal processing schemes, passive acoustic methods can be effectively deployed and are favored due to the low cost of the microphone sensors, with an additional attraction that they can be mounted *ex-situ*. Interpretation of the signals required a model or means of correlation with process parameters. A simple example of a passive acoustic method is to listen to individual grains arriving in a pile, and any subsequent avalanche as the structure becomes unstable. In such a simple case, the particle arrival rate and the exact initiation of local or bulk collapse and its duration can often be sensed. The use of acoustic methods coupled with DEM is yet to be widely reported although several measurements of these phenomena have been noted.

For more complex processes containing granular materials, systematic investigations have been applied to 'fingerprint' slurry flow in pipelines and to classify the performance of hydrocyclones.[14–17] Using a 190 kHz piezoelectric acoustic sensor, emissions generated within a 125 mm hydrocyclone separator

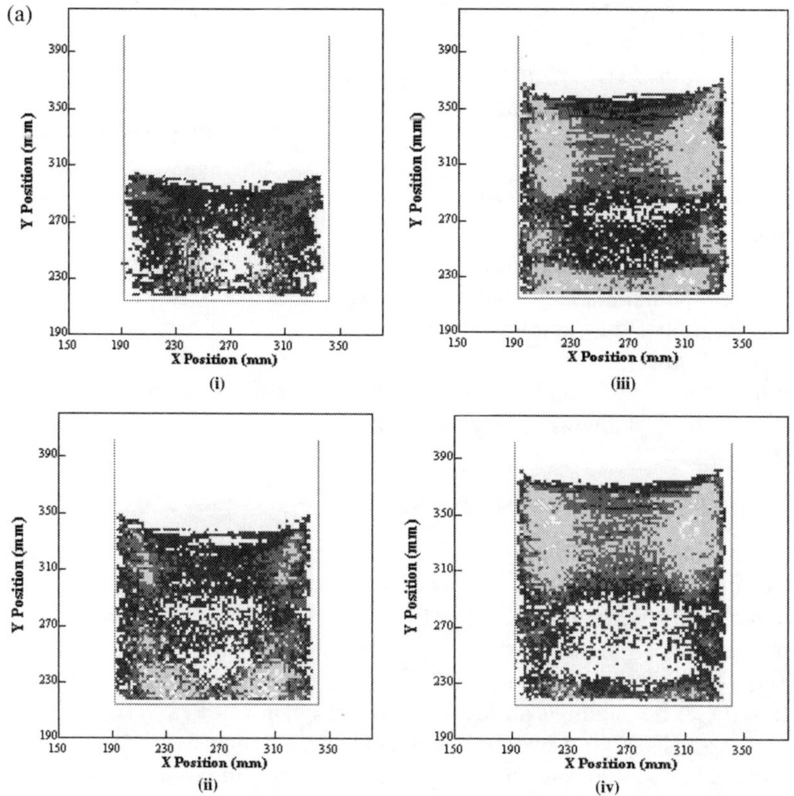

Figure 4a   *Example of occupancy and fill information as a function of mill stirring rate (200, 400, 600, 800 rpm). This together with detailed velocity maps can be obtained to compare with distinct element simulation*

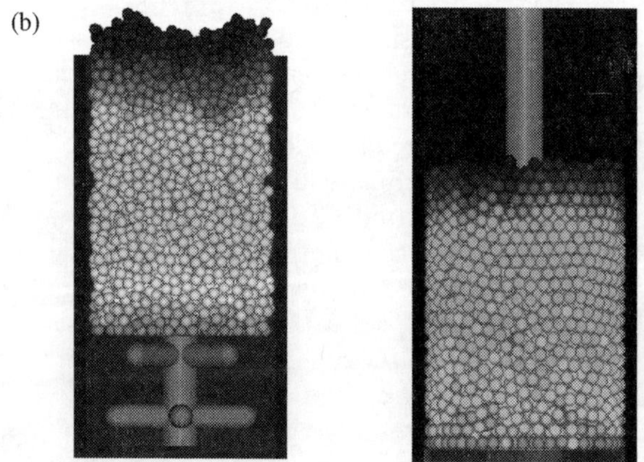

Figure 4b   *Example of occupancy and fill information – distinct element simulation*

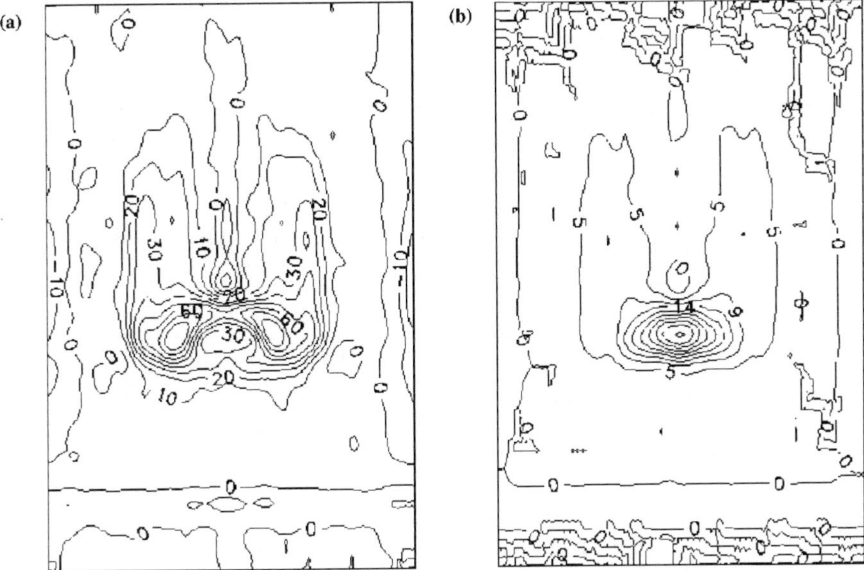

Figure 5 *Use of positron emission tracking to map the occupancy of particles of specific size in a stirred tank to quantify segregation and degree of mixing. Shown here is the effect of coarsening the size distribution on the concentration contours on reducing the proportion of coarse particle (600–710 microns) from 80% (left) to 20% (right) in a suspension of fine particles (150–210 microns) for constant total volume fraction and stirring rate[11]*

and a 44.5 mm diameter steel pipeline handling slurries of fine silica particles were collected for both spectral analysis and modeling purposes. The power density spectra of the collected signals were calculated by performing Fast Fourier Transformation (FFT) using appropriate windowing methods.[16]

Figure 6 and Figure 7 illustrate typical acoustic spectral characteristics for

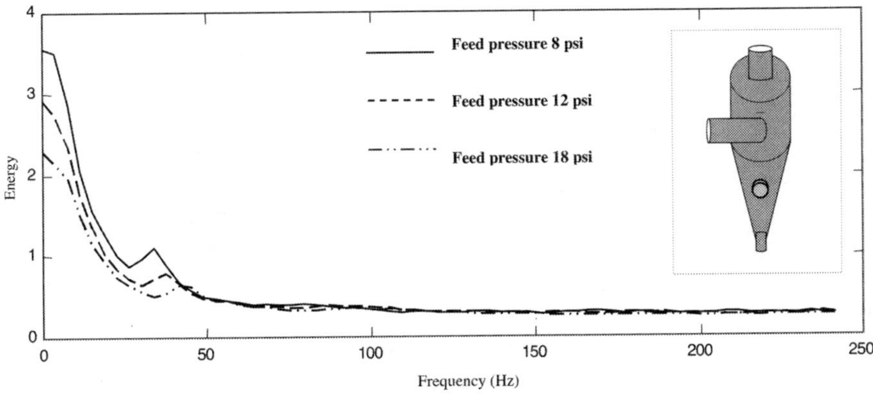

Figure 6 *Effect of feed pressure on spectral characteristics of acoustic emissions at 10wt% solid concentrations in a hydrocyclone investigation*

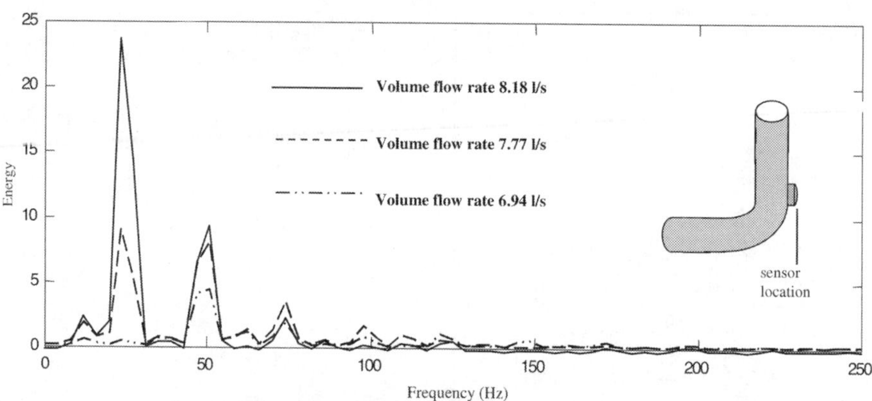

Figure 7 *Effect of volume flow rate on spectral characteristics of acoustic emissions at 30wt% solid concentration in pipeline flow investigation*

each process. It can be seen that in both case studies, the amplitude of the acoustic spectrum is sensitive to changes in certain key process parameters. The figures show variations due to feed flow/pressure, with further detailed analysis of the results indicating that the spectral behaviour of the acoustic emission signal is also closely related to the hydrodynamic features of the flow involved in each process. In the hydrocyclone investigation, the spectral characteristic of acoustic signals appears to be a combined manifestation of both turbulent and swirling flow structures inside the hydrocyclone.[15] In the pipeline flow monitoring, it can be shown that the pump-driven pulsation of the slurry, modulated by a number of flow parameters in the vicinity of the sensor mounting location, has been a major cause for the periodic occurrence of spectral peaks in the spectrum. Variations in these pulses can therefore be used to monitor the process. A multivariate stepwise regression technique can be applied to the experimental data, in order to quantify the relationship between process parameters and the statistical and spectral characteristics of the acoustic signals. It was demonstrated that for each case studied, predictions of less than 3% comparative errors were readily achieved for many of the parameters being monitored, including the hydrocyclone feed pressure, underflow concentration, solids concentration, volumetric flowrate, and others (Figure 8).

2.4 Forward Light Scattering

Amid a variety of optical-based sensing methods, fine particle size analysis using forward light scattering (i.e. laser diffraction or small angle light scattering) has been the established practice for laboratory-scale quality control, with a burgeoning role for real-time process monitoring. It evaluates size based on the principle that particles scatter incident light at an angle (θ) inversely proportional to their length scales. The light intensity (I) is measured by an array of detectors placed at various angles, whose signals are then compared with patterns predicted using an optical model to construe a Particle Size Distribu-

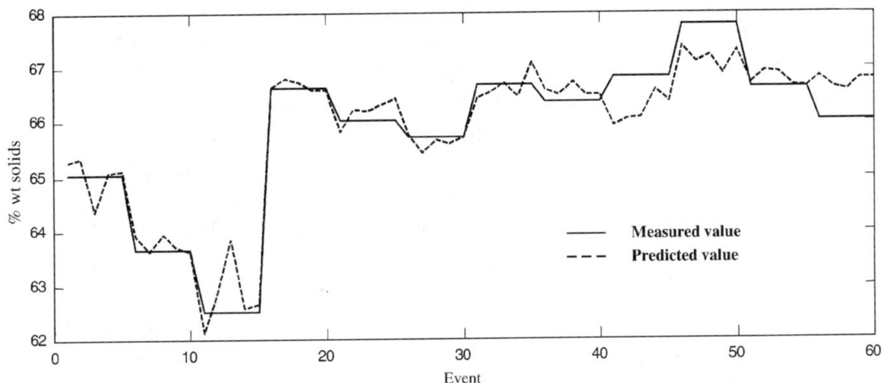

Figure 8 *Comparison of the model predictions against the actual values for hydrocyclone underflow solid concentrations*

tion (PSD) through a deconvolution procedure. Most commercial apparatus use a built-in model based on the Mie theory to calculate a volume-based PSD from sub-micron to millimetres range (typically 0.02–2000 µm), despite the theory's assumption of spherical shape. The PSD should correspond adequately to the actual distribution unless the particles are highly non-spherical (e.g. very elongated or crystalline). The Mie theory is not restricted to particles of specific optical properties or size, as concerns regarding the strong interactions between light and matter are addressed by this approach. It thus has a more universal applicability than models such as Rayleigh approximation that is valid only for small particles with low refractive index, or Fraunhofer diffraction that estimates size based on projected area for large, opaque particles.[18]

Several advantages of laser diffraction for monitoring pneumatic flows are its non-destructive/intrusive nature, rapid acquisition rate, no prerequisite for calibration, and its excellent reproducibility. It measures diffraction from all particles; therefore reducing size bias that is problematic in of particle counting or image analysis, provided the sample is representative and properly distributed. The main drawback is the need for a relatively low dosage of materials to minimize multiple scattering (that otherwise results in overestimating the presence of small particles), since most deconvolution procedures assume that light is scattered only once. The ideal obscuration spans between 0.5 and 5% (around 4–10 g) for dry powder, with even lower limits for finer particles as data quality is dependent on the number of scatterers passing through the laser beam.[19] This could pose a significant obstacle for in-process measurement with little opportunity for sample conditioning. Algorithms to account for multiple scattering[20] can be incorporated to allow slightly more flexible concentrations (up to 25% by volume). It is possible to install this instrument for direct in-line analysis if the powder conveyed is sufficiently dilute and well distributed, such as in the exit points to unit operations like jet mills or gas atomisers for metal powders production.[21] Otherwise, accompanying units to regulate the amount of samples taken from the main flow are often compulsory. It could be a relatively simple air venturi assembly as adopted by Malvern 'Insitec' that disperses powder from a static sample flute connected to a single or several fixed sampling points.[21] A more intricate approach by Sympatec 'Mytos & Twister' employs a dynamic 'finger' to collect samples by moving in a spiral track down the centre of a process pipe.[22] Industrial analysers based on forward scattering principles have been successfully implemented for on-line monitoring in print toner[21] and cement manufacturing,[23] enabling better optimisation through direct adjustment of parameters and curtailing of stabilisation time whenever there are changes in the operation.

A further benefit with this technique is the possibility of obtaining additional data from the intensity patterns. The magnitude of the scattering wave vector, q, is a function of the scattering angle θ,

$$q = \frac{4\pi n_o}{\lambda_o} \sin\left(\frac{\theta}{2}\right) \tag{1}$$

where n_o and λ_o are the refractive index of dispersion medium and the wavelength of incident light in vacuum, respectively. Only light scattered from particles separated by distances smaller than $1/q$ would be in phase, while the rest would be incoherent. Accordingly q could act as a probe for regions of size $1/q$ when agglomerates are present, by direct measurement of the time-averaged angular light intensity, $I(q)$, at different q-values. Using the approach of Rayleigh–Gans–Debye (RGD), the scattering body is modeled as a collection of non-interacting entities; $I(q)$ is therefore considered as a product of a form factor, $P(q)$ that signifies intensity from a single component (or particle), and the structure factor, $S(q)$, which describes additional intensity due to particle configuration in an assembly. At certain q-range (where one is effectively looking at a section much larger than primary particles and away from the edge of agglomerate), $P(q)$ is relatively constant so that $I(q)$ is simply due to the structural effects. For a fractal geometry, this leads to the classic expression[24]

$$I(q) \propto S(q) \propto q^{-d_F} \tag{2}$$

where d_F is a mass fractal dimension that indicates the space-filling ability (or compactness) of an object. The use of light scattering to obtain structure information has been employed predominantly in fundamental studies; still, it is a compelling tool for on-line characterisation such as in aerosol-based ultrafine particles production where the cluster structure is a chief parameter of interest. Although the RGD approximation is restricted to particles with low refractive indexes, this procedure can be used when there is little optical contrast between sample and background, and when methods like image analysis simply become too impractical.[25] Recent theoretical approach, utilising Mie instead of Rayleigh scattering, promises to extend the validity for a wider variety of particle size and opacity.[26] Other statistics that can be summarised include the equivalent solid content through intensity data at $q \to 0$,[25] and the average density of agglomerates from apparent volume fraction, estimated based on the extinction of transmitted light.[27]

For large, opaque particles that can be characterised via their projected areas, particle shape also influences the diffraction patterns mainly in the azimuthal direction, whereas their size is reflected in the radial intensity profile. Quantitative shape information can potentially be attained by measuring the azimuthal light intensity, and transforming them via cross-correlation.[28] Novel invention of a robust metal-oxide semiconductor sensor that improves the sensing capability of regular detectors likewise opened up the possibility of achieving simultaneous detection of both size and shape for on-line uses.[29]

2.5 Scanning Laser Microscopy

In contrast to the laser diffraction technique, scanning laser microscopy relies on the *duration* rather than intensity of signals generated as the illuminating light is reflected back by individual particles. For this reason, it can be used on dense suspensions for most materials (save for purely transparent objects that behave as specular scatterers) in any two-phase dispersed systems (solid-liquid, liquid-liquid, or solid-gas) without further dilution. The technology is available commercially in the form of Focused Beam Reflectance Measurement (FBRM®) probes from Lasentec. The basic set-up consists of a laser focused through a lens rotating at a fixed tangential velocity, and projected through a sapphire window, thus allowing the light focal point to move along a circular path and detect particles flowing past the window. The back-scattered light, generated when the beam intersects a particle, is then captured by a photo detector inside the probe (Figure 9).

Particles ranging from 0.5 μm to 2.5 mm can be measured in suspensions with virtually no limiting maximum concentration. The focal point can be adjusted to catch reflected light if there is a possibility of the laser path being blocked by other particles. Although not as imperative for size interpretation as in small angle light scattering, the optical properties of both solute and solvent do govern signal strength. If the particles are not strong isotropic scatterers; or in case of liquid-liquid dispersions, have a low relative refractive index, the detector may recognise pulses generated only from the perimeter rather than the entire body that could lead to over-counting and under-sizing.[30] Other factors affecting data quality include the probe tip, which should be positioned to ensure that sample flows towards its window, and sufficient agitation rate to provide rigorous mixing whilst avoiding air bubbles formation that could distort measurement.[31]

Due to the fast scanning speed (from 2 m/s and up to 10 m/s with an air-bearing lens system), the particles are relatively stationary compared to incident radiation. The time taken for the light to be reflected from one edge of a particle to the opposite edge, multiplied by the scan speed, is thus equal to the distance in straight line between any random two points on the object's boundary. This measured parameter is known as a chord length (Figure 10).

During a period of measurement, the number of particles is counted based on

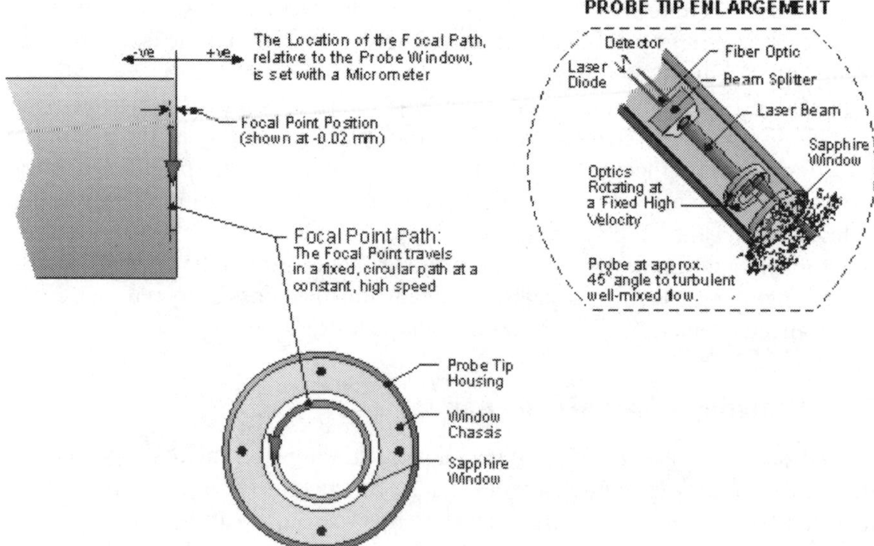

Figure 9   *Configuration of a Lasentec FBRM® probe (http://www.lasentec.com/ fbrm.html)*

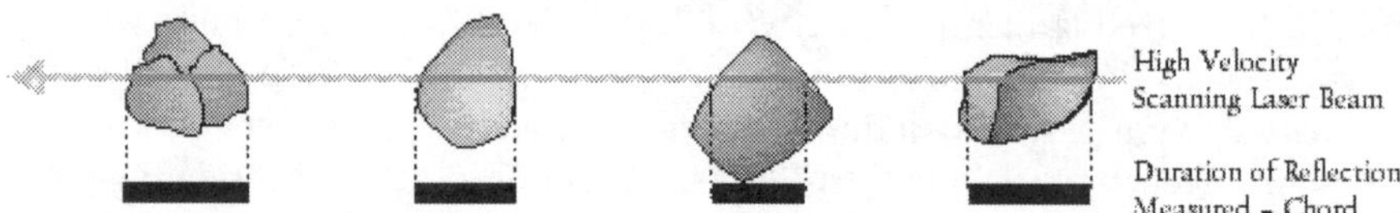

Figure 10   *'Chord length' measurement of an object at random orientations*

pulses generated by individual particles as they pass through the focal point, giving a chord length number distribution to represent the sample PSD. Accordingly, this technique does not presuppose any shape when analysing particles or agglomerates, while apparatus that rely on forward scattering principles generally utilise Mie model for spheres to translate intensity patterns into size distributions. Given that particles are scanned at arbitrary orientations, the Chord Length Distribution (CLD) should adequately represent the actual size distribution if the particles are close to spherical, and less so when there is an increasing irregularity or difference in dimensions between the major and minor axes. In such cases, different weightings to the chord length distribution may improve the comparison to actual PSD, although they have to be employed with caution due to the systematic errors associated from deriving these distributions.[32] By the same token, the estimated chord length distributions are not exactly practical for quantitative validation since size data are often expressed in diameters. Furthermore, the chord length itself has to be regarded as apparent, rather than representation of physical size, due to the influence of factors such as particle shape, orientation, optical properties, and process conditions on its measurement.[31] Algorithms such as probability apportioning and finite element methods to predict diameter distributions from CLD may be applied with limited accuracy, based on the assumption that each measured chord length is obtained independently.[33]

Scanning laser microscopy (as FBRM®) is beginning to be recognized as the direct monitoring device of choice in process industries, particularly those involving crystallization, granulation, and dissolution, with emerging applications for flocculation and emulsion. Because of its quick response rate, effectively no specific sampling requirement, and adaptability to harsh conditions, the technology at its current status is adapted mostly for detecting dynamic changes *in-situ* or in pipelines, rather than to yield accurate size information. It permits the tracking of particle and/or agglomerate populations so that the effect of altering parameters, such as addition of binders or increasing solids concentration, can be quantified. An example of this was the probe implementation onto 'white pans' reactors that produce white sugar from thickened beets juice at the American Crystal Sugar Company in Crookston, Minnesota. The formation of small crystals (between 0–50 μm) that fluctuated in size during boiling was observed. This was detrimental to the overall efficiency since they had to be flushed with excess water. By monitoring the chord length distribution trends of both product and feeds in real time, the culprit could be identified as the 3% periodic spike in solids concentration of input liquor that caused less than optimum crystallization.[34] The chord lengths and particle counts could also serve to 'fingerprint' the endpoints for batch-wise operations, e.g. high shear granulation to bind drug substance into a filler matrix. This would allow process variables to be selected accordingly with live data, as opposed to routine trial and error with off-line mesh evaluation.[35] The outputs that this technique generated can potentially be used as tools to control and automate polymer dosing and mixing rate in flocculation,[36–37] as well as agglomeration of suspensions in fluidised beds.[38]

References

1. R. A. Williams and M. S. Beck in "Tomographic Techniques for Process Design and Operation", ed. M. S. Beck, E. Campogrande, M. A. Morris, R. A. Williams, and R. C. Waterfall, Computational Mechanics Publications, Southampton, 1993, 490.
2. R. A. Williams and S. J. Peng in "Controlled Formation of Particles, Droplets, and Bubbles", ed. D. J. Wedlock, Butterworth-Heinemann, Oxford, 1994, 327.
3. R. A. Williams, J. Gregory, M. Elimelech and X. Jia, *Particle Aggregation and Deposition Processes: Measurement, Modelling, and Simulation*, Butterworth-Heinemann, Boston, 1998.
4. R. A. Williams and X. Jia in "Hydrocolloids – Part 2 Fundamentals and Applications in Food, Biology, and Medicine", ed. K. Nishinari, Elsevier Science, 2000, 19.
5. K. L. Ostrowski, S. P. Luke, M. A. Bennett and R. A. Williams, *Powder Tech.*, 1999, **102**, 1.

6. A. Jaworski and T. Dyakowski, *Powder Tech.*, 2002, **125**, 279.
7. D. Neuffer, A. Alvarez, D. H. Owens, K. L. Ostrowski, S. P. Luke and R. A. Williams, Proceedings of the 1st World Congress on Industrial Process Tomography, Buxton, 1999, 71.
8. K. L. Ostrowski, S. P. Luke, M. A. Bennet and R. A. Williams, *Chem Eng. J.*, 1999, **3524**, 1.
9. T. A. York, I. Evans, Z. Pokusevski and T. Dyakowski, *Sensors and Actuators A: Physical*, 2001, **92**: 103, 74.
10. D. J. Parker, R. N. Forster, P. Fowlers and P. S. Taylor, *Nucl. Instrum. & Methods*, 2002, **A477**, 540.
11. S. L. McKee, D. J. Parker and R. A. Williams in "Frontiers in Industrial Process Tomography", ed. D. M. Scott and R. A. Williams, AIChE and Engineering Foundation, New York, 1995, 249.
12. J. Conway-Baker, R. W. Barley, R. A. Williams, X. Jia, J. Kostuch, B. McLoughlin and D. J. Parker, *Min. Eng.*, 2002, **15**, 53.
13. T. Folkestad and K. S. Mylvaganam, *Oil & Gas Journal*, 1990, **27**:8, 33.
14. R. Hou, A. Hunt and R. A. Williams, *Minerals Engineering*, 1999, **11**:11, 1047.
15. R. Hou, A. Hunt and R. A. Williams, *Powder Technology*, 1999, **106**, 30.
16. R. Hou, *Acoustic Monitoring of Particulate Flows*, Ph.D. Thesis, University of Exeter, UK, 2000.
17. R. Hou, A. Hunt and R. A. Williams, *Powder Tech.*, 2002, **124**, 287.
18. H. C. van de Hulst, *Light Scattering by Small Particles*, Wiley, New York, 1969.
19. R. Jones, *G. I. T. Lab. J.*, 2002, **2**, S.46.
20. E. D. Hirleman, *Part. Part. Syst. Charact.*, 1988, **5**, 57.
21. G. M. Crawley, *Powder Metallurgy*, 2001, **44**:4, 304.
22. M., Puckhaber, S. Röthele and W. Witt, *Powder Handling. Proc.*, 1998, **10**:4, 416.
23. A. R. Godek, GCL: N&S American Cement, 2000, 37.
24. C. M. Sorensen in "Handbook of Surface and Colloid Chemistry" ed. K. S. Birdi, CRC Press, New York, 1997, 533.
25. G. C. Bushell, Y. D. Yan, D. Woodfield, J. Raper and R. Amal, *Adv. Coll. Int. Sci.*, 2002, **95**, 1.
26. A. Thill, S. Lambert, S. Moustier, P. Ginestet, J. M. Audic and J. Y. Bottero, *J. Coll. Int. Sci.*, 2000, **228**, 386.
27. P. T. Spicer, S. E. Pratsinis, J. Raper, R. Amal, G. Bushell and G. Meesters, *Powder Tech.*, 1998, **97**, 26.
28. C. Heffels, D. Heitzmann, E. D. Hirleman and B. Scarlett, *Part. Part. Syst. Charact.*, 1994, **11**, 194.
29. Z. Ma, H. G. Merkus and B. Scarlett, *Powder Tech.*, 2001, **118**, 180
30. R. G. Sparks and C. L. Dobbs, *Part. Part. Syst. Charact.*, 1993, **10**, 279.
31. P. J. Dowding, J. W. Goodwin and B. Vincent, *Coll. Surf. A: Physicochem. Eng. Aspects*, 2001, **192**, 5.
32. A. Tadayyon and S. Rohani, *Part. Part. Syst. Charact.*, 1998, **15**, 127.
33. P. A. Langston, A. S. Burbidge, T. F. Jones and M. J. H. Simmons, *Powder Tech.*, 2001, **116**, 33.
34. J. Sobolik, Proceedings of the Lasentec Users Forum, Charleston, South Carolina, 2002, M–2–143P, Rev A.
35. M. Menning, M. Cheng and T. Ju, Proceedings of the Lasentec Users Forum, Barcelona, Spain, 2001, M-2–122P, Rev A.
36. R. A. Williams, S. J. Peng and A. Naylor, *Powder Tech.*, 1992, **73**, 75.

37. P. Fawell, Proceedings of the Lasentec Users Forum, Charleston, South Carolina, 2002, M–2–146P, Rev A.
38. L. Mörl and J. Drechsler, Proceedings of the Lasentec Users Forum, Orlando, Florida, 2000, M–2–007P, Rev A.

CHAPTER 10

Fluidization of Fine Powders

J. ZHU

Powder Technology Research Centre, Department of Chemical and
Biochemical Engineering, The University of Western Ontario, London,
Ontario, N6A 5B9, Canada

Nomentclature

A	Hamaker coefficient
F_{van}	van der Waals force
H	separation distance
G	Vibration number
R	radius of spherical particles
U_g	superficial gas velocity
U_{amf}	apparent minimum fluidization velocity
U_{dis}	disruptive fluidization velocity
U_{mb}	minimum bulling velocity
U_{mf}	minimum fluidization velocity
θ	vibration angle

1   Introduction

Fluidization occurs when particulate materials are suspended in an upflowing
fluid phase. It is often the preferred mode of operation if a fluid phase and a
particulate phase (powder) need to be brought into good contact for physical or
chemical processing, due to its special characteristics: uniform and extensive
gas-solid contact, good solids mixing leading to uniform temperature throughout the bed, high mass and heat transfer rates, easy solids handling, etc. The
particles that are suitable for fluidization range from sub-microns to several
millimetres, with increasing difficulty in fluidization as particle size goes to the
two extremes. In 1973 Geldart[1] classified gas-solid fluidization into four groups
according to their fluidization behaviour as influenced by particle size and density (see Figure 1): Group A is called aeratable powder which can be fluidized well
and experiences a period of uniform bed expansion when the gas velocity is
increased beyond the minimum fluidization velocity. As shown in Figure 1,

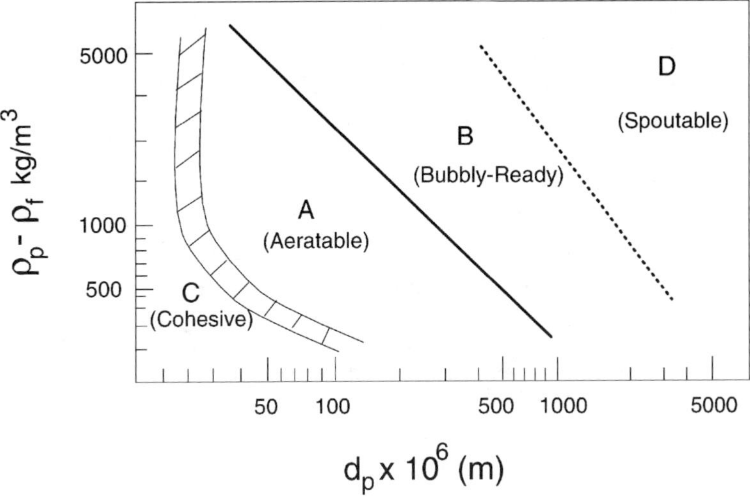

Figure 1 *Geldart powder classification*

Group A particles are normally in the range of 25–40 µm to about 150–200 µm. Group B is the bubbly-ready or sand-like powder, usually in the range of 150–200 µm to about 700–900µm, which is larger than Group A and goes directly to bubbling fluidization as soon as the gas velocity reaches the minimum fluidization velocity. Group D is also called spoutable powder, which is larger than both Groups A and B powders and sometimes tends to experience some instability when fluidized. For the 700–900 µm to several millimetre Group D powders, spouted bed is a more suitable contact mode than a regular fluidized bed. Group C powder is very fine, smaller than 25–40 µm, and usually difficult to fluidize due to the strong interparticle forces. Therefore, it is also called cohesive powder.

The cohesive nature of Group C powder comes from the fact that when the particle size becomes smaller, the relative magnitude of the interparticle forces increases significantly in relation to the gravitational and drag forces exerted on the particles. Such strong interparticle forces make the individual particles cling to each other and therefore form agglomerates, which can lead to severe agglomeration, and bed channelling or even complete defluidization. Therefore, Group C powder was generally considered not fluidizable and was not the subject of much study in the field of fluidization until the last two decades. In recent years, however, Group C particles have been widely used in new advanced materials and chemical industries due to their special characteristics.[2] With their high specific surface area, for instance, smaller primary particles commonly lead to a better quality of final product in the ceramic industry or in powder metallurgy.[3] Finer paint powders also lead to significant improvement on the coating finish in the powder coating industry. Aerogel powders can provide very high surface area for catalytic chemical reactions.[4] Moreover, fine and ultra-fine powders are also of increasing importance in the pharmaceutical, plastics, and food industries.[5,6]

The lower limit of the Group C powders have not been clearly defined and

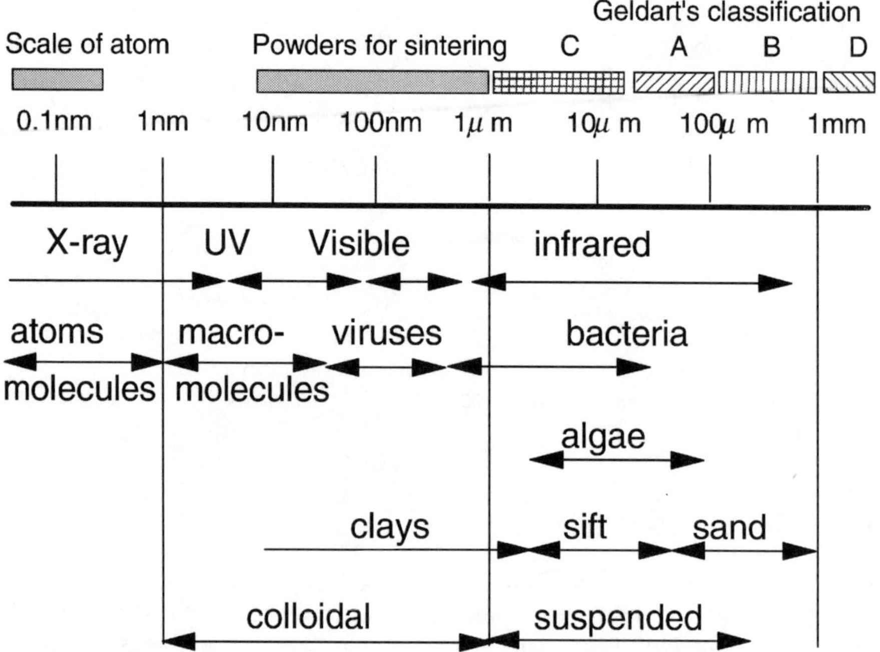

Figure 2 *Powder size classification with other typical particular matters and with radio-magnetic wave length*

Geldart's initial classification implies it goes to the molecular level. As shown in Figure 2, they may be further divided into micron particles (>1 μm), submicron particles (50–100 nm to 1 μm) and nanometer particles (<50–100 nm).[7]

Group C powders have often been referred to as cohesive powders, fine powders, or ultrafine fine powders in comparison with Group A powders which are sometimes also called fine powders. For clarity, fine and ultrafine powders (or particles) are considered synonyms in this chapter and both are referred to as Group C powders, where the term cohesive powder is used to stress the cohesive nature of the powders whenever necessary.

2   Interparticle forces

There are three types of interparticle forces, the Van der Waals force, the capillary force and the electrostatic force.[8] Van der Waals force is a collective term taken to include the dipole/dipole, dipole/non-polar and non polar/non-polar forces arising between molecules.[9] It is the dominant interaction force between particles in a powder as well as in a fluidized bed, including both the forces between the fluidized particles and between the particles and the wall of the apparatus in which the fluidized bed is generated. This force always exists and is usually the largest interparticle force among the three types. Van der Waals force only becomes noticeable when the particles come sufficiently close, e.g., 0.2 to 1 nm and when the particles are small enough, e.g. 10 μm or less.[8] Van der Waals force may be understood by imagining an instantaneous picture of molecules possessing different electronic configurations, giving them a dipolar character. This temporary situation will act on the neighbouring molecules rendering these also dipolar. As a result thereof and as a consequence of the general attraction between dipoles, molecules attract each other, even when they are apolar singly. Van der Waals force may be estimated using the following equation for spherical particles:[10]

$$F_{Van} = \frac{AR}{12H^2} \tag{1}$$

Surface roughness, geometrical structure and possible deformation of the individual particles can significantly change the Van der Waals force.

Electrostatic force can occur by tribo-electric charging or by the formation of a potential difference when particles of different work functions are brought into contact. The resulting Coulomb attraction make the powder adhesive. Capillary force comes from the fluid condensation in the gap between the particles in close contact, resulting in additional or liquid bridging force among particles. It should be noted that a higher capillary force comes at the expense of the electrostatic force, which diminishes with the increase of moisture.

In a recent review paper Seville et al.[9] suggested sintering as another interparticle force. While this force is qualitatively different from the other three forces discussed above, material migration due to diffusion, viscous flow or some other mechanism or combination of mechanisms can gradually lead to sintering, especially under high temperature.

3 Fluidization Characteristics of Fine Particles

The basic characteristic of fine particles is usually described as cohesive and difficult to fluidize due to strongly interparticle forces.[1,11–15] Very few Group C powders can fluidize on their own, especially those under 10 μm. As the particles form agglomerates, fluidization is difficult and can sometimes be impossible especially when the particle size becomes smaller. The powder bed often exhibits severe channelling and rat-holing, with most of the bed areas not fluidized at all, rather than a smooth fluidization. Such phenomena completely defeat the purpose of fluidization given the very poor gas-solid contact. Some particles can form fairly stable agglomerates through self-agglomeration, often at higher gas velocities, which then behave more like Group A particles and form good fluidization of the agglomerates.[16,17] While it does seem to provide a good fluidization, such fluidization of the agglomerates can only be useful for increasing solids handlability but not to create good gas-solid contact required for many chemical and physical processes. Some other particles may be fluidized with the addition of other energy or materials (collectively called fluidization aids), such as mechanical vibration, e.g. see Mori.[18] Fluidization aids will be the subject of the next section.

3.1   Channelling

Channelling is one of the most important fluidization characteristics of fine particles and happens more often at low superficial fluidization velocity. Due to the cohesive nature of the bed, upflowing gas has difficulty to fluidize the fine particles, but pushes its way nearly vertically up through voids extending from the distributor to the bed surface. These vertical channels may move across the bed with time, leaving the other regions of the bed defluidized. There are also small cracks in the bed which drain into these vertical channels.[1,11,19,20] Those cracks can be in any orientation, but may have a somewhat greater preference to form horizontally than vertically. With increasing gas velocity, larger channels, also called rat-holes, can form for some extremely cohesive particles.[17,18,21] This phenomenon arises because the interparticle forces are noticeably greater than the other forces (e.g., drag and gravity) the fluid can exert on the particles.[22]

3.2   Agglomeration

Particles inside a fluidized (or partially fluidized or defluidized) bed of Group C powder may exist in three forms: single particles, natural agglomerates,[23–25] and fluidized agglomerates. Fine particles tend to form spherical agglomerates due to cohesiveness when kept in a heap, stored in a vessel, or while being transferred (i.e. wherever relative motion exists between particles). These agglomerates, called natural agglomerates, also called initial or first agglomerates, are generally friable and light in structure, and possess relatively close size.[21] When fluidized, these so-called natural agglomerates undergo reorganization with respect to the member particles, or are fragmented into smaller agglomerates or even discrete particles. Another type of agglomerates, called the fluidized agglomerates or secondary agglomerates, also forms when the bed is fluidized.[16,21,23–26] In conclusion, there are three states of agglomerations for fine particles: unagglomerated single particles (newly produced), natural agglomerates (when stored), and fluidized agglomerates (when fluidized).[21]

Not much is known about the mechanism of agglomeration other than it is mainly due to the strong interparticle forces. Baerns reported agglomeration phenomena and attributed it to interparticle forces of fine powders.[11] It was further shown that the fluidization behaviour of fine particles mainly depends on the properties of the agglomerates formed when fluidized.[16,23,27] For example, the bed of particles enters into self-agglomerating fluidization when stable and fluidizable agglomerates are formed. This is usually the case for smaller particles where the relative interparticle force is much larger.[14,16,17,22,28,29] These more stable agglomerates behave like larger Group A powders and thus fluidize well. On the other hand, when the interparticle forces are not as strong, the agglomerates formed are not strong enough as stable pseudo (and larger) particles and thus can not be easily mobilized to allow good fluidization. This leads to partial fluidization and normally happens with the intermediate sized Group C powders. For larger Group C powders, unstable lumps or chunks of particles ranged up to the size of the vessel may form, leading to channeling, e.g. see Leu and

Huang.[30] For all fine particles, the fluidized bed almost always channels at lower velocities, when there are no fluidization aids present. For particles with less interparticle forces, the bed remains channelling or partially fluidized, and less stable agglomerates or lumps of particles are formed during fluidization.

For the intermediate and larger Group C powders, proper use of fluidization aids can significantly reduce the interparticle forces, to make most of those powders fluidizable as single particles or as very small agglomerates. To control agglomerate formation, it is important to understand the properties of agglomerates. Besides the species of particles, agglomerate density and size are two of the most important properties of agglomerates. Agglomerate density is the function of bulk density of particles and is also affected by the undergoing process of packing and storage.[31–34] Agglomerate size is dependent not only on the size of particles, but on the dynamic equilibrium between agglomerate formation and agglomerate dissolution resulting from the configuration of the vessel and the operating parameters.[21,35,36]

3.3   Onset of Pseudo-minimum Fluidization—Bed Disruption

As the lowest limit of operating gas velocity for a fluidized bed, minimum fluidization velocity is another important parameter of particle fluidization. However, no clear minimum fluidization velocity can be identified for most Group C powders. As shown by data obtained in our Powder Technology Research Centre (Figure 3), as a result of channelling and agglomeration, any meaningful measurements of conventional U_{mf} (minimum fluidization velocity) and U_{mb} (minimum bubbling velocity) through the typical pressure drop versus gas velocity relationship become virtually impractical because the bed pressure drops are non-reproducible and most of the times also not equal to the weight of particles per unit cross-sectional bed area.[14,15,37]

For those fine powders that can fluidize by themselves, the minimum gas velocity that can make the particles fluidize is at a velocity much higher than the anticipated minimum fluidization velocity. Such a characteristic gas velocity is called the apparent or pseudo-minimum fluidization velocity, U_{amf}, or the disruptive velocity U_{dis}.[16,17] During this process (from the anticipated minimum fluidization to the apparent minimum fluidization), fine particles undergo agglomeration and it is the agglomerates formed that start to fluidize at the disruptive velocity. This pseudo-minimum fluidization velocity depends on both the properties of the particles and the agglomerates formed thereto. Such results have been obtained by using, for example, Cu/Al_2O_3 and Ni/Al_2O_3 aerogels,[16] Fe_2O_3/SiO_2 aerogels,[36] submicron Ni, Si_3N_4, and Al_2O_3 particles[17] and very dense tungsten carbide powders.[23]

Fine particles, which do not form fluidizable agglomerates, are also fluidizable when certain external force is introduced, such as vibration. For example, the results in Figure 3 obtained from our Research Centre clearly show the fluidization of fine $CaCO_3$ particles with the aid of mechanical vibration. In such situations, there exists a measurable pseudo-minimum fluidization velocity which is closely related to the external aids used, as well as the particle

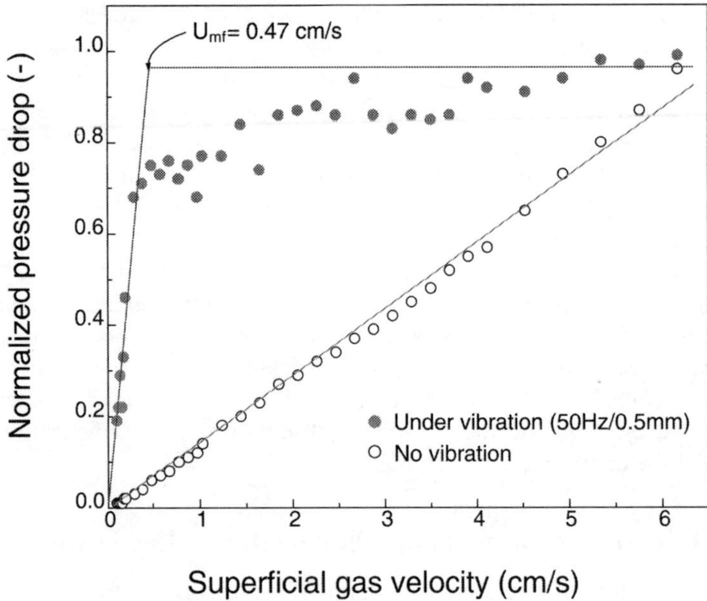

Figure 3 *A typical pressure gradient versus gas velocity curve for Group C powder (5.5 μm CaCO, fluidized with and without mechanical vibration)*

properties. For example, the minimum fluidization velocity to fluidize such particles under vibration depends not only on the particle properties but also on the intensity and frequency of the vibration.[18,38]

Currently, the pseudo-minimum fluidization velocities of fine particles can only be obtained by experiment and there is no universal formula to predict them due to the lack of a good correlation between the properties of fine particles and the minimum fluidization behaviour. The minimum fluidization velocity calculated from conventional equations derived for large particles and air are unsuitable for fine particles due to the formation of agglomerates.

3.4   Bed Expansion and Pressure Drop

Bed expansion of particles is one of the most important fluidization characteristics of powders and reflects the extent of particle suspension by fluidizing gas. A higher bed expansion generally indicates a better fluidization with more gas contained in the particulate phase, resulting in better gas-solid contact. Pressure drop across a fluidized bed is related directly to the fluidization states of particles, because the pressure drop should equal the total weight of particles per unit cross-sectional area if the particles are fully fluidized. The extent to which the bed pressure drops is close to the total weight per unit area, gives another indication of good fluidization, and therefore can also be used to evaluate the difficulty in fluidizing fine particles. Due to the complexity of fluidization and the great differences in bed expansion of fine particles of different size ranges, it is best to discuss bed expansion and pressure drop separately for several size ranges.

3.4.1 *Particles of Several Microns and Above*

For particles in this range, interparticle forces lead to cracking and channeling and most particles do not fluidize consistently over time at low velocity. Only partial fluidization is observed.[15,20] However, the bed mostly still exhibits a certain extent of bed expansion, at least under high gas velocities, with defluidized regions normally forming from the distributor where cracks and channels form.[30] The volume of the dead region decreases with increasing particle size and superficial gas velocity. Away from the defluidization region, bubbles may start to form at higher velocity and increase in size as they rise towards the top of the bed. These bubbles disrupt the cracks and channels, thus maintaining the top bed in a fluidized or semi-fluidized state. This kind of particle is often temporarily fluidized at higher superficial velocity but such state is unstable. Attempts to sustain such particles in a very stable fluidized state for a longer period of time often fail[20] without the aid of other external methods.

The partial fluidization results in a sizeable pressure drop across the bed but the presence of cracks and their developments make the pressure measurement unpredictable. The reproducibility of these measurements is poor because each experiment produces a fresh set of cracks, with their "density", inclinations and tortuosities that determine the pressure drop changing each time. This constant "tussle" between the interparticle cohesive forces which tend to build a defluidized zone and the hydrodynamic force which give rise to bubbling leads to fluctuations in the bed pressure drop and bed height.

3.4.2 *Particles Less than 0.5 Microns or Smaller*

These particles easily form natural agglomerates. When fluidized, they also tend to form large agglomerates, i.e. fluidized agglomerates. For these particles, bed expansion shows two typical situations. One is that the bed expands suddenly after the bed structure is disintegrated at the disruptive velocity and then the bed height increases homogeneously with increasing velocity. The other is that the bed expands stage-wise with increasing gas velocity. At low velocity the bed remains static or channelling and no clear expansion is observed.[16,17,21,24,39]

Except for some particles with strong electrostatics, like ZrO_2,[17] stable fluidization of this kind of particle can always be achieved eventually and stable pressure drop is obtainable at higher velocity than U_{dis}. In the initial stage, pressure drop across the bed, corresponding to bed expansion described above, may increases suddenly or stage-wise with increasing gas velocity, see Chaouki et al.[16] and Wang et al.[21] After the initial channels disappear, the pressure drop increases gradually with increasing gas velocity, like group A particles, until it reaches a stable value.[17]

3.4.3   Particles with Intermediate Sizes

Particles between the two ranges described above show some transitional features. Most particles with wide size distribution tend to form non-uniform agglomerates during fluidization at velocities higher than the disruptive velocity. As a result, there is a fixed-bed region of large agglomerates at the bottom, a fluidized region of smaller agglomerates in the middle and a dilute-phase region of even smaller agglomerates further up in the fluidized bed.[21,30,40] Beyond a certain gas velocity, the bed expansion depends on the height of the fluidizable agglomerate layer, with bed height increasing with increasing gas velocity. Some other particles with narrow size distribution such as $CaCO_3$ tend to form uniform large agglomerates which are not strong enough to be mobilized. When the bed is vigorously "shaken" under high gas velocity, the structure may be disintegrated into smaller agglomerates. However, the bed still remains essentially in a fixed bed state and no expansion is clearly observed for these particles.[17,20,21,40,41]

For such particles with transitional properties, the situation with pressure drops across the bed becomes more complicated. For most cases, the pressure drop across the bed fluctuates with time but gradually approaches a constant value. At different superficial velocities, this "final" pressure drop may be less than the weight of particles due to partial channelling, or greater than the weight of particles due to friction and adhesion with the wall.[1,17,20,21,30] Partial fluidization at the top and channeling or stationary at the bottom both result in less pressure drop.

3.5   Further Classification of Group C Powders

As discussed before, there are some different behaviours for Group C powders of different size ranges. Different research groups have proposed several classifications:

Mori et al.[18] developed a vibro-fluidized bed to fluidize Group C particles. Based on their experimental observation, fine particles in a wide range of fine particles down to the submicron level could be classified into three subgroups with respect to their vibro-fluidizability:

- The first group includes powders which are fairly easily fluidized and give large bed expansion and elutriation;
- The second group includes very cohesive powders such as $MgCO_3$ and $CaCO_3$ which are still difficult to fluidize even subject to strong mechanical vibration;
- The third group includes common fine powders which are fluidized under bubbling conditions similar to Group A powders in the vibro-fluidized bed and their elutriation can be easily controlled.

Similar behaviours are observed by Leu and Huang[30] when studying fluidization of cohesive powders in a sound wave vibrated fluidized bed.

Fluidization behaviours of fine particles without external forces were observed carefully by Zhao et al.[40] and Wang et al.[21] and the combination of their observations led to the following subgrouping:

- Type 1: Channelling and slugging always take place due to the appreciable cohesive forces between fine particles and between agglomerates. Experiments showed that these channels or plugs could be broken and stable fluidization could be obtained by external forces, such as tapping and vibration of the bed.
- Type 2: Agglomerates with sizes similar to Geldart group A particles are formed when they are fluidized. The fluidization behaviour of these agglomerates is similar to that of Group powders as observed by Chaouki et al.[16] and Morooka et al.[17] In general, Type 2 particles are much less than 1 μm and belong to the class of non-metallic oxides, and they form stable agglomerates rather than exist singly.
- Type 3: Large agglomerates with sizes similar to Geldart Groups B or D particles are formed when fluidized, such as while fluidizing $CaCO_3$ as observed by Morooka et al.[17] The fluidization behaviours of these over-sized agglomerates are similar to those of the Geldart Groups B or D powders. Many of these powders are metallic and ionic compounds, and are less than 5 μm. These particles can form natural agglomerates but these agglomerates are generally not stable.
- Type 4: Particles of transition characteristics, with smaller agglomerates in the upper region of a fluidized bed and larger agglomerates in the lower region, some of which may even be defluidized. The disparate sizes of these agglomerates may generally reach some steady and relatively uniform value after repeated solids recirculation. As a result, relatively homogeneous fluidization may eventually ensue. These particles are usually oxides and range in sizes from several microns to tens of microns.

The fluidization state obviously also changes with gas velocity. Observing the different states of fluidization in a conical column while fluidizing NiO/Al_2O_3 cryogels under reducing operating velocity, Venkatesh et al.[42] proposed the following different states of fluidization:

- State A – completely mixed fluidization under high gas velocity (U_g = 0.44 m/s): At this velocity, the bed attained a state of intense solids mixing. A slight bed expansion was also observed. There was no observable agglomeration in the homogeneous fluidized bed
- State B – core-annulus segregative fluidization under intermediate velocity ($0.33 \leq U_g \leq 0.44$ m/s): Completely mixed fluidization did not exist below 0.44 m/s and gave place to agglomeration of a fraction of the bed. Spontaneous formation of a cylindrical core of the agglomerates was observed in this state of flow regime.
- State C – post-fluidization condition ($U_g \leq 0.33$ m/s): Upon further reduction of the gas velocity, the segregated bed entered the packed condition. The dimensions of the constituent particles were apparently the same as in

State B. The bed pressure drop decreased with decreasing fluid velocity in this state, whereas the bed pressure drops remained constant in both States A and B flow regimes.

Given the complicated nature of the fine powder fluidization and relatively small amount of research work, the above classifications are far from conclusive and should only be regarded as initial work.

4   Aids In Fluidizing Fine Powders

Given the importance of fine powders in modern industries, and since the main difficulty associated with their fluidization is the strong interparticle forces, many measures have been taken to overcome the interparticle forces, and hence improve the fluidization. Those measures are collectively called fluidization aids. Most of those measures include the introduction of extra energy into the system, to help break up the agglomerates, such as mechanical and acoustic vibration, mechanical stirring, and magnetic and electrical field disturbance. Other measures approach the problem by reducing the surface force of the particles, such as surface adsorption and modifications, or adding larger or finer particles as flowing agents. Some of the commonly used methods are discussed below.

4.1   Mechanical Vibration

Mechanical vibration can be applied through vibrating the entire fluidization column or only vibrating some internals inside the bed. Very often, the frequency and amplitude of the vibrating motor can be adjusted to achieve the best effect for a given system. Mori et al.[18] developed a vibro-fluidized bed to fluidize Group C powders. The fluidized bed and its windbox are fixed on the upper surface of a rectangular vibro-stand which is supported on the base by four springs. A pair of vibro-motors are fitted on two side walls of the vibro-stand. The amplitude of the vibration can be changed by changing the angle of the impetus plates of the vibro-motors. The frequency is controlled by an electrical controller and the vibration angle can be varied by changing the fitting angle of the vibro-motors. They found that a wide range of fine particles down to submicron level can be fluidized fairly well at a relatively lower gas velocity and there exists an optimum vibration frequency for each powder. The authors also discussed applications of vibro-fluidized beds in drying, solvent removal, humidification and dry mixing. Dutta and Dullea[43] explained how vibration can enhance the fluidization quality of cohesive powders and attributed such effects to simultaneously increasing the bed pressure drop and bed expansion and decreasing the elutriation loss. Jaraiz et al.[29] further developed a theory to estimate the particle to particle cohesive force from pressure drop versus expansion data of packed vibrated beds of very fine particles subject to a gentle upflow of gas, which can eventually help estimate the extent of vibration required to aid the fluidization.

Examples from our studies on the effect of mechanical vibration are given in

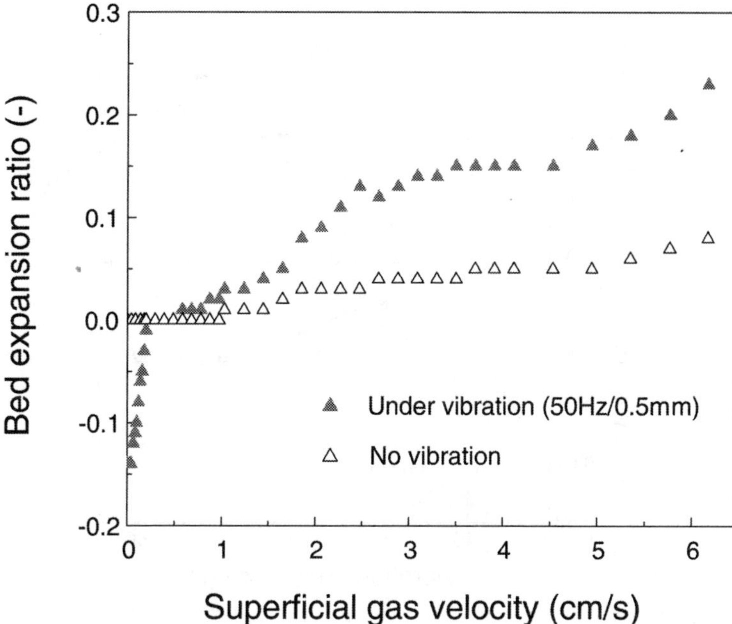

Figure 4   *A typical bed expansion for Group C powder (5.5 μm CaCO, fluidized with and without mechanical vibration)*

Figures 3 and 4, where the vibration is shown to both increase the fluidization quality and the bed expansion for 5.5 μm $CaCO_3$. Without mechanical vibration, Figure 3 shows that the bed is hardly fluidized and Figure 4 shows little bed expansion. With vibration, however, both figures indicate that the bed is well fluidized. Figure 5 further shows the effect of vibration frequency and vibration angle, suggesting that a 45° vibration angle and a frequency around 20Hz provide the best aid to fluidization with this type of particle.

The intensity of the vibration is often described by the vibration number, G, (e.g. see Marring et al.[38] and Noda et al.[44]) defined as the ratio of the maximum vibration acceleration to the acceleration due to gravity. By means of an empirical relation valid for their experimental conditions used, the vibration intensity required for good fluidization was determined versus the unconfined yield strength of the lightly consolidated powder batches.[38] The data of pressure drop and bed expansion under vibration was used to evaluate interparticle forces and the linear relation between bed pressure drop and particle coordination number was verified experimentally by Jaraiz et al.[29]

4.2   Acoustic Wave Vibration

As early as 1955, Morse[45] discovered that the fluidizability of cohesive powders can be improved by the application of acoustic fields by putting loud-speakers at the bottom of a fluidized bed. Applying low frequency and high intensity sonic energy markedly improved the fluidization of non-fluent, fine-grained powders

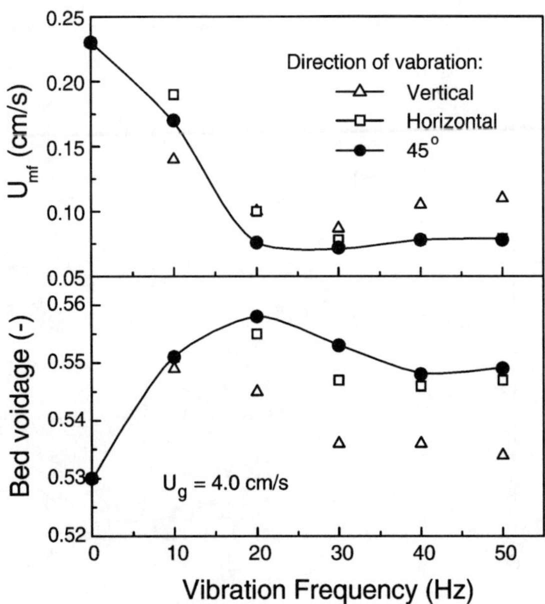

Figure 5   *Effect of vibration frequency and vibration angle on minimum fluidization velocity and on bed expansion (29 µm glass beads)*

with channelling and stagnation. Later, sound was used by Montz et al.[46] to remove micron-sized solid contaminants adhering to the surfaces of larger particles. Nowak et al.[47] used a speaker powered by an audio amplifier as the source of vibration and observed that the fluidization quality of cohesive powders is very good when low frequency (lower than 120 Hz) acoustic energy of sufficient intensity (great than 100 dB) is introduced. The minimum fluidization velocity is decreased and the fluidization quality and heat transfer are improved with the acoustic energy. Maximum effect was observed at acoustic resonance. Chirone et al.[27,48] and Russo et al.[49] published a series of papers in which they described experimental and theoretical studies on the influence of acoustic waves on fluidization of cohesive powders. They reported that beds of 11 µm-sized zeolitic catalyst, 8 µm-sized ash and 5 µm-sized talc could be homogeneously suspended in gas streams when operated in an acoustic field. It was also found that channel-free, homogeneous beds of a cohesive 0.5–45 µm catalyst powder were obtained provided the operation was carried out with appropriate combinations of bed weight and acoustic field intensity and frequency. High-intensity sound also reduced elutriation rates of fine particles from beds. Furthermore, their experimental findings indicate that acoustic fields used were not able to reduce catalyst aggregates down to the single particle level but were sufficiently powerful to break up clusters into sub-agglomerates made of a few coarse particles densely packed with hundreds of small particles. Leu et al.[30] also found similar positive influences of sound wave vibration on fluidizing eight types of Group C powders in a fluidized bed. However, they found that not all of these particles can be completely fluidized and their fluidization behaviour also depends on the static bed height. Experiments by Chirone et al.[48] further demonstrated that application of an acoustic field does not bring beds of very small pigments made of 0.16 μm A-Hr Tioxide and 0.30 μm Tr-92 Tioxide to a state of uniform fluidization. Severe channelling is observed with these solids throughout the range of superficial gas velocities tested.

Figures 6 and 7 show the experimental results from our Powder Technology Research Centre on 0.7 μm TiO_2 powder with the introduction of an acoustic wave from the bottom of the bed. Acoustic vibration is clearly seen to enhance the fluidization, leading to greater bed pressure drop and higher bed expansion. The measurements of sound pressure level above the fluidized bed further reveals that more acoustic energy is absorbed by the particle bed when fluidized at higher gas velocity. This suggests that more acoustic energy is absorbed by the fluidized bed for the improvement of the fluidization quality with increasing gas velocity.

Temkin[50] suggested the following mechanism of sound propagation in powder beds: when sound propagates through a suspension, it is attenuated by the particle-wave interaction, such as the oscillation of the particles and the scattering of the incident wave, while the energy of the sound wave decays quickly with the distance of propagation in the dense bed of particles. That's one reason why the sound wave vibrated bed is often a shallow bed.[30] The generating mechanism of impact sound between two particles and attenuation of sound in a suspension of particles have been investigated (e.g. see Temkin[50]). However, the quantitative relationships between the properties of acoustic waves and powders in this process still need further research.

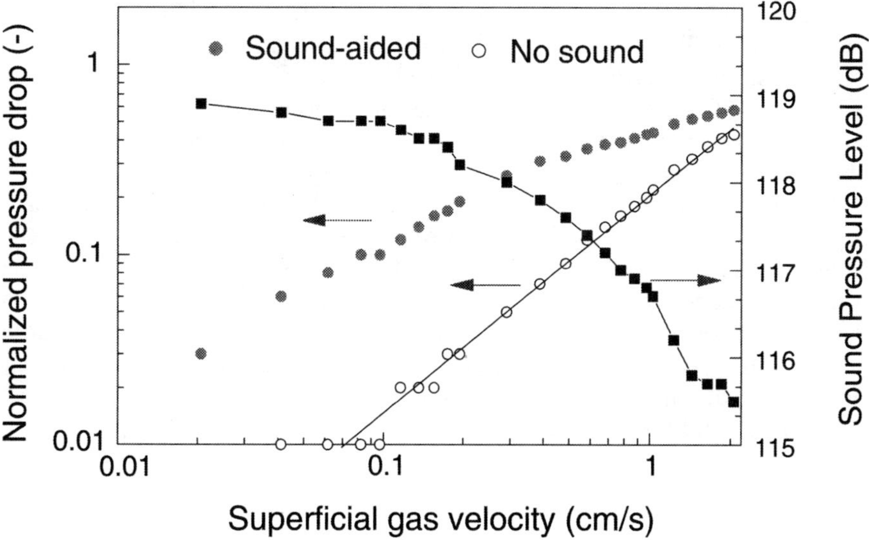

Figure 6   *Effect of acoustic vibration on the pressure drop across the bed and the attenuation of the sound waves (0.7 μm TiO₂)*

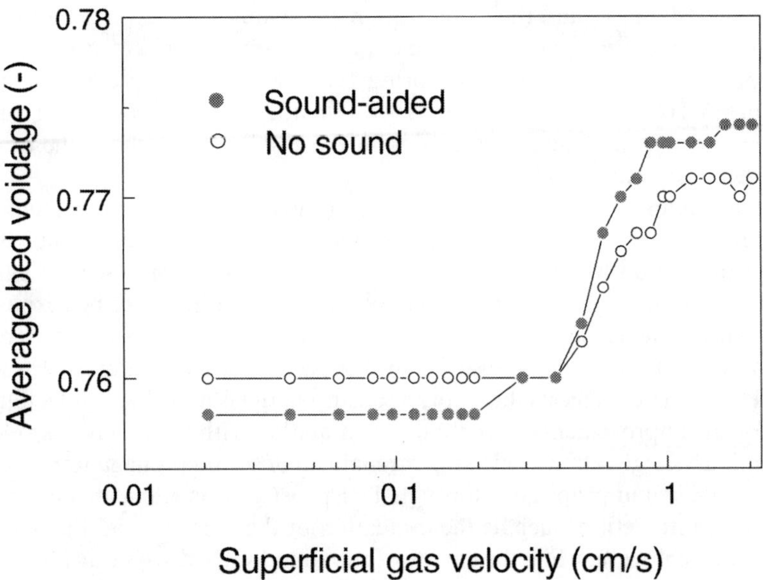

Figure 7   *Effect of acoustic vibration on the bed expansion (0.7 μm TiO$_2$)*

4.3   Mechanical Stirring and Gas Pulsation

Mechanical stirring has been found to be effective for promoting the fluidization of cohesive powders.[51] Brekken et al.[52,53] fluidized cohesive flour and starch by mechanical stirring. They also investigated the effects of stirring speed, gas velocity and bed depth on bed expansion and pressure drop. Slow agitation was confirmed to be capable of breaking down channels and enabling good quality particulate fluidization in beds of resin and silica powders.[54] More recently, mechanical agitation has been considered as a good choice in fluidizing fine particles by Nezzal et al.[55] However, it was further found that fluidization quality of very cohesive particles could not be improved if the rotation speed is higher than 100 rpm.[55]

Pulsating the fluidizing air to the fluidized bed acts in a similar manner as mechanical stirring and can also improve gas-solid contact, e.g. see Massimilla et al.[56] and Alfredson and Doig[57]. Evens et al.[58] reported a vortex fluidized bed with four secondary tangential air injections on the wall to aid the fluidization of large but cohesive particles (1mm in diameter). Recently, our research group has also invented a rotating fluidized bed, which alternates the fluidizing gas entrance to the fluidized bed through a sintered cylindrical distributor, which can not only aid the fluidization but also effectively prevent the particles from sticking to the inner bed wall.

4.4 External Magnetic Forces

Magnetic fluidization has received much attention since 1959 and an extensive review on this general topic has been provided by Liu et al.[59] and more recently by Hristov.[60] However, only two papers could be found on fluidizing Group C powders in a magnetic field. The first paper is by Zhu and Li,[61] who investigated the fluidization behaviour of Group C powders in an axial uniform magnetic field. They observed that particle chains were formed in the fluidized bed and channels and bubbles were eliminated effectively in the magnetic field. They believed that the improvement of fluidization quality owes to the formation of a chain like structure. They also studied the fluidization mechanism carefully and developed a mathematical model. The other experimental work was by Lu and Li,[62] where binary particles of $CaCO_3$ (0.37 μm) and Fe_2O_3 (0.8 μm) were fluidized well in a rotating magnetic field. It is also found that magnetic particles (0.8μm Fe_2O_3, 3.5μm Fe and 3.7μm Fe_3O_4) in a rotating magnetic field can display four kinds of particle motions: vibrating, forming rotating chain, moving around the beaker wall and remaining still.

In contrast to mechanical vibration, magnetic field disturbance improves the fluidizability of fine powders through the control of magnetic particles used in the fluidized bed. Those magnetic particles interact with fine particles and help prevent fine particles from forming large agglomerates or break the agglomerates after forming chains. The key problem with a magnetic fluidized bed is its difficulty of scale-up.

4.5 Addition of Coarser or Finer Particles as Flow Conditioners

Coarse particles in the bed of fine particles, if mobilized, can act as an additional agitator to aid the fluidization of fine particles.[63] For example, Kato et al.[64,65] developed a 'powder-particle' fluidized bed to specially fluidize fine particles by feeding the fine powders continuously into a bed in which coarse particles (Group B) are being fluidized. The presence of coarse particles acts as a turbulence promoter. This technique has been utilized in various processes, such as the continuous drying of fine particle-water slurry,[66] and catalytic pyrolysis of coal and of plant biomass.[67] Addition of coarser particles on the elutriation rate of Group C fine powders have been investigated.[68–70] Similar works by Liu and Kimura[39] in investigating possibilities of the direct use of fine silicon particles in nitrification of silicon particles, showed that entrainment of silicon particles (3.7 μm) with the addition of large particles (330 μm Si or 370 μm SiN) was controlled in the acceptable range.

The addition of finer particles can also improve the fluidization quality of Group C fine powders. As pointed out by Visser,[8] a most important parameter determining the Van der Waals attraction between particles is the separation distance. So, any means of enlarging the separation distance will substantially reduce the adhesive force and consequently improve the fluidization behaviour of fine particles. Fredrickson[71] showed that starch was made fluidizable by incorporating 0.1 wt.% of very fine tricalcium phosphate. Fried and Wheelock[72] also found that incorporation of a small amount of submicron silica (0.5–3 wt.%) in flour would make the flour less agglomerative and thereby render it fluidizable.

Steeneken at al.[73] discussed the influence of Sipernat 22S, a precipitated spray dried silica, on the mechanical properties of potato starch powder. Addition of a very small amount of such flow aid turned the cohesive starch into a free flowing powder. Degussa Corporation reported in 1984 the effectiveness of Aerosil 200 (12nm) as a flow agent in the grinding of sulfur and found that the flowability decreases when the percentage of Aerosil exceeds an optimum. Dutta and Dullea[74] mixed a small amount of highly dispersed fluidizing aids, γ aluminum oxide (29nm) or Aerosil 200 (12nm, with a charge – 7.5E-8 C/kg) into Group C powders and found a significant reduction in the cohesiveness of the powders.

It is well known that the addition of finer particles may improve the fluidization and handlability of fine Group C powders: the mechanism behind such improvement was not clear. Research work in our research centre has revealed the reason for such marked improvements and provided guidelines on how to find and manufacture the most effective finer additives to aid the fluidization of Group C powders.[75]

The above method of adding (coarser or finer) particles into Group C powders to improve their fluidization has a great potential in industrial fluidized beds since such methods are easier to apply and to scale up than most of the other fluidization aids.

4.6   Fluidization Using Different Gases and through Surface Modification

Adsorption of gas(es) on the surface of fine particles may also lead to a change of cohesiveness and therefore the fluidization quality. It has been found that due to gas adsorption on particle surfaces, fluidization behaviours of group A particles are strongly dependent on the type of fluidization gas used[76–78] and agglomeration happens when switching fluidizing gases.[79,80] Because there exists greater interparticle forces between fine particles compared with group A particles, gas adsorption undoubtedly has more influence on the fluidization behaviour of Group C fine particles. For example, Geldart et al.[14] found that the expansion of Group C powders increases with increasing gas viscosity by using different gases. Furthermore, the cohesiveness of powders was found to further increase when the pressure of the gas is raised.[77,78] However, since the results of gas adsorption on particle surfaces are different due to the differences of molecular polarity of adsorbed gases, the situation is by far not clear. Our recent work[81] has indicated that low molecular weight gas molecules may have a high potential to be absorbed on the surface of fine particles and therefore can enhance the fluidization.

Other modifications, either chemical or physical, may also changes the fluidizability of fine powders. Since surface modification has no effect on the particle properties but only on the van der Walls forces, it should be very attractive in the material and other industries.

5   Effects of Temperature on Fine Powder Fluidization

Applications of fine powder fluidization are often carried out under high temperatures, several hundred to more than one thousand degree centigrade.[41,64,82] Apparently, the hydrodynamics obtained under ambient conditions may not always hold true under high temperature operating conditions. Unfortunately, studies on the effects of temperature on the fluidization of fine particles are often carried out for specific processes of application, so that we can only provide some limited information here for reference purpose:

With increasing temperature from ambient to 973K, Morooka et al.[17] found that apparent minimum fluidization velocity and the size of agglomerates in the bed of Si_3N_4 particles all decreased. An observation on the fluidizing behaviour of ultra-fine particles of SiO_2 (7 nm) and MgO (13 nm) at elevated temperature up to 800°C by Ushiki et al.[25] demonstrated that large fluidized agglomerates are formed due to the increasing interparticle forces, and higher gas velocity is required for maintaining fluidization at high temperature. Ushiki further noticed a uniformly expanded fluidized bed was formed at an intermediate temperature when MgO particles are fluidized. Using a dimensionless fluidized agglomerate size, they found that this dimensionless size for silica powder increased with reducing initial agglomerate size and with rising temperature, but the same dimensionless size for magnesia powder decreased with rising temperature in the range below about 350°C and then increased more rapidly than that of silica powders at a temperature higher than 350°C. Recently, similar phenomena were observed by Lettieri et al.[83] In their experiments, fluidization of fine silica particles compared to larger (Group A) silica was carried out at temperature up to 650°C. It was found that the bed expansion and the deaeration time of the fine particles decreased with increasing temperature, both indicating less ideal fluidization with increasing temperature.

Apparently, different effects of temperature were reported in different cases. Two factors associated with the increase of temperature are the decreased gas density and increased viscosity. While the effects of those two factors are well predicted for the fluidization of Groups A and B powders, the reasoning is far from that simple for Group C powders, since the effect of temperature on the interparticle forces are not necessarily clear. More studies need to be carried out to accumulate more data before any firm conclusion can be made.

6   Applications/Potential Applications

Fine particle materials have attracted a great deal of attention recently due to their potential in industrial applications. Since most fine powders can be fluidized with the aid of external forces and/or additives, the utilization of fluidized beds has provided an opportunity for the development of new handling technology for fine powders. Several selected applications with the fluidization of fine powders are reviewed here.

6.1   Production of Fine Particles

Fluidized bed reactors can be used to produce fine particles. These processes often use existing promoters in the bed or reactive precursors in-situ with the reactant gas as the fluidizing gas. For example, a fluidized bed reactor has been applied to obtain aluminium nitride (AlN) crystallites smaller than 0.2 μm by nitriding aluminum particles. In this process, high-purity (99.99%) aluminum powder of 15 μm was agitated in a fluidized bed at ambient temperature, and a nitrogen steam was fed to entrain the powder into the freeboard region, which was kept at 1423–1823 K. An N_2/NH_3 gas mixture was also introduced into the freeboard. The residence time in the heated area was several seconds, and the conversion reached 100% at 1523K with 5 mol% NH_3.

Fluidized beds are also used in the chemical synthesis of nanophase materials such as mixed metal oxides and salts (see Whyte et al.[84]). They reported that the fluidization process can preserve the structural uniformity of the final product.

A Plasma-activated Spout/Fluidized Bed (PSFB) was proposed by Horio et al.[85] to produce fine alloy particles. In their system, a plasma jet was introduced upward from the conical bottom of the bed, which was operated at atmospheric pressure. Plasma flame was also injected horizontally by Cavadias and Amouroux[86] from a horizontal nozzle installed in the upper part of the fluidized bed. However, designs to improve contact between plasma gas and bed particles remain for further studies.

6.2   Modification of Powder Surface and Introduction of Specific Functions

To deposit a thin, adherent and uniformly dispersed coating onto the individual particles by Chemical Vaporization Deposition (CVD), each element of substrate surface in the dispersed heterogeneous system must be brought into contact with the coating precursor at uniform concentration and temperature. All the conditions are met in a fluidized bed reactor, provided that the particles inside the fluidized bed are uniformly fluidized. The combination of conventional fluidized bed technology with standard CVD methods has proven to be effective to coat particles. Early efforts in this area include the coating of uranium pellets with carbon[87] and the dispersion of boron carbide in a tungsten-carbide-cobalt matrix.[88] More recently, Toda and Kato[89] coated ultrafine magnesia particles with silica using a fluidized bed CVD technology, to improve acid resistance, dispersion property and sintering behaviour. Mixed tetraethyloxysilane-ethanol-water vapour was introduced in the fluidized bed with air or nitrogen as the carrier gas, and uniform coating of non-crystalline silica was achieved at 473–673°C. Kaae[90] coated particles with mixed carbide phase. Temperature and concentration gradients were virtually eliminated by making the substrate a heated bed of particles fluidized by a stream of inert gas. Metal precursors were introduced into the beds by vapour addition to the fluidizing gas or by direct injection of liquid into the bed to apply adherent coatings of aluminum on powdered mica and powdered nickel (see Sanjurjo et al.[91]). In their experiment, titanium precursors newly generated in situ by the reaction between HCl and metallic titanium were also successfully used to coat powdered mica. This CVD process has also been extended to other areas such as the production of composite particles.[92] The uniformity of deposit distribution inside the agglomerates has been found to be improved with decreasing reaction temperature and agglomerate size, and with increasing internal void fraction and the size of the primary particles.[93] Fluidized beds have also been applied to process fine particles for advanced materials such as the growth of diamonds, decarbonization of SiC, and production of polycrystalline silicon, etc., as summarized in more detail by Morooka et al.[7]

6.3 Fine Particles as Catalyst or Flow Aids

The smaller the particles, the larger the specific surface area. Fluidization of fine particles, for example, aerogel, makes the use of smaller powdered catalyst of high efficiency possible. It is well known that aerogel catalysts are produced by sol-gel methods associated with supercritical drying. The resulting catalysts, in the form of simple or mixed oxides and supported metals have very large surface areas and large pore volumes. Their excellent resistances to heat treatments allow them to be used for many types of catalytic reactions up to 450–500°C.[4] Aerogel catalysts, for example, $Cu-Al_2O_3$ aerogel catalyst in the selective conversion of cyclopentadiene into cyclopentene,[82] and $Ni-SiO_2$ aerogel catalyst in toluene hydrogenation into methylcyclohexane,[41] show in general greater activities and selectivity. Their stability with time on stream is also remarkable.[4]

Fine particles can also be used as flow aids. For example, fine powders have been used to aid the feeding of fine coal, sawdust and biomass.[94] A small fraction (0.1–5.0 wt%) of fines added to other particles, such as cohesive coal powder, can significantly reduce their tensile strength and increase the fluidity.[95] However, Kono et al.[96] also reported that there is a saturation point for the addition of Group C additives, when Group C additives begin to segregate and precipitate from the homogeneous aerated emulsion phase of the main particles.

The addition of fine particles in a fluidized bed process has also been shown to lead to increased yield. For example, Yates and Newton[97] found that increased fines in a reactor bed resulted in an increase in chemical conversion with only a slight decrease in selectivity. They explained that the increase in conversion is due to more gas passing through the emulsion phase, indicating a better fluidization.

If cohesion forces can be controlled, fluidization processes will become quite attractive for making iron powder for powder metallurgy processes.[98] Direct iron ore reduction was once excluded from fluidized processes because of the strong cohesion forces, but may now make a comeback, when more studies are carried out to understand the reason behind the cohesive forces of fluidized metal particles at high temperature.[98]

6.4 Pharmaceutical Applications

Recent development in fine particle fluidization technology has led to its application in the pharmaceutical industry. A review on particulate design for pharmaccutical powders was recently published by Jono et al.,[99] where the requirements and difficulties of fluidizing Group C pharmaceutical powders are discussed.

Mehta et al.[100] did some experiments on the coating method of tablets and fine particles. All experiments were done using fluidized bed technology and two systems were tested, an aqueous coating and an organic coating. Jones[101] also looked at coating techniques in a fluidized bed. To meet the requirement of specialized dosage forms, using a fluidized bed enables the particle (or tablet) to be coated evenly with little bridging between particles.[101] Jones also suggested that more research needs to be done on the effects of moisture and heat content in fluidized bed processes since appropriate moisture will dispel electrostatic charges but too much will lead to agglomeration.

Pulmonary drug delivery is another area where the knowledge of fine powder fluidization and handling have found and will find more applications, since the drug particles must be in the 1–5 μm range in order for them to be effectively inhaled. Another challenge is the very small dosage required for each inhalation (10 μg to 1 mg per dose). The combination of the above two makes it next to impossible to precisely meter and dispense such a small quantity of drug into the small blisters for dry powder inhalers, so that large amount of Group A particles are often mixed with the Group C drug particles to make them flow better and also to increase the volume of each dosage. However, this is not ideal since the Group A carriers mostly land in the mouth together with many Group C drug particles. With patented fine powder fluidization technologies developed in our Powder Technology Research Centre,[81,102,103] we have been able to meter and dispense a quantity as low as 20 μg of pure Group C drug particles, with an accuracy of 3% or better, without having to use any carrier particles. A new inhaler has also been developed that delivers only pure drug and at very high efficiencies which are 2–3 folds higher than those currently available on the market.

6.5 Powder Coating

Powder paint coating is a process where paint powder is directly applied to part surfaces with the aid of electrostatic charging. It is superior to liquid paint coating in that (1) it eliminates the use of petroleum-based solvents that are both expensive and harmful to the environment, and (2) it allows overspray paint powder to be recycled and reused. However, only coarse paint powders of 30–60 microns have been used so far for powder coating since finer powders make the powder extremely difficult to flow. The main problem with coarse paint powders is that they cannot form a smooth enough layer on the surface, so that the coating finish is not very uniform after curing. With our new technology on fine powder fluidization, we are able to nicely fluidize Group C paint powders in the range of 10–20 μm, the application of which has resulted in superb surface finishes that are comparable with liquid coating. This new technology has now been commercialized.

References

1. D. Geldart, 'Types of gas fluidization', *Powder Technol.*, 1973, **7**, 285–297.
2. M. Elimelech, J. Gregory, X. Jia and R. A. William, *Particle Deposition and Aggregation*, Butterworth-Heinemann, Oxford, 1995.
3. J. Bridgwater, 'Particle technology', *Chem. Eng. Sci.*, 1995, **50**, 4081–4089.
4. G. M. Pajonk, 'Aerogel catalysts', *Applied Catalysis*, 1991, **72**, 217–266.
5. A. M. Mehta, M. J. Valazza and S. E. Abele, 'Evaluation of fluidized bed processes for enteric coating systems', *Pharmaceutical Technology*, 1986, **10**, 46, 48–51, 53, 55, 56.
6. K. Wetterlin, 'Turbuhaler: A new powder inhaler for administration of drugs to the airways', *Pharm. Technol.*, 1988, **5**, 506–508.
7. S. Morooka, T. Okubo and K. Kusakabe, 'Recent work on fluidized bed processing of fine particles as advanced materials', *Powder Technol.*, 1990, **63**, 105–112.
8. J. Visser, 'An invited review – Van der Waals and other cohesive forces affecting powder fluidization', *Powder Technol.*, 1989, **58**, 1–10.
9. J. P. K. Seville, C. D. Willett and P. C. Knight, 'Interparticle Forces in fluidization: a review', *Powder Technol.*, 2000, **113**, 261–268.
10. H. C. Hamaker, 'The London-Van der Waals attraction between spherical particles', *Physica (Utrecht)*, 1937, **4**, 1058–1072.
11. M. Baerns, 'Effect of interparticle adhesive forces on fluidization of fine particles', *Ind. & Eng. Chem. Fund.*, 1966, **5**, 508–516.
12. K. Sugihara and S. Ono, 'Gavanomagnetic properties of graphite at low temperature', *J. Phys. Soc. Japan*, 1966, **21**, 631–637.
13. D. Geldart, 'The effect of particle size and size distribution on the behaviour of gas-fluidized beds', *Powder Technol.*, 1972, **6**, 201–209.
14. D. Geldart, N. Harnby and A. C. Y. Wong, 'Fluidization of cohesive powders', *Powder Technol.*, 1984, **37**, 25–37.
15. D. Geldart and A. C. Y. Wong, 'Fluidization of powders showing degrees of cohesiveness I: bed expansion', *Chem. Eng. Sci.*, 1984, **39**, 1481–1488.
16. J. Chaouki, C. Chavarie and D. Klvana, 'Effect of interparticle forces on the hydro-dynamic behaviour of fluidized aerogels', *Power Technol.*, 1985, **43**, 117–125.
17. S. Morooka, K. Kusakabe, A. Kobata and Y. Kato, 'Fluidization state of ultrafine powders', *J. Chem. Eng. of Japan*, 1988, **21**, 41–46.
18. S. Mori, A. Yamamoto, S. Iwata, T. Haruta and I. Yamada, 'Vibro-fluidization of Group-C particles and its industrial applications', *AIChE Symp. Ser.*, 1990, **86** (276), 88–94.
19. D. Geldart and A. C. Y. Wong, 'Fluidization of powders showing degrees of cohesiveness II: experiments on rates of de-aeration', *Chem. Eng. Sci.*, 1985, **40**, 653–661.
20. S. R. Iyer and L. T. Drzal, 'Behavior of cohesive powders in narrow-diameter fluidized bed', *Powder Technol.*, 1989, **57**, 127–133.
21. Z. L. Wang, M. Kwauk and H. Z. Li, 'Fluidization of fine particles', *Chem. Eng. Sci.*, 1998, **53**, 377–395.
22. O. Molerus, 'Interpretation of Geldart's Type A, B, C and D powders by taking into account interparticle cohesion forces', *Powder Technol.*, 1982, **33**, 81–87.

23. A. W. Pacek and A. W. Nienow, 'Fluidization of fine and very dense hard metal powders', *Powder Technol.*, 1990, **60**, 145–158.

24. B. Hua, L. M. Hu, C. Z. Li and H. Z. Li, 'Process of collapse and expansion of ultrafine powder fluidized bed', *J. of East China Univ. of Sci. and Technol.* (Chinese), 1994, **20**, 290–294.

25. H. Ushiki, 'Non-uniformity space time concept in condensed matters', *Kobunshi*, 1995, **44**, 726–731.

26. M. Horio, A. Saito, K. Unou, H. Nakazono, N. Shibuya, S. Shima and A. Kosaka, 'Synthesis of diamond particles with an acetylene circulating fluidized bed', *Chem. Eng. Sci.*,1996, **51**, 3033–3038.

27. R. Chirone, L. Massimilla and S. Russo, 'Bubble-free fluidization of a cohesive powder in an acoustic field', *Chem. Eng. Sci.*, 1993, **48**, 41–52.

28. K. Rietema, 'Powders, What are they?', *Powder Technol.*, 1984, **37**, 5–23.

29. E. Jaraiz, S. Kiruma and O. Levenspiel, 'Vibrating beds of fine particles: estimation of interparticle forces from expansion and pressure drop experiments', *Powder Technol.*, 1992, **72**, 23–30.

30. L. P. Leu and C. T. Huang, 'Fluidization of cohesive powders in a sound waves vibrated fluidized bed', *AIChE Symp. Ser.*, 1994, **90** (301), 124–141.

31. K. Nishii, Y. Itoh, N. Kawakami and M. Horio, 'Pressure swing granulation, a novel binderless granulation by cyclic fluidization and gas compaction', *Powder Technol.*, 1993, **74**, 1–6.

32. H. O. Kono, L. Richman, J. Su and D. Smith, 'Characterization of flow properties of very fine powders at ambient and elevated temperature [a novel experimental and theoretical approach]', *AIChE Symp. Ser.*, 1996, **93**, 141–146.

33. T. Zhou and H. Li, 'Estimation of agglomerate size for cohesive particles during fluidization', *Powder Technol.*,1999, **101**, 57–62.

34. W. Pietsch, 'Readily engineer agglomerates with special properties from micro- and nanosized particles', *Chem. Eng. Progress*, 1999, **95**, 67–81.

35. H. O. Kono, C. Matsuda and D. C. Tian, 'Agglomeration cluster formation of fine powders in gas-solid two phase flow', *AIChE Symp. Ser.*, 1990, **86** (276), 72–77.

36. H. Li, R. Legros, C. H. M. Brereton, J. R. Grace and J. Chaouki, 'Hydrodynamic behavior of aerogel powders in high velocity fluidized beds', *Powder Technol.*, 1990, **60**, 121–129.

37. R. J. Dry, M. R. Judd and T. Shingles, 'Two-phase and fine powders', *Powder Technol.*, 1983, **34**, 213–223.

38. E. Marring, A. C. Hoffmann and L. P. B. M. Janssen, 'The effect of vibration on the fluidization behavior of some cohesive powders', *Powder Technol.*, 1994, **79**, 1–10.

39. Y. D. Liu and S. Kimura, 'Fluidization and entrainment of difficult-to-fluidize fine powder mixed with easy-to-fluidize large particles', *Powder Technol.*, 1993, **75**, 189–196.

40. G. Y. Zhao, C. W. Zhu and V. Hlavacek, 'Fluidization of micron-size powders in a small-diameter fluidized bed', *Powder Technol.*, 1994, **79**, 227–235.

41. D. Klvana, J. Chaouki, M. Repellin-Lacroix and G. M. Pajonk, 'New method of preparation of aerogel-like materials using a freeze-drying process', *Journal of Physique, Colloque*, 1989, (C4, Proc. Int. Symp. Aerogels-ISA 2, 2nd, 1988).

42. R. D. Venkatesh, J. Chaouki and D. Klvana, 'Fluidization of cryogels in a conical column', *Powder Technol.*, 1996, **89**, 179–186.

43. A. Dutta and L. V. Dullea, 'Effects of external vibration and the addition of fibers on the fluidization of a fine powder', *AIChE Symp. Ser.*, 1991, **87** (281), 38–46.

44. K. Noda, Y. Mawatani and S. Uchida, 'Flow pattern of fine particles in a vibrated

fluidized bed under atmospheric or reduced pressure', *Powder Technol.*, 1998, **99**, 11–14.

45. R. D. Morse, 'Sonic energy in granular solid fluidization', *Ind. Eng. Chem.*, 1995, **47**, 1170–1175.

46. K. W. Montz and J. K. Beddow, 'Adhesion and removal of particulate contaminants in a high-decibel acoustic field', *Powder Technol.*, 1988, **55**, 133–140.

47. W. Nowak and M. Hasatani, 'Fluidization and heat transfer of fine particles in an acoustic filed', *AIChE Symp. Ser.*, 1993, **89** (296), 137–149.

48. R. Chirone and L. Massimilla, 'Sound-assisted aeration of beds of cohesive solids', *Chem. Eng. Sci.*, 1994, **49**, 1185–1194.

49. P. Russo, R. Chirone, L. Massimilla and S. Russo, 'The influence of the frequency of acoustic waves on sound-assisted fluidization of fine particles', *Powder Technol.*, 1995, **82**, 219–230.

50. S. Temkin, *Elements of Acoustics*, John Wiley & Sons, New York, 1981, p. 455.

51. T. M. Reed III and M. R. Fenske, 'Effects of agitation on gas fluidization of solids', *Ind. Eng. Chem.*, 1955, **47**, 275–282.

52. R. A. Brekken, E. B. Lancaster and T. D. Wheelock, 'Fluidization of flour in a stirred, aerated bed I. General fluidization characteristics', *Chemical Eng. Progress, Symposium Series* 1970, **66** (101), 81–90.

53. R. A. Brekken, E. B. Lancaster and T. D. Wheelock, 'Fluidization of flour in a stirred, aerated bed II. Solids mixing and circulation', *Chemical Eng. Progress, Symposium Series*, 1970, **66** (105), 277–284.

54. K. Godard and J. F. Richardson, 'Use of slow speed stirring to initiate particulate fluidization', *Chem. Eng. Sci.*, 1969, **24**, 194–195.

55. A. Nezzal, J. F. Large and P. Guigon, 'Fluidization behavior of very cohesive powders under mechanical agitation', *Fluidization VIII*, Engineering Foundation, New York, 1999, 77–82.

56. L. Massimilla, G. Volpicelli and G. Raso, 'Pulsing gas fluidization beds of particles', *Chem. Eng. Progr. Ser.*, 1966, **62**, 63–70.

57. P. G. Alfredson and I. D. Doig, 'Behavior of pulsed fluidized beds: II Bed Contraction', *Transaction of the Institute of Chemical Engineers*, 1973, **51**, 241–245.

58. P. L. Evans, J. P. K. Seville and R. Clift, 'Mobilization of cohesive particles in a novel spouted bed', *Fluidization VI*, Engineering Foundation, New York, 1989, 253–260.

59. Y. A. Liu, R. K. Hamby and R. D. Colberg, 'Fundamental and practical developments of magnetofluidized beds: A review', *Powder Technol.*, 1991, **64**, 3–42.

60. J. Y. Hristov, 'Fluidization in a magnetic field is more than 40 years old. It is time to strike a balance before the new century', *Recent Progress en Genie des Procedes*, 2000, **14**, 251–260.

61. Q. S. Zhu and H. Z. Li, 'Study on magnetic fluidization of Group C powders', *Powder Technol.*, 1996, **86**, 179–185.

62. X. Lu and H. Li, 'Fluidization of $CaCO_3$ and Fe_2O_3 particle mixtures in a transverse rotating magnetic field', *Powder Technol.*, 2000, **107**, 66–78.

63. T. Zhou and H. Li, 'Effect of adding different size particles on fluidization of cohesive particles', *Powder Technol.*, 1999, **102**, 215–220.

64. K. Kato, T. Haruta, A. Koshinuma, I. Kanazawa and T. Suguhara, 'Decarbonization of silicon carbide-carbon fine particles mixture in a fluidized bed', in *Fluidization VI*, *AIChE*, New York, 1989, p. 351.

65. K. Kato, T. Takarada, N. Matsuo, T. Suto and N. Nakagawa, 'Residence time distribution of fine particles in a powder-particle fluidized bed', *Inter. Chem. Eng.*, 1994, **34**, 605–610.

66. N. Nakagawa, K. Ohsawa, T. Takarada and K. Kato, 'Continuous drying of a fine particles-water slurry in a powder-particle fluidized bed', *J. Chem. Eng. of Japan*, 1992, **25**, 495–501.

67. T. Takarada, T. Tonishi, Y. Fusegawa, N. Nakagawa and K. Kato, 'Catalytic pyrolysis of coal in a powder-particle fluidized bed', Fluid. VII, *Proc. Eng. Found. Conf. Fluid, 7th*, 1992, 515–523.

68. X. X. Ma, Y. Honda, N. Nakagawa and K. Kato, 'Elutriation of fine powders from a fluidized bed of a binary particle-mixture', *J. Chem. Eng. of Japan*, 1996, **29**, 330–335.

69. X. X. Ma and K. Kato, 'Effect of interparticle adhesion forces on elutriation of fine powders from a fluidized bed of a binary particle-mixture', *Powder Technol.*, 1998, **95**, 93–101.

70. D. Taneda, H. Takahagi, S. Aoshika, N. Nakagawa and K. Kato, 'Elutriation of fine particles in a powder-particle fluidized bed', *Kagaku Kogaku Ronbunshu*, 1998, **24**, 418–424.

71. R. E. C. Fredrickson, *U.S. Patent* 3,003,894, 1961.

72. E. M. Fried and T. D. Wheelock, 'Fluidized bed characteristics of wheat flour', *Chem. Eng. Prog. Symp. Ser.*, 1966, **62** (69), 114.

73. P. A. M. Steeneken, A. J. J. Woortman, A. H. Gerritsen and H. Poort, 'The Influence of flow conditioners on some mechanical properties of potato starch powder', *Powder Technol.*, 1986, **47**, 239–247.

74. A. Dutta and L. V. Dullea, 'A comparative evaluation of negatively and positively charged submicron particles as flow conditioners for a cohesive powder', *AIChE Symp. Ser.*, 1990, **86** (276), 26–40.

75. J. Zhu and H. Zhang, 'Fluidization Additives to Fine Powders', *U.S. Patent*, applied, July 2002.

76. D. Geldart and A. R. Abrahamsen, 'Homogeneous fluidization of fine powders using various gases and pressures', *Powder Technol.*, 1978, **19**, 133–136.

77. H. W. Piepers, E. J. E. Cottaar, A. H. M. Verkooijen and K. Rietema, 'Effects of pressure and type of gas on particle-particle interaction and the consequences for gas-solids fluidization behavior', *Powder Technol.*, 1984, **37**, 55–70.

78. H.-Y. Xie, 'The role of interparticle forces in the fluidization of fine particles', *Powder Technol.*, 1997, **94**, 99–108.

79. K. Rietema and J. Hoebink, 'Transient phenomena in fluidized beds when switching the fluidizing agent from one gas to another', *Powder Technol.*, 1977, **18**, 257–265.

80. T. Kai and T. Takahashi, 'Formation of particle agglomerates after switching fluidizing gases', *J. Chem. Eng. of Japan*, 1997, **43**, 357–362.

81. J. Zhu, J. R. Grace and S-Y. Jiao, 'Novel Fluidization Aids', *U.S. Patent*, 6,212,794, 2001.

82. J. Chaouki, C. Chavarie, D. Klvana and G. Pajonk, 'Study of selective hydrogenation of cyclopentadieneon fluidized copper/alumina aerogel', *Canadian Journal of Chem, Eng.*, 1986, **64**, 440–446.

83. P. Lettieti, J. G. Yates and D. Newton, 'On the fluidization of some industrial catalysts at high temperature: the effect of interparticle forces', *AIChE Symp. Ser.*, 1999, **321**, 100–105.

84. J. R. Whyte, B. H. Kear, L. E. McCandlish and B. K. Kim, 'Preparation of nanophase WC-Co composite powders in a fluidized bed reactor', *AIChE Symp. Ser.*, 1992, **88**, 116–121.

85. M. Horio, R. Hanaoka and M. Tsukada, 'A plasma-spout/fluidized bed as a new powder processing reactor', *Proceeding of Japanese Symposium on Plasma Chemistry*, 1988, **1**, 213–218.

86. S. Cavadias and J. Amouroux, 'Synthesis of nitrogen oxides in plasma reactors', *Bulletin de la Societe Chimique de France*, 1986, **2**, 147–158.

87. J. Guilleray, R. L. R. Lefevre, M. S. T. Price and J. P. Thomas, *Proc. 5ᵗʰ Int. Conference on Chemical Vapor Deposition, The Electrochemical Society*, NJ, 1975, 727–748.

88. G. Arthur and M. P. Johnson, Proc. 5ᵗʰ Int. Conf. On Chemical Vapor Deposition, The Electrochemical Society, NJ, 1975, 509–522.

89. Y. Toda and A. Kato, 'Silica coating of ultrafine magnesia', *Ceramics International*, 1989, **15**, 161–166.

90. J. L. Kaae, 'Codeposition of compounds by chemical vapor deposition in a fluidized bed of particles', *Ceramic Eng. and Sci. Proceedings*, 1988, **9**, 1159–1168.

91. A. Sanjurjo, B. J. Wood, K. H. Lau, G. T. Tong, D. K. Choi, M. C. H. Mckubre, H. K. Song and N. Church, 'Titanium-based coating on copper by chemical vapor deposition in fluidized bed reactors', *Surface and Coating Technology*, 1991, **49**, 110–115.

92. S. Morooka, A. Kobata and K. Kusakabe, 'Rate analysis of composite ceramic particles production by CVD reactions in a fluidized bed', *AIChE Symp. Ser.*, 1991, **87** (281), 32–37.

93. B. Golman and K. Shinohara, 'Modeling of CVD coating inside agglomerate of fine particles', *J. Chem. Eng. of Japan*, 1998, **31**, 103–110.

94. D. S. Scott and J. Piskorz, 'Low rate entrainment feeder for fine solids', *Ind. Eng. Chem. Fundam.*, 1982, **21**, 319–322.

95. H. O. Kono, C. C. Huang, M. Xi and F. D. Shaffer, 'The effect of flow conditioners on the tensile strength of cohesive powders structure', *AIChE Symp. Ser.*, 1989, **85** (270), 44–48.

96. H. O. Kono, C. C. Huang, E. Morimoto, T. Nakayama and T. Hikosaka, 'Segregation and agglomeration of Type C powders from homogeneously aerated Type A–C powder mixtures during fluidization', *Powder Technol.*, 1987, **53**, 163–168.

97. J. G. Yates and D. Newton, 'Fine particle effects in a fluidized bed reactor', *Chem. Eng. Sci.*, 1986, **41**, 801–806.

98. T. Mikami, H. Kamayi and M. Horio, 'The Mechanism of de-fluidization of iron particles in a fluidized bed', *Powder Technol.*, 1996, **89**, 231–238.

99. K. Jono, H. Ichikawa, M. Miyamoto and Y Fukumori, 'A review of particulate design for pharmaceutical powders and their production by spouted bed coating', *Powder. Technol.*, 2000, **113**, 269–277.

100. A. M. Mehta, M. J. Valazza and S. E. Abele, 'Evaluation of fluid-bed processes for enteric coating systems', *Pharm. Technol.*, 1986, **10**, 46, 48–51, 53, 55–56.

101. D. M. Jones, 'Factors to consider in fluid-bed processing', *Pharm. Technol.*, April 1985, 50–62.

102. J. Zhu, J. R. Grace and N. Pourkavoos, 'Dispensing of Ultra-Fine Particles in a Reproducible Manner', *U.S. Patent*, 6,183,169, 2001.

103. J. Zhu, J. Z. Wen, Y. Ma and H. Zhang, 'Apparatus for Volumetric Metering of Small Quantity of Powder from Fluidized Beds', *U.S. Patent*, applied, 2001.

CHAPTER 11

The Kinetics of High-Shear Granulation

G. K. REYNOLDS, C. F. W. SANDERS, A. D. SALMAN and M. J. HOUNSLOW

Particle Products Group, Department of Chemical and Process Engineering, University of Sheffield, Sheffield S1 3JD

Nomenclature

b	Breakage function or binder quantity
B	Birth rate
C	Compaction rate
d	Particle/granule diameter
D	Death rate
e	Coefficient of restitution (ratio of incident particle momentum to rebound momentum)
f	Arbitrary function
h	Viscous layer thickness
h_a	Characteristic surface roughness length scale
k_v	Shape factor
l	Particle/granule diameter
m	Granule mass
M	Tracer mass
n	Particle/granule number
n_{eq}	Number of size intervals
N	Particle/granule number
S	Selection rate constant
St_v	Viscous Stokes' number (see Equation 1)
St_v^*	Critical viscous Stoke's number (see Equation 2)
t	Time
u	Particle/granule size (volume)
u_0	Initial velocity
v	Particle/granule size (volume)
w	Moisture content

| W | Wetting rate |
| x | Mass fraction of component A |

Greek letters		*Superscripts*		*Subscripts*	
β	Aggregation rate constant or kernel	agg	Aggregation	0	of size distribution
β_0	Time dependent aggregation rate constant	A	Aggregation	1	of tracer distribution
ε	Granule porosity	break	Breakage	c	critical
Φ	Binder volume fraction	B	Breakage	i	of size class i
μ	Binder viscosity			L	liquid/binder
ρ	Granule density				
Ψ	Collision efficiency (see Equation 29)				

1   Introduction

Granulation is a process used in many industries. It is used for improving the properties of fine powders in terms of handling, transport and mixing. Granulation typically involves size enlargement of fine powders through mixing with a liquid phase, termed a binder. High shear granulation specifically achieves this through intense mechanical agitation with an impeller, and often an additional device termed a chopper. Granules produced from a high shear granulation process are typically dense and relatively strong and spherical.

Successful modelling of high shear granulation requires understanding the properties and behaviour of the particulate material within the granulator and the rates of the processes that change these properties. Properties affecting granule behaviour within the granulator can include size and composition. Composition can include different primary particles, binder content and porosity. Rate processes in the system can include nucleation, aggregation and breakage.

In approaching the modelling of granulation, it becomes important to select the granule properties that dominate these rate processes and determine at what length-scale these properties should be considered. For example should we consider granule properties on a size-averaged basis, or do we need to consider properties on a single granule or even intra-granule basis.

Once the length scale of the model is decided, it is then necessary to obtain kinetic information of the dominant mechanisms in the process. This is often limited by what can be realistically measured from experimental work.

This chapter introduces and reviews various modelling approaches at different length scales for the prediction of granulation under high shear conditions.

2   Micro-scale Approaches

Micro-scale approaches aimed at predicting the coalescence rate of granules within a given system were first attempted by Ouchiyama and Tanaka.[1,2] They attempted to define the probability of coalescence. Coalescence was determined based on the tensile stress exerted at the area of contact due to the moment of the separating forces. Applying Hertzian theory of elastic bodies and a relation for plastic bodies, they obtain an equation for the probability of successful coalescence. This approach was the first to attempt to provide a physical basis in the definition of coalescence rates and granule interaction. It incorporates granule strength and deformability, but fails to take account of surface liquid effects. Some parameters used in the formulation can be calculated from the granule properties, but some require experimental determination, adding further burden in using this approach.

Ennis *et al.* present the next attempt at a micro-scale approach to the determination of granule coalescence.[3] The basis for their approach is the argument that only colliding granules with insufficient energy to rebound will coalesce. They assume that binder is uniformly distributed over the surface of spherical, mono-sized granules. Collisions are elastic, with viscous forces dominating over negligible capillary contributions. These assumptions and a consideration of the force balance lead to a viscous Stokes' number, with granule density ρ, granule diameter d, initial velocity u_0, and binder viscosity μ:

$$St_v = \frac{8\rho du_0}{9\mu} \tag{1}$$

They then define a critical Stokes' number, above which coalescence does not occur, incorporating the coefficient of restitution e, the viscous layer thickness h, and taking account of surface asperities in terms of some characteristic length scale, h_a:

$$St_v^* = \left(1 + \frac{1}{e}\right)\ln\left(\frac{h}{h_a}\right) \tag{2}$$

This approach manages to capture the importance of the dynamic viscous liquid bond between particles within granulating systems, but the assumptions made simplify the model to an idealised situation.

Increasing numbers of theoretical approaches to coalescence have been developed more recently. Typically these models make use of an energy or force balance to predict whether a collision between two granules will lead to a rebound or coalescence. With a more theoretical basis, these models are much more dependent on knowledge of the collision environment and the properties of the colliding granules. This leads to the requirement for further experimental data in order to determine the overall aggregation rate constant. Work in this area includes that by Moseley and O'Brien,[4] Simons *et al.*,[5] Tardos *et al.*,[6] Lian *et al.*,[7] Seville *et al.*,[8] Thornton and Ning,[9] and Liu *et al.*[10]

Experimental work to determine the properties and behaviour of colliding granules is also being conducted. Iveson and Litster studied granule consolidation experimentally.[11,12] They developed an experimental methodology to study the formulation properties on granulation behaviour. They also conducted experiments using cylindrical pellets made from glass ballotini and a variety of binders (water, surfactant solutions and glycerol).[13] These were impacted on a flat surface by being dropped from heights ranging from 10 to 30cm. By measuring the deformation area after impact, they calculated the granule yield stress. They concluded that there are three energy dissipation mechanisms affecting impact deformation. These are attributed to inter-particle friction, capillary and viscous forces. Prediction of the binder content effect on the yield stress is not possible unless the balance of these forces is known. Fu *et al.* studied the deformation and rebound behaviour of wet granules (i.e. granules with liquid binder) under impact conditions.[14] The main focus of the study was to record the coefficient of restitution by measuring the impact and rebound speed on a semi-infinite rigid target. A wide variety of wet granules were manufactured by varying production time, binder to solid ratio, and primary particle size with an average granule size of 6.5mm. The effect of these parameters on the coefficient of restitution was interpreted. This experimental work is a significant step towards understanding the behaviour of wet granules to improve the modelling of granule interaction.

A micro-scale approach to modelling breakage contains a much smaller body of literature than aggregation. However, more recently breakage is being considered increasingly important in certain systems.

3   1-D Population Balance Models

A population balance is a statement of continuity, typically for a particulate system, which describes how the number population varies with time and space. In particular for granulation processes a one-dimensional population balance model will be a statement of continuity for the change of a size distribution of particles with time. The one-dimensional term for this type of population balance refers to its concern with just one particle property, typically size usually expressed as volume. The key assumption to note, inherent in all one-dimensional population balances, is that the variable of concern, typically size, is the only variable that has a significant effect on the kinetics of the rate process. Changes in all other properties, such as binder content and porosity in granular systems, are considered to have no impact on the rate process. Whether or not this is an appropriate assumption will be also be considered below.

A population balance can be considered analogous to energy and mass balances that are generally involved in the design of process engineering unit operations. Instead of balancing energy or mass over a given system, number is the focus. Conceptually and in the case of granulation processes, this leads to it being necessary that the change in the number of granules within a system needs to be accounted for through mechanisms that create and destroy individual granules. These mechanisms can be generalised as nucleation, coalescence and breakage. Nucleation is the creation of a particle or granule of a given finite size for incorporation into a population balance, leading to an increase in the number of particles in a system. Hounslow explains that in modelling granulation, it can be allowed that the very fine primary powder is not considered particulate, and nucleation is the mechanism by which grains of this powder clump together to create a new granule.[15] Effectively the modelled granulation process occurs within a continuum of fine primary particles. In this case it is also necessary to allow that size enlargement of granules by acquisition of primary particles may occur. In the granulation literature this process is termed layering. Aggregation is the process by which several discrete granules combine to form a single conglomerate, leading to a decrease in the number of granules in a system. Finally breakage leads to a net increase in the number of granules in a system, as one granule is lost and substituted for the many fragments it creates.

The application of population balances to process engineering problems first began being used in the 1960s. Ramkrishna[16] attributes Valentas *et al.*[17] as the first to consider population balance analysis of breakage and coalescence processes in dispersed phase systems in chemical engineering literature. At a similar time, Hulbert and Katz derived expressions for the creation and loss of crystals within a supersaturated solution.[18] They derived a birth term $(B(v,t))$ for the creation of crystals of size v at a time t by aggregation of two crystals of initial sizes u and v-u:

$$B_{\mathrm{agg}}(v,t) = \frac{1}{2}\int_0^v \beta(u,v-u,t)\, n(u,\,t)\, n(v-u,t)du \tag{3}$$

The aggregation or coalescence rate is represented by β. This can be described as a second order rate constant, or kernel, as it takes account of the size of the two aggregating bodies. They also derived a death term $(D(v,t))$ for the loss of particles of size v at time t through aggregation with another particle:

$$D_{\mathrm{agg}}(v,t) = n(v,t)\int_0^\infty \beta(u,v,t)n(u,t)du \tag{4}$$

These terms of the creation and destruction of crystals due to a coalescence mechanism can also be used in other processes. In high-shear granulation, granules can coalesce leading to a change in the number of particles described by these terms.

For batch aggregation processes, Randolf and Larson expressed the overall population balance equation as:[19]

$$\frac{\partial n(v,t)}{\partial t} = B - D \tag{5}$$

In this balance, $n(v,t)$ is the number density function of particles of volume v. This number density function changes by the net effects of birth terms (B) that create new particles of size v and death terms (D) that remove particles of size v. Conceptually this statement of continuity forming a population balance is straightforward. However, in applying and using Population Balance Equations

(PBE's), the difficulty arises in determining the birth and death terms and quantifying their rates.

In granulation processes, the main mechanisms leading to change in the number density function can be categorised as nucleation (i.e. creation of new granules from primary particles and binder), aggregation (i.e. granules coming together to form larger conglomerates), or breakage (i.e. granules decomposing into two or more fragments). A review of approaches to including these mechanisms in population balances follows.

3.1 Nucleation

As stated above, nucleation in high shear granulation can be considered the process by which grains of primary powder clump together (usually with the aid of a binder) in order to create a distinct granule. Recently a lot of work has focussed on the mechanisms of nucleation in granulation (for example Litster *et al.*[20]). However, there is little information available on the quantifying of nucleation rates in granulation. Due to the lack of suitable treatment of nucleation, it is usually not included in population balance modelling of high shear granulation processes, in favour of an initial size distribution taken after the initial 'mixing and nucleation' time.

3.2 Aggregation

Historically a binary aggregation rate constant or 'kernel', β, has been used to represent the rate of aggregation of two particles of given sizes (see above description of Hulbert and Katz[18]). Sastry developed this approach further by dividing β into two parts:[21]

$$\beta(u,v,t) = \beta_0(t)\beta(u,v) \tag{6}$$

The first size independent term, β_0, can be considered to account for operating and formulation parameters (such as impeller speed and binder ratio). The second term expresses the effect of size on the rate of aggregation of two particles. This decomposition allows for the selection of size-dependent aggregation rate relationships with a more physical basis, without needing to include parameters specific to the granulation process being studied. The dependencies based on the particular process can be then incorporated into β_0 to allow an overall aggregation rate to be obtained. This approach of obtaining aggregation rates has been widely used, and a number of theoretical and empirical terms have been developed for the size-dependent term, described as follows.

Table 1 shows a list of a number of physically based aggregation kernels. The simplest kernel is the size-independent kernel, where it is assumed there is no size dependence on aggregation rate. In this case every granule has an equal opportunity to aggregate with another. Smoluchowski,[22] in his study of colloidal systems, derived the perikinetic and orthokinetic kernels. The orthokinetic kernel, derived for systems under shear is particularly appropriate for high-shear

Table 1   *Summary of physical aggregation kernels*

Kernel ($\beta(u,v)$)	Name	Basis	Reference
1	Size independent	Size independent	
$u + v$	Sum	Coagulation of rain drops	Golovin (1963)[23]
$(u^{1/3} + v^{1/3})^3$	Orthokinetic	Shear	Smoluchowski (1917)[22]
$(u^{1/3} + v^{1/3})(u^{-1/3} + v^{-1/3})$	Perikinetic	Random/Brownian motion	Smoluchowski (1917)[22]
$(u^{1/3} + v^{1/3})^2 \cdot \sqrt{\left(\dfrac{1}{u^2} + \dfrac{1}{v^2}\right)}$	EME	Equipartition of momentum	Hounslow (1998)[24]
$(u^{1/3} + v^{1/3})^2 \cdot \sqrt{\left(\dfrac{1}{u} + \dfrac{1}{v}\right)}$	EKE	Kinetic theory of gases – equipartition of kinetic energy	Hounslow *et al.* (2001)[25]

granulation processes. It can also be seen that the shear kernel of Smoluchowski can be approximated by the sum kernel of Golovin.[23] Aggregation between similar sized particles gives some discrepancy, but this reduces with more disparate particles. Similarly the size independent kernel can be used to approximate the perikinetic kernel. These particular kernels also show bias towards large-large particle interactions. Hounslow presents a kernel based on the kinetic theory of gases.[24] In this case it is assumed that each granule is subject to the same random fluctuation impulses so that the random component of translational momentum will be constant. From this the random component of velocity will be inversely proportional to the granule mass. The equipartition of momentum energy kernel is shown in Table 1. Hounslow *et al.* also use a kernel based on the kinetic theory of gases.[25] In this kernel it is assumed that the random fluctuations of kinetic energy of the particles is constant. The equipartition of kinetic energy kernel is also shown in Table 1.

In addition to the physically based kernels described, empirical kernels have also been developed and used to model aggregation in granulation processes. The product kernel is a simple kernel where the rate depends on the product of the sizes of the two aggregating particles (a variation on Golovin's sum kernel):

$$\beta(u,v) = uv \tag{7}$$

Kapur developed an empirical relation with constants a and b that would need to be determined for a given system:[26]

$$\beta(u,v) = \frac{(u + v)^a}{(uv)^b} \tag{8}$$

Sastry proposed an alternative empirical kernel, which has some similarity to the equipartition of kinetic energy kernel (Table 1).[27]

$$\beta(u,v) = (u^{2/3} + v^{2/3})\left(\frac{1}{u} + \frac{1}{v}\right) \tag{9}$$

Adetayo and Ennis developed a semi-empirical kernel to model a variety of contradictory experimental observations found in drum granulation processes.[28] This kernel is size independent for particles with a size less than a critical 'cut off' value. The critical size is based on consideration of deformation theory and micro-level Stokes analysis (see section on micro-scale approaches). This is a significant step towards unifying the micro-scale approach to modelling coalescence rates with the traditional approach to developing the population balance model.

3.3 Breakage

Breakage in granulating systems has been observed experimentally, but it is not until recently that breakage has been incorporated into the modelling of these processes. Annapragada and Neilly,[29] note qualitatively that the modelling of only aggregation terms in batch granulation processes is not sufficient to describe the system.

Breakage kinetics can be described by two functions. The first function, the selection rate constant, $S(v,t)$, describes the rate of breakage of particles of a given size. The second function, the breakage function, $b(u,v)$, describes the sizes of the fragments from the breaking particle. As for aggregation, the birth and death terms due to breakage can be defined for inclusion in the population balance equation. The death rate due to breakage can be written as:

$$D_{\text{break}}(v,t) = S(v,t)n(v,t) \tag{10}$$

The birth rate is the weighted sum of the death rates of larger particles that produce fragments of size v:

$$B_{\text{break}}(v,t) = \int_{v}^{\infty} b(v,u)D(u,t)du = \int_{v}^{\infty} b(v,u)S(u,t)n(u,t)du \tag{11}$$

These terms can be added to the birth and death terms in the population balance equation in order to include breakage processes.

Pearson *et al.* conducted experimental studies of breakage using tracers in high-shear granulation.[30] They developed this work (Hounslow *et al.*[25]) to determine the selection rate constant, $S(v,t)$, and the breakage function $b(u,v)$. From their work, they found the selection rate constant to be dependent on granule age (i.e. time), but not granule size. They approximated the breakage function, $b(u,v)$, by using two overlapping log-normal distributions based on the distribution of fragment sizes in tracer studies.

3.4 Solution of 1-D Population Balances

Combining the aggregation and breakage terms discussed, we have a population balance similar to:

$$\frac{\partial n(v,t)}{\partial t} = \frac{1}{2}\int_0^v -\beta(u,v-u,t)n(u,t)n(v-u,t)du - n(v,t)\int_0^\infty \beta(u,v,t)n(u,t)du$$

$$+ \int_v^\infty b(v,u)S(u,t)n(u,t)du - S(v,t)n(v,t) \tag{12}$$

Typically the solution of a population balance equation requires what is termed the inverse problem. This is where the kinetics are determined based on experimental data giving the particle size distribution. This is frequently reduced to a fitting problem and in many cases a number of expressions with different rate constants may give a good fit. In this case it is not immediately clear which kernels are appropriate for a given process, and mechanisms cannot be inferred from the solution to the inverse problem. Smit *et al.* provide a technique for discriminating between aggregation models.[31,32] They note that mathematically gelling kernels (those capable of predicting gelation) are often used to model aggregating processes where this phase transition does not take place. Smit *et al.* also state that some kernels can be eliminated if they cause an explosion in the third or sixth moment of the size distribution.[33] However it is also noted by Ilievski *et al.*[34] and Adetayo and Ennis[28] that even if a good fit is found there is no certainty that it is the best fit or has any physical basis. Pearson *et al.* also note that in comparisons between the analytic solutions for the size-independent and sum kernels and experimental data, both can provide a good fit.[35] The size-independent kernel results in β_0 being a strong function of time. Conversely the sum kernel results in a weak time dependence for β_0. They state that no definite conclusions can be made concerning the dependence of granule growth rate on size and time from a single experiment. This confounding of size and age adds a layer of complexity to the inverse problem. Population balance equations can only be analytically solved for a couple of very specific cases, notably size-independent aggregation, and aggregation using the sum kernel. For most granulation processes considering anything more elaborate, numerical techniques are required. Ramkrishna cites a number of techniques for solving population balance equations.[16,36] In general the most frequently used numerical techniques for solving one-dimensional population balance equations in granulation processes fall into two broad categories. Nicmanis and Hounslow describe these as firstly finite-element methods/methods of weighted residuals in which the solution is approximated as linear combinations of basis functions over a finite number of subdomains (termed 'elements').[37] The second category is finite difference methods/discretised population balances, in which the population balance equation is approximated by differencing schemes (Hounslow *et al.*,[15] Hounslow,[38] Hill and Ng,[39] and Hounslow *et al.*[39]

3.5　Example of a 1-D Population Balance Model

An example of the use of 1-dimensional population balance models (Hounslow *et al.*[25]) to describe experimental high shear granulation data including breakage is summarised. Breakage was studied by introducing coloured tracer granules of a given sieve cut into placebo batches in order to observe the subsequent redistribution of coloured tracer. Two 1-dimensional discretised population balance models were derived to describe the evolution of the granule size distribution and the tracer mass distribution:

$$\frac{dN_i}{dt} = B_{0,i}^A(t) - D_{0,i}^A(t) + B_{0,i}^B(t) - D_{0,i}^B(t) \tag{13}$$

$$\frac{dM_i}{dt} = B_{1,i}^A(t) - D_{1,i}^A(t) + B_{1,i}^B(t) - D_{1,i}^B(t) \tag{14}$$

The aggregation terms for the granule size distribution and the tracer mass distribution are given respectively:

$$B_{0,i}^A(t) - D_{0,i}^A(t) = N_{i-1}\sum_{j=1}^{i-2} 2^{j-i+1}\beta_{i-1,j}N_j + \tfrac{1}{2}\beta_{i-1,i-1}N_{i-1}^2 - N_i\sum_{j=1}^{i-1} 2^{j-i}\beta_{i,j}N_j - N_i\sum_{j=i}^{n_{eq}}\beta_{i,j}N_j \tag{15}$$

$$B_{1,i}^A(t) - D_{1,i}^A(t) = M_{i-1}\sum_{j=1}^{i-2} 2^{j-i+1}\beta_{i-1,j}N_j + N_{i-1}\sum_{j-1}^{i-2} 2^{j-i+1}\beta_{i-1,j}M_j + N_i\sum_{j=1}^{i-1} (1 - 2^{j-i})\beta_{i,j}M_j$$

$$+ \beta_{i-1,i-1}N_{i-1}M_{i-1} - M_i\sum_{j=1}^{i-1} 2^{j-i}\beta_{i,j}N_j - M_i\sum_{j=i}^{n_{eq}}\beta_{i,j}N_j \tag{16}$$

The breakage terms for the granule size distribution and the tracer mass distribution are given respectively:

$$B_i^B - D_i^B = -S_iN_i + \sum_{j=i}^{n_{eq}} b_{i,j}S_jN_j \tag{17}$$

$$B_{1,i}^B - D_{1,i}^B = -S_iM_i + \sum_{j=i}^{n_{eq}} 2^{i-j}b_{i,j}S_jM_j \tag{18}$$

The equipartition of kinetic energy kernel was found to give the best fit to experimental data when compared with the size independent kernel and the shear kernel (see Table 1). A refined breakage model was used comprising of the sum of two truncated log-normal distributions, which was based on the normalised tracer distribution one minute after the addition of tracer granules from a size range of 1.0–1.18 mm:

$$b(x,l) = 0.484 b_{f,1}(x,l) + 0.516 b_{f,2}(x,l) \tag{19}$$

where $b_{f,1}$ and $b_{f,2}$ are of the form given as follows with $\bar{l}_{gv} = 160$ µm, $\sigma_g = 1.70$ (mode 1), and $\bar{l}_{gv} = 3.8$ mm, $\sigma_g = 2.66$ (mode 2), respectively:

$$b_f(x,l) = \left(\frac{l}{x}\right)^3 \frac{\dfrac{1}{x\sqrt{\pi/2}\,\ln\sigma_g}\exp\left[-\left(\dfrac{\ln x/\bar{l}_{gv}}{\sqrt{2}\,\ln\sigma_g}\right)^2\right]}{1 + erf\left[\dfrac{\ln l/\bar{l}_{gv}}{\sqrt{2}\,\ln\sigma_g}\right]} \tag{20}$$

Experimental observations suggested the breakage rate was very high initially, but subsequently decayed rapidly. This was modelled by an exponential decay in the selection rate constant:

$$S(t,l) = S_A \exp\left(-\frac{t-480}{20}\right) \tag{21}$$

Very little dependence of β_0 with time was found and this was described using the following relation:

$$\beta_0(t) = \beta_A + \beta_B(t-480) \text{ kg/s} \tag{22}$$

An integral fitting technique was utilised to determine S_A, β_A and β_B, giving the following:

$$S(t,l) = 0.025\exp\left(-\frac{t-480}{20}\right) \tag{23}$$

$$\beta_0(t) = 1.30 \times 10^{-9} - 6 \times 10^{-13}(t-480) \tag{24}$$

These parameters were used to model several experimental batches where different sized tracer granules were introduced. One set of results is shown in Figure 1. It was found that this model was capable of describing granule size distributions and tracer mass distributions simultaneously with great accuracy.

The inclusion of breakage in this model is a significant step in the modelling of high shear granulation, where breakage is considered to be an important process. The breakage process included in this model also removes the unsatisfactory need to have time as a correlating variable. This approach is often used in exclusively aggregating models, despite the fact that time dependent rate constants are normally evidence of some other unaccounted time dependent physical property. The true kinetics of the system and a real rate-based understanding can only be found when that physical property is identified. This again leads to the disadvantage of 1-dimensional or single phase population balance models

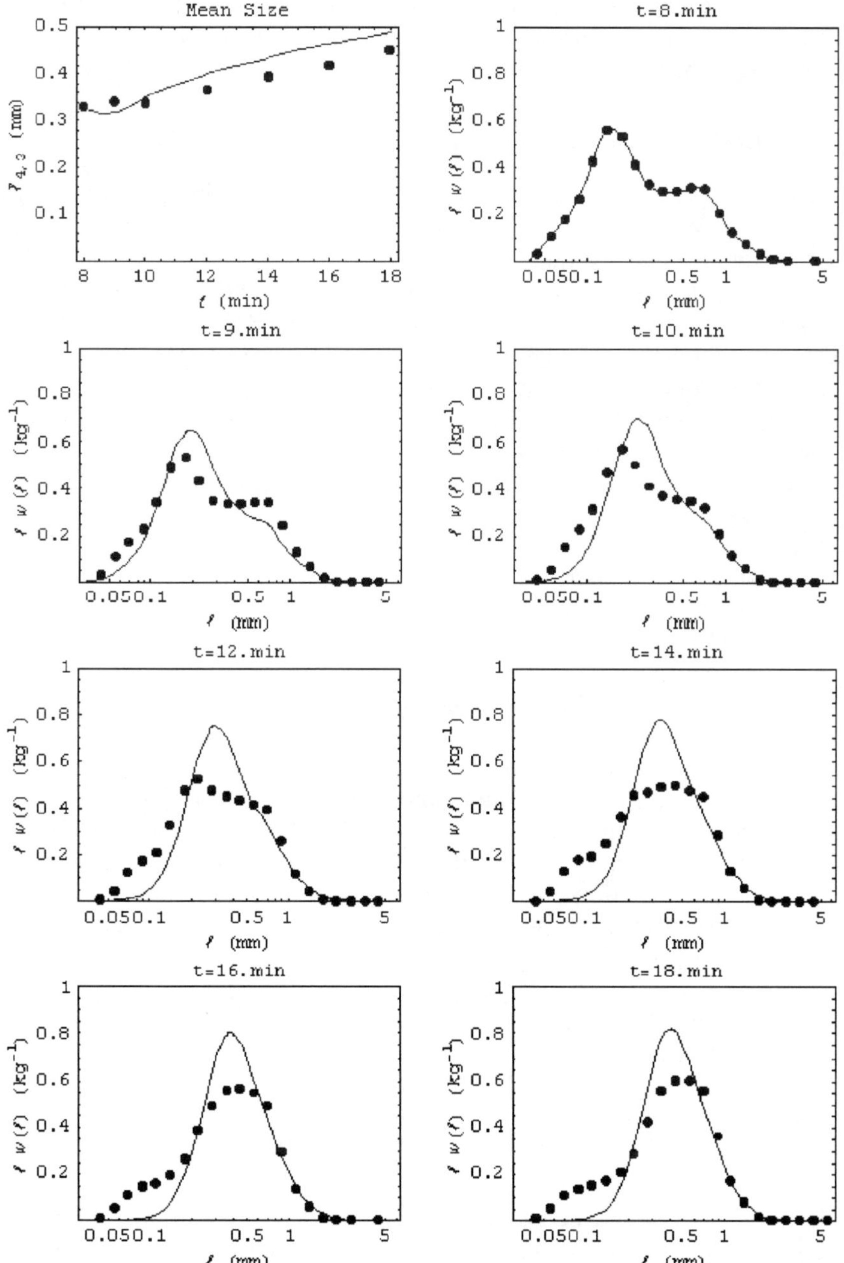

Figure 1 *Simulated (line) and experimental (circles) granule-size distributions for the 1.1mm tracer granule experiment using the kinetics of Equations 23 and 24. The first figure shows the 4,3 mean size as a function of time. The remaining figures show the mass GSD as it evolves over time*

(Hounslow *et al.*[25] Reproduced with permission of the American Institute of Chemical Engineers. Copyright 2001 AIChE, all rights reserved.)

to providing a physically based model of high shear granulation processes. The need for inclusion of further granule properties beyond size becomes more important as the physical basis is developed.

4   Multi-Dimensional Population Balance Models

Single phase population balance models describe the distribution of the size of the granules. With this level of definition these models can have size dependent aggregation or breakage rates. This is sufficient in a system where the collision rate is purely size dependent, or if the efficiency of the collision (i.e. likelihood of two colliding granules sticking) is purely size dependent. However much of the micro-scale modelling of high shear granulation (Section 2) has shown that this efficiency is also dependent on other granule properties.

The rate of granule collision is determined by the equipment and operating variables (e.g. impeller speed) and the size, mass and shape of the particles.

The efficiency of granule coalescence is a function of properties including:

- the size
- the amount of binder
- the properties of the binder
- the porosity
- the impact velocity
- the deformation/deformability/elasticity

Inclusion of these properties in the model, will allow more physics and micro-scale research (e.g. Ennis *et al.*[3] and Simons *et al.*[5]) to be linked into the population balance. 'Writing' multi-dimensional population balance equations is straightforward, however solving them is much more difficult. There are no useful analytical solutions, and numerical solutions consume a lot of computational time. By inclusion of more granule properties, it is to be expected that rate constants, which in reality depend on these properties, can be more accurately determined.

Multiple phase PBE's can easily be written, for example this four phase model suggested by Iveson:[11]

$$\frac{\partial n(m,\varepsilon,w,x,t)}{\partial t} = B(m,\varepsilon,w,x,t) - D(m,\varepsilon,w,x,t) + C(m,\varepsilon,w,x,t) + W(m,\varepsilon,w,x,t) \quad (25)$$

The model describes the mass, binder content, porosity and amount of two solid components in time. B and D are the birth and death terms due to coalescence, C is the compaction term, W the wetting term, m the granule mass, ε the granule porosity, w the moisture content, x the mass fraction of component A and t the time. The compaction term uses an empirical model to describe the average porosity with time.

This model has not been solved. Another way to describe multiple phases in granulation is with the use of Monte Carlo methods (Ramkrishna[16,36]). Monte

Carlo models use probability functions to describe events. After a random number is generated, the probability function determines which events occur.

Lee and Matsoukas used a model with constant number of granules to avoid difficulties of the population getting too small or too large, because of aggregation or breakage.[40] Their model describes aggregation and binary breakage. It shows good comparison with other models, but has not yet been compared with experimental data.

4.1　Use of Multi-dimensional Population Balance Equations

An often used method to account for more properties of the granules (in addition to size) in the model, is to make the rate constants time dependent. In these type of models, the time dependence can be used as just a fitting parameter to describe the experimental data, or based on assumptions or measurements of granule properties changing in time. However particles do not carry a stopwatch around. Time is not a particle property (Hounslow *et al.*[25]).

Time dependent models can account for changing properties in time (like for example consolidation of the granules), but they cannot distinguish between granules that have different properties, but identical age. (The granules can have a different size, but not, for example, a different binder content).

To compare a real multi phase population balance model with experiments, all granule properties that are in the model should be measured. Alternatively, all important properties should be determined, measured and then incorporated into a model. An initial step is to consider a two phase model that describes a two dimensional distribution function:

$$f(u,b) \tag{26}$$

Where f is no longer just a function of size (u) but also of another property, for example the binder content (b). However a two dimensional distribution function contains considerably more data than a one dimensional function. A discretized one dimensional function has u points, whereas a discretized two dimensional function has $u \times b$. All that data not only needs to be modelled, but also measured experimentally in order to verify the model. That means, in the case of the dimensions being size and binder content, measuring the size and binder content of enough single granules to obtain the full population distribution. To simplify this, the assumption can be made that all granules of a certain size (u) have the same amount of binder (b). In this phased average case there will only be two one dimensional distributions ($n(u)$ and $b(u)$):

$$n(u) = \int_0^u f(u,b)\, db \tag{27a}$$

$$b(u) = \rho \int_0^u b \cdot f(u,b)\, db \tag{27b}$$

This decreases computational needs and the experimental data can be obtained, for example, by sieving to determine the size distribution ($n(u)$) and by measuring the amount of binder in each sieve fraction ($b(u)$).

The first model that describes more than one measured granule property is the two phase model of Hounslow, Pearson and Instone that describes the granule size distribution and the distribution of a tracer in a high shear granulator.[25] A size cut of granules with this tracer was added after a certain granulation time, after which the granulation process was continued and modelled. These experiments were done to learn more about the breakage during the process.

Another use of this model has been described by Biggs.[41] In this work both the Granule Size Distribution (GSD) and the Binder Size Distribution (BSD) of two different high shear granulation experiments are described by the model. The first experiment uses a formulation with calcium carbonate and melted poly-ethylene glycol as binder. The second set of experiments uses a pharmaceutical formulation of lactose, starch and hydroxy propyl cellulose in water as a binder. In this model not only are two phases described, but also the aggregation rate is dependent on the binder content in the granules: a wet granule is more likely to stick than a dry granule.

This model is the first population balance model where the aggregation rate is not just size dependent, but dependent on the composition of the granules. By assuming that granules and particles are spherical and that all granules in one size class have the same binder fraction, the discretized tracer model (Hounslow, Pearson and Instone[25]) can be used. The tracer model was expanded to allow aggregation kinetics to depend on binder fraction:

$$\beta(u,v) = \beta_0 \times \psi(u,v) \times \beta^* (u,v) \tag{28}$$

where the aggregation rate is a function of the aggregation rate constant (β_0), a size dependent part ($\beta^*(u,v)$) and the efficiency of collision (ψ) which is a function of the binder content of two colliding particles according to a step function represented numerically by:

$$\psi_{i,j} = \frac{(\max(\Phi_{l,i},\Phi_{l,j})/\Phi_c)^{48}}{1 + (\max(\Phi_{l,i},\Phi_{l,j})/\Phi_c)^{48}} \tag{29}$$

where $\Phi_{l,i}$ and $\Phi_{l,j}$ are the binder volume fractions of the two colliding granules and $\psi_{i,j}$ is the efficiency of the collision between granules i and j. Scott *et al.* suggested that when two granules collide, at least one of the granules must have a binder content above some critical value.[42] If one of the granules contains more than the critical volume fraction Φ_c, then the efficiency is 1, and the collision is successful. If neither of the particles contains enough liquid, then the efficiency is 0. In Figures 2 and 3 it can be seen that for both formulations the model describes the experimental data. Also in Figures 2 and 3 are the binder fractions:

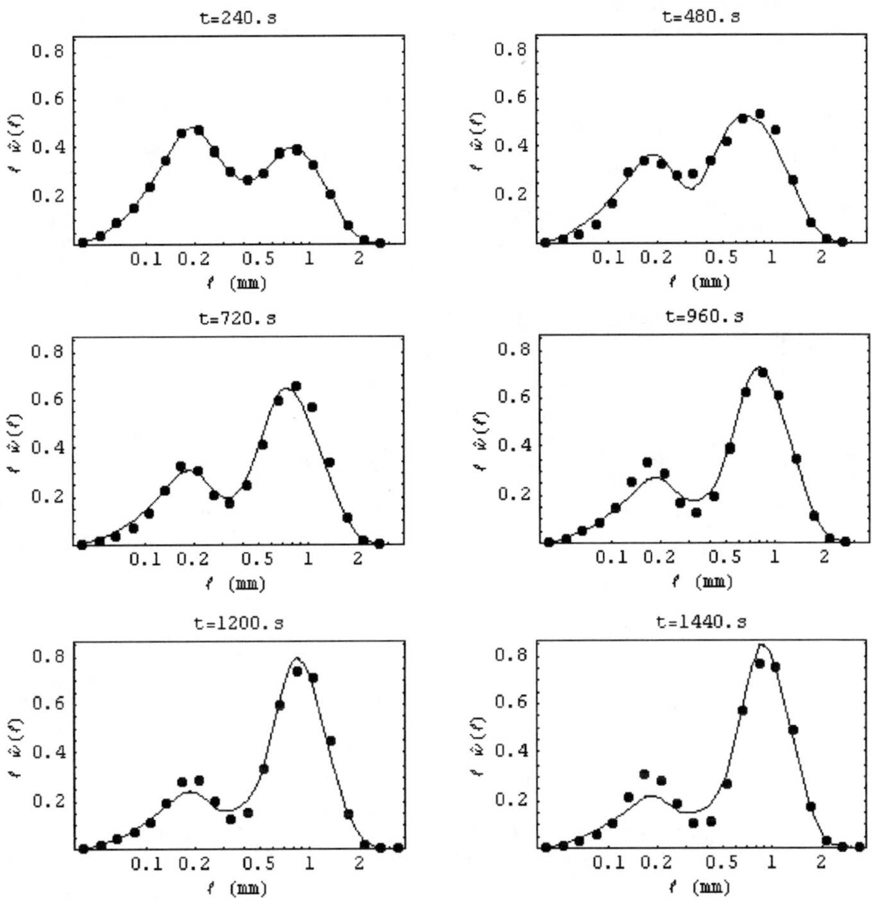

Figure 2a　*The GSD (Granule Size Distribution) for granulation of calcium carbonate ("•" experiment, "—" model)*

$$\Phi_{i,t} = \frac{M_{i,t}/\rho_L;}{k_v \cdot l^3 \cdot N_{i,t}} \tag{30}$$

with k_v, the shape factor of the granules, assumed to be ($\pi/6$). Small granules contain less binder than large granules and therefore are less likely to stick together than large (wet) granules. The values of the model parameters are given in Table 2.

Table 2　*Model parameters*

	Granulation of Calcium Carbonate	*Granulation of Lactose and Starch*
β_0	$3.4\ 10^{-10}$	$4.3\ 10^{-10}$
Φ_c	0.22	0.18
SSE	0.016	0.066

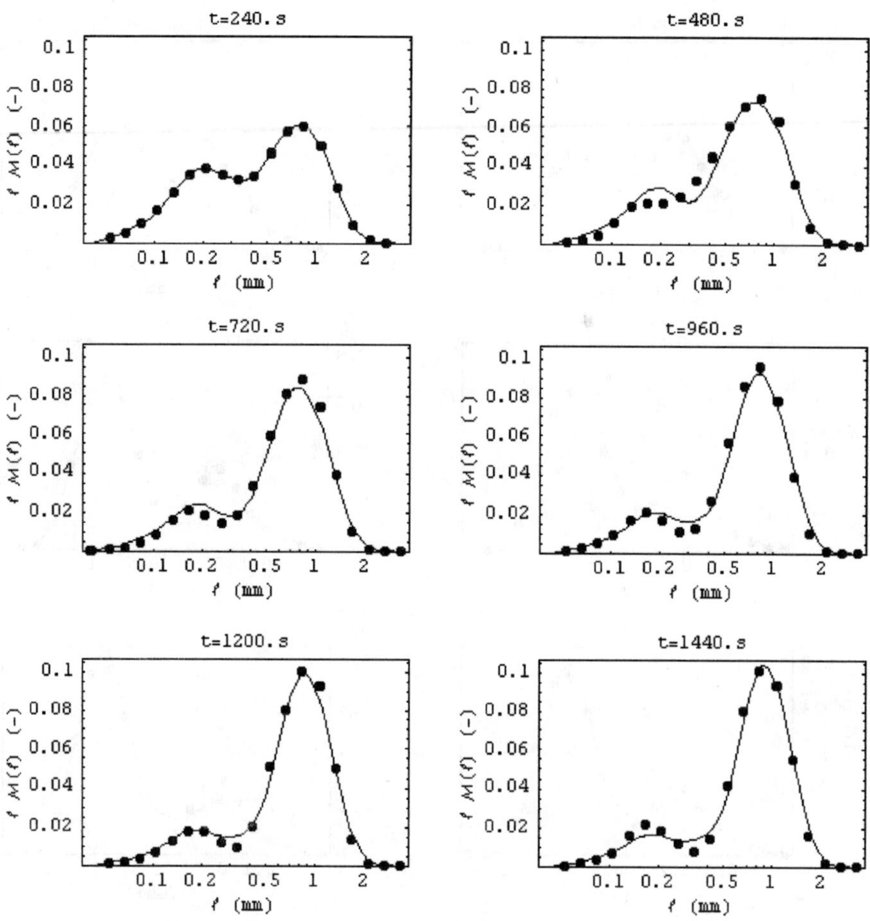

Figure 2b   *The BSD (Binder Size Distribution) for granulation of calcium carbonate ("•" experiment, "—" model)*

The Sum of Squared Error (SSE) between experimental and model values is given by:

$$SSE = \sum_t w_t^n \left(\sum_l w_l^n (N_{t,l} - \tilde{N}_{t,l})^2 \right)$$

$$+ \sum_t w_r^m \left(\sum_l w_l^m (M_{t,l} - \tilde{M}_{t,l})^2 \right) \tag{31}$$

This model shows that with only one rate constant and a critical binder content it is possible to describe both the GSD and BSD during a whole granulation process.

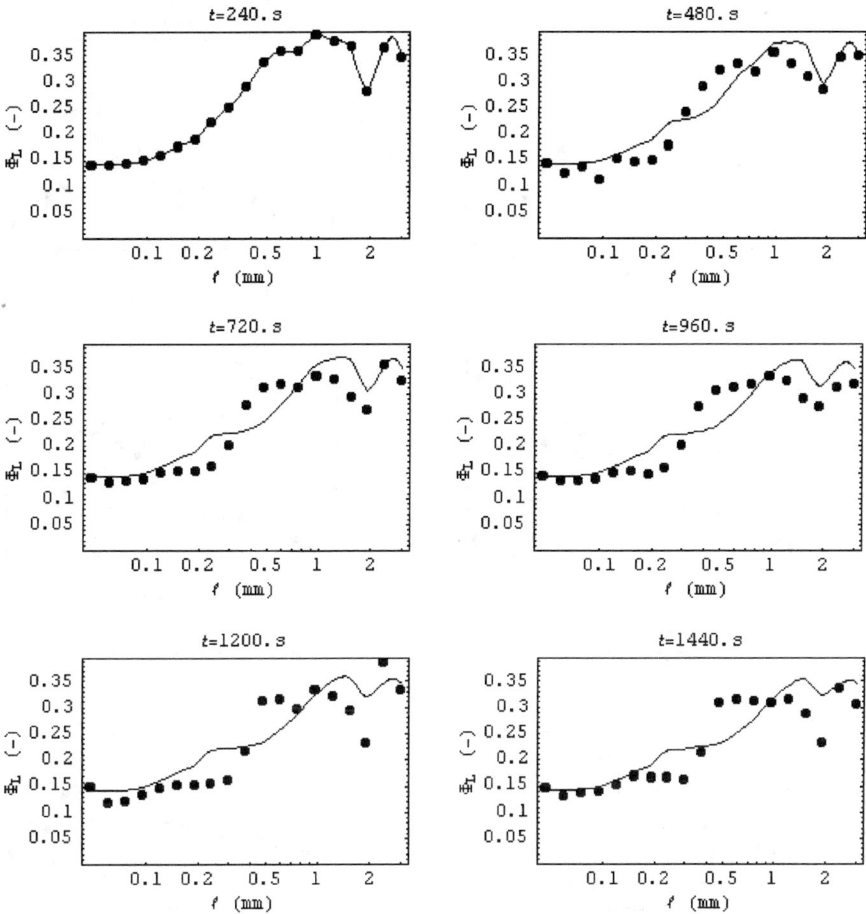

Figure 2c *The Binder Fraction for granulation of calcium carbonate ("•" experiment, "—" model)*

5   Conclusions

Population balance models are the natural framework in which to describe the kinetics of granulation, relating as they do changes in the distributions of the products, for example the size distribution, to the properties of the product, for example binder content. In this paper the focus has necessarily been in setting up the conservation statements that comprise the population balance, but the matters of significance are the kinetics of the process as conveyed by the rate constants, β, β_0, S etc. The purpose of the work presented here is to reduce the description of the behaviour of a granulating system to a description of these rate constants.

We have shown that the application of 1-D population balance models necessarily requires that the rate constants depend on time – which is altogether

  The Kinetics of High-Shear Granulation

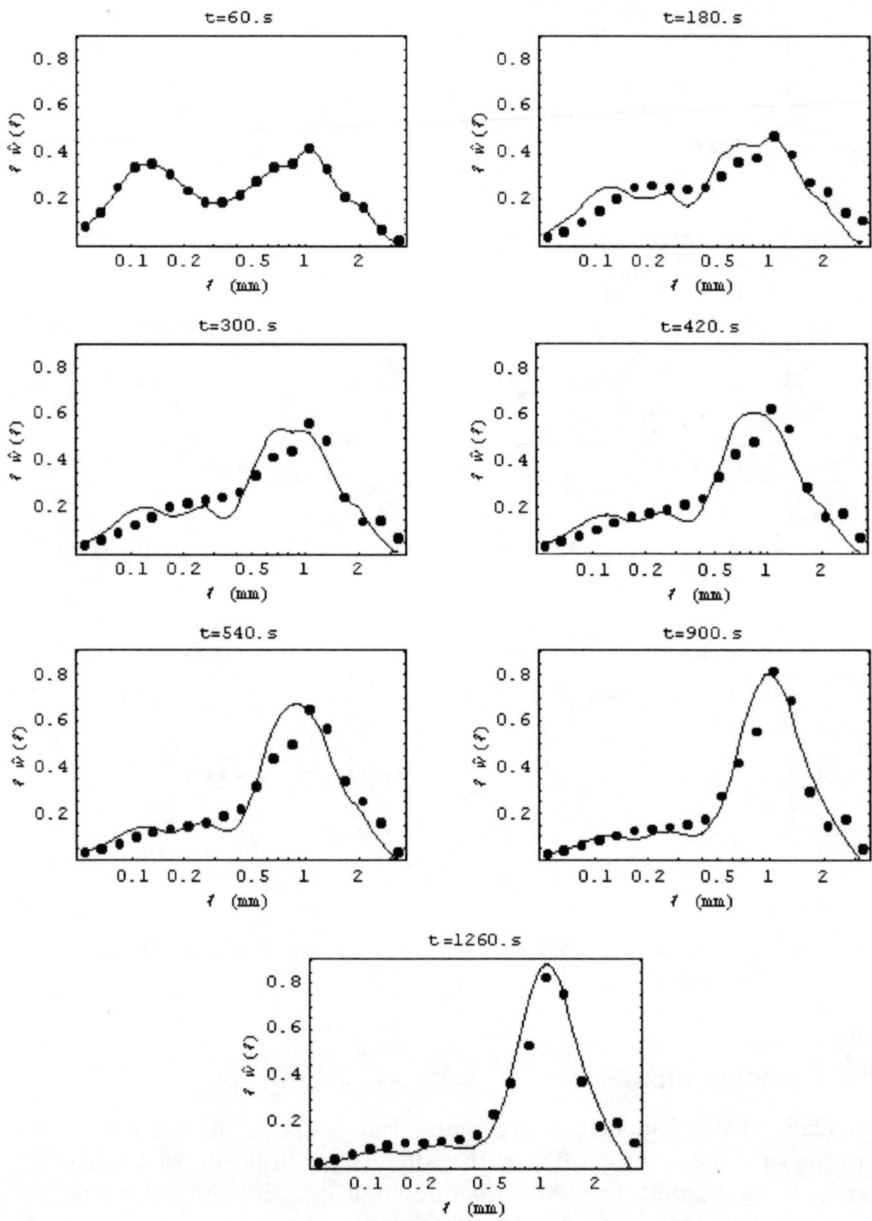

Figure 3a *The GSD (Granule Size Distribution) for granulation of lactose and starch ("•" experiment, "—" simulated results)*

unsatisfactory. The solution to this problem, of using 2-D models, so that the rate constants can depend on the composition of the granules, would be enormously demanding. We propose instead to use multiple 1-D models that allow the average properties of granules of a certain size to be represented. This model

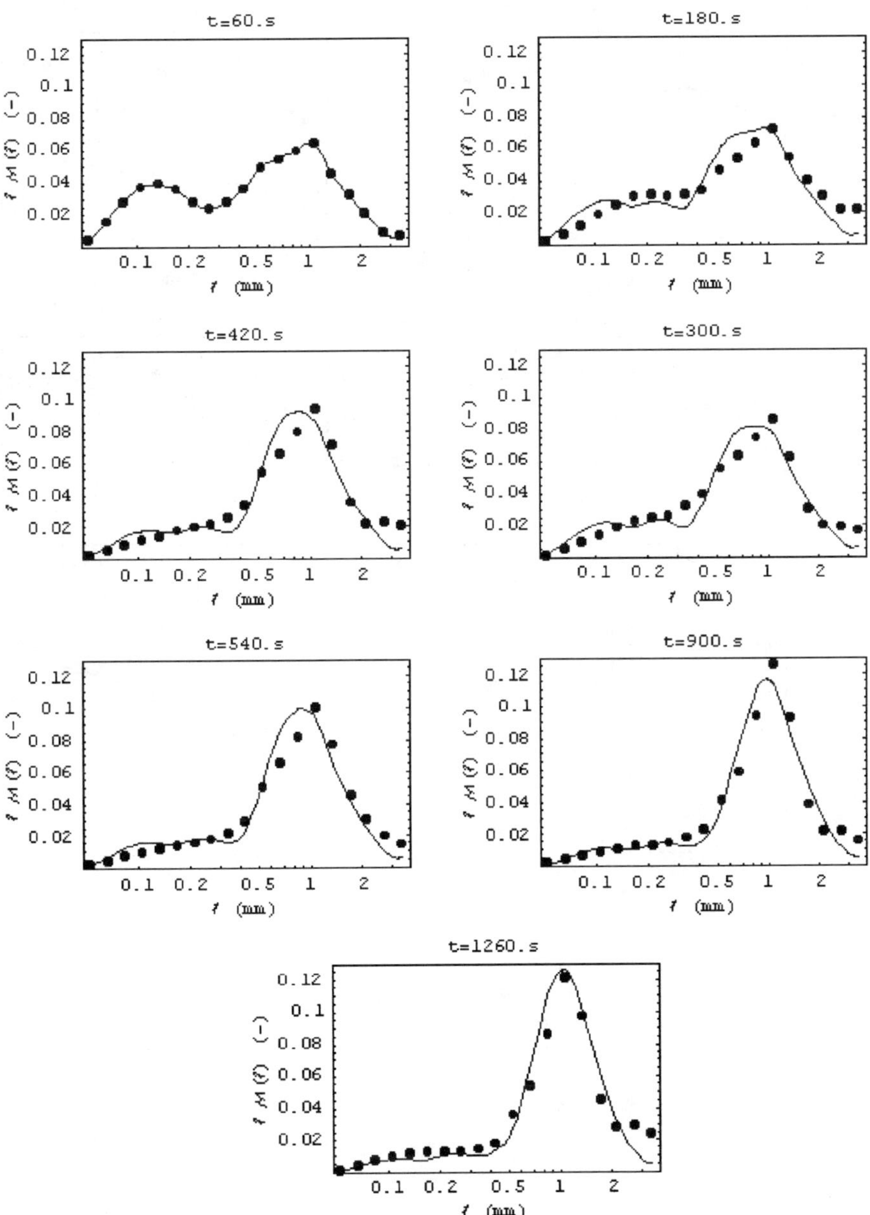

Figure 3b   *The BSD (Binder Size Distribution) for granulation of lactose and starch ("•" experiment, "—" simulated results)*

reduces the size-enlargement kinetics to a rate constant and a critical binder content. In other words, we can describe this phase of a granulation process by means of two numbers, meeting our objective of describing the kinetics of granulation.

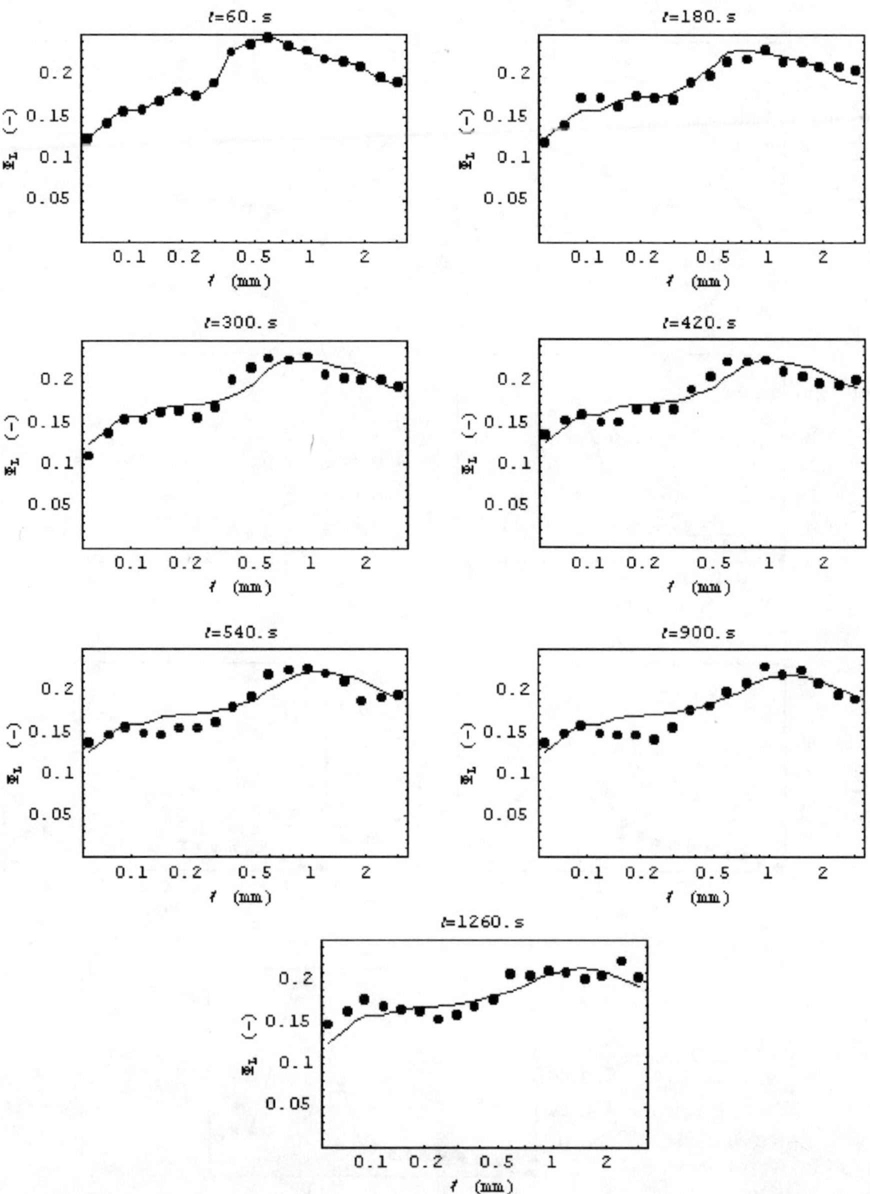

Figure 3c *The Binder Fraction for granulation of lactose and starch ("•" experiment, "—" simulated results)*

Much, however, remains to be done. Our description of breakage is very limited, and we have as yet no description of nucleation. We believe that the methods we have described here can be applied to these other mechanisms and that it is possible to develop true kinetic models for granulation.

References

1. N. Ouchiyama and T. Tanaka, "The probability of coalescence in granulation kinetics", *Ind. Eng. Chem. Process Des. Dev.*, 1975, **14**, 286.
2. N. Ouchiyama and T. Tanaka, "Kinetic analysis and simulation of batch granulation", *Ind. Eng. Chem. Process Des. Dev.*, 1982, **21**, 29.
3. B. J. Ennis, G. Tardos and R. Pfeffer, "A microlevel-based characterization of granulation phenomena", *Powder Technol.*, 1991, **65**, 257.
4. J. L. Moseley and T. J. O'Brien, "A model for agglomeration in a fluidised bed", *Chem. Eng. Sci.*, 1993, **48**, 3043.
5. S. J. R. Simons, J. P. K. Seville and M. J. Adams, "Mechanisms of agglomeration", *6th Int. Symp. Agglomeration*, Nagoya, Japan, 15–17 Nov. 1993, 117.
6. G. I. Tardos, M. Irfan Khan and P. R. Mort, "Critical parameters and limiting conditions in binder granulation of fine powders", *Powder Technol.*, 1997, **94**, 245.
7. G. Lian, C. Thornton and M. J. Adams, "Discrete particle simulation of agglomerate impact coalescence", *Chem. Eng. Sci.*, 1998, **53**, 3381.
8. J. P. K. Seville, H. Silomon-Pflug and P. C. Knight, "Modelling of sintering in high temperature gas fluidisation", *Powder Technol.*, 1998, **97**, 160.
9. C. Thornton and Z. Ning, "A theoretical model for the stick/bounce behaviour of adhesive, elastic-plastic spheres", *Powder Technol.*, 1998, **99**, 154.
10. L. X. Liu, J. D. Litster, S. M. Iveson and B. J. Ennis, "Coalescence of deformable granules in wet granulation processes", AIChE J., 2000, **46**, 529.
11. S. M. Iveson, J. D. Litster and B. J. Ennis, "Fundamental studies of granule consolidation Part 1: Effects of binder content and binder viscosity", *Powder Technol.*, 1996, **88**, 15.
12. S. M. Iveson and J. D. Litster, "Fundamental studies of granule consolidation Part 2: Quantifying the effects of particle and binder properties", *Powder Technol.*, 1998, **99**, 243.
13. S. M. Iveson and J. D. Litster, "Liquid-bound granule impact deformation and coefficient of restitiution", *Powder Technol.*, 1998, **99**, 234.
14. J. Fu, A. D. Salman and M. J. Hounslow, "Experimental study on impact deformation and rebound of wet granules", *World Congress on Particle Technology 4*, Sydney, Australia, 21–25 July 2002, No. 113.
15. M. J. Hounslow, R. L. Ryall and V. R. Marshall, "A discretized population balance for nucleation, growth and aggregation", *AIChE J.*, 1988, **34**, 1821.
16. D. Ramkrishna, *Population Balances: Theory and Applications to Particulate Systems in Engineering*, 1st ed., Academic Press, 2000.
17. Kenneth J. Valentas, Olegh Bilous and Neal R. Amundson, "Analysis of breakage in dispersed phase systems", *Ind. Eng. Chem. Fund.*, 1966, **5**, 271.
18. H. M. Hulbert and S. Katz, "Some problems in particle technology", *Chem. Eng. Sci.*, 1964, **19**, 555.
19. A. D. Randolf and M. A. Larson, *Theory of particulate processes*, 1st ed., Academic Press, 1971.
20. J. D. Litster, K. P. Hapgood, J. N. Michaels, A. Sims, M. Roberts, S. K. Kameneni and T. Hsu, "Liquid distribution in wet granulation: dimensionless spray flux", *Powder Techol.*, 2001, **114**, 32.
21. K. V. S. Sastry and D. W. Fuerstenau, "Mechanisms of agglomerate growth in green pelletization", *Powder Technol.*, 1973, **7**, 97.
22. M. Smoluchowski, "Mathematical theory of the kinetics of the coagulation of colloidal solutions", *Zeitschrift fur Phusikalische Chemie*, 1917, **92**, 129.

23. A. M. Golovin, "The solution of the coagulation equation for raindrops, taking condensation into account", *Soviet Physics-Doklady*, 1963, **8**, 191.
24. M. J. Hounslow, "The population balance as a tool for understanding particle rate processes", *KONA Pow. & Part.*, 1998, **16**, 180.
25. M. J. Hounslow, J. M. K. Pearson and T. Instone, "Tracer studies of high-shear granulation: II. Experimental results", *AIChE J.*, 2001, **47**, 1984.
26. P. C. Kapur, "Kinetics of granulation by non-random coalescence mechanism", *Chem. Eng. Sci.*, 1972, **27**, 1863–1869.
27. K. V. S. Sastry, "Similarity size distribution of agglomerates during their growth by coalescence in granulation or green pelletization", *Int. J. Miner. Process.*, 1975, **2**, 187.
28. A. A. Adetayo and B. J. Ennis, "Unifying approach to modelling granule coalescence mechanisms", *AIChE J.*, 1997, **43**, 927.
29. A. Annapragada and J. Neilly, "On the modeling of granulation processes: a short note", *Powder Technol.*, 1996, **89**, 83.
30. J. M. K. Pearson, M. J. Hounslow and T. Instone, "Tracer studies of high-shear granulation: I. Experimental results", *AIChE J.*, 2001, **47**, 1978.
31. D. J. Smit, M. J. Hounslow and W. R. Paterson, "Aggregation and gelation – I. Analytical solutions for CST and batch operation", *Chem. Eng. Sci.*, 1994, **49**, 1025.
32. D. J. Smit, M. J. Hounslow and W. R. Paterson, "Aggregation and gelation – III. Classification of kernels and case studies of aggregation and growth", *Chem. Eng. Sci.*, 1995, **50**, 849.
33. D. J. Smit, M. J. Hounslow and W. R. Paterson, "A fundamental test of the suitability of aggregation kernels", *6th Int. Symp. Agglomeration*, Nagoya, Japan, 15–17 Nov. 1993, 52.
34. D. Ilievski, E. T. White and M. J. Hounslow, "Agglomeration mechanism identification case study: Al(OH)3 agglomeration during precipitation from seeded supersaturated caustic aluminate solutions", *6th Int. Symp. Agglomeration*, Nagoya, Japan, 15–17 Nov. 1993, 93.
35. J. M. K. Pearson, M. J. Hounslow, T. Instone and P. C. Knight, "Granulation Kinetics: The Confounding of Particle Age and Size", *World Congress on Particle Technology* 3, 1998, 86.
36. D. Ramkrishna, "The status of population balances", *Rev. Chem. Eng.*, 1985, **3**, 49.
37. M. Nicmanis and M. J. Hounslow, "Finite-element methods for steady-state population balance equations", *AIChE J.*, 1998, **44**, 2258.
38. M. J. Hounslow, "A discretized population balance for continuous systems at steady state", *AIChE J.*, 1990, **36**, 106.
39. P. J. Hill and K. M. Ng, "New discretization procedure for the breakage equation", *AIChE J.*, 1995, **41**, 1204.
40. K. Lee and T. Matsoukas, "Simultaneous coagulation and break-up using constant-N Monte Carlo", *Powder Technol.*, 2000, **110**, 82.
41. C. A. Biggs, C. Sanders, A. C. Scott, A. W. Willemse, A. C. Hoffmann, T. Instone, A. D. Salman and M. J. Hounslow, "Coupling granule properties and granulation rates in high-shear granulation", *Powder Technol.*, 2003, **130**, 162.
42. A. C. Scott, M. J. Hounslow and T. Instone, "Direct evidence of heterogeneity during high-shear granulation", *Powder Technol.*, 2000, **113**, 205.

CHAPTER 12

Dynamics of Particles in a Rotary Kiln

D. M. SCOTT and J. F. DAVIDSON

Department of Chemical Engineering, University of Cambridge, Pembroke Street, Cambridge CB2 3RA, UK

Nomenclature

A	cross sectional area of granular bed, m^2
C_A	constant defined in equation (10)
C_B	constant defined in equation (10)
D	diameter of cylinder, m
Fr	Froude number, $= \omega^2 R/g$
g	acceleration due to gravity, m/s^2
h	bed depth, m
L	length of cylinder, m
$L_{ellipse}$	axial length of ellipse, m
n	rotational speed, rev/s; also, mean number of avalanches to traverse the active layer
Q	volumetric flow rate of granular material, m^3/s
Q_d	$Q \sin \gamma / n D^3 \tan \beta$
R	radius of cylinder, m
T_d	$\tau n D \tan \beta / L \sin \gamma$
t	time, s
t_{13}	cycle time, s
t_{12}	avalanche time, s
t_b	breakthrough time, s
x	distance from discharge, m
y	h/R
β	angle of inclination of cylinder, degrees
γ	angle of repose, degrees
γ_d	dynamic angle of repose, degrees
γ_s	static angle of repose, degrees

319

ϕ	half of angle subtended at axis of cylinder by bed, rad
θ	angle of inclination of bed surface to cylinder axis in no-flow situation, degrees
σ	standard deviation of residence time, s
τ	mean residence time, s
ω	rotational speed, rad/s

1 Introduction

Rotary kilns are widely used in the processing of granular solids in the chemical and metallurgical industries, in which they are used to perform operations such as mixing, drying, heating and gas-solid reactions. An industrial kiln typically consists of a refractory-lined steel cylinder, with a diameter of 2–4 m, and a ratio of length to diameter greater than 20. The cylinder is inclined by a few degrees to the horizontal, and rotates slowly, some minutes per revolution. Granular material is continuously fed into the top end. The particles roll down, forming a moving bed of granular material, whose surface, as seen in a cross section normal to the axis of the kiln, is inclined to the horizontal at an angle close to the angle of repose. Some kilns have lifters, but the work described in this chapter is for those kilns without lifters. An example is in the manufacture of TiO_2, a white pigment used principally in paints, paper and plastics, which is the most important pigment in the world, with expected significant growth. There has been substantial investment in environmental improvement. Manufacture by the sulphate process, which accounts for more than half the production, uses rotary kilns. The gas is heated in a combustor, and enters the lower end of the kiln at around 1000 °C. The feed to the kiln is wet solids at ambient temperature, and as it passes through the kiln, it is first dried, and then heated to around 950 °C, so that the material transforms from the anatase to the desired rutile form, which transformation is exothermic.

The motion of the bed of granular solid in the transverse plane takes a number of distinct forms, as described by Henein *et al.*[1] At very low rotational speeds slipping may occur, where the granular bed retains its form as a body of particles, but slips at the kiln wall. At slightly higher rotational speeds there is slumping or avalanching: this is a quasi-periodic motion in which the bed rotates with the kiln wall in rigid body rotation, until an avalanche breaks off; the avalanche is mostly from the upper part of the free surface. When the avalanche stops, the inclination of the bed surface is less than at the start of the avalanche. The period between avalanches falls with increasing rotational speed, eventually giving the rolling mode in which there is a continuous stream of granules falling down the free surface. At higher rotational speeds centrifugal effects are important, and the bed surface is curved. In industrial kilns, the typical modes are avalanching and rolling, which may occur simultaneously at different axial positions along the kiln.

This chapter describes work done in Cambridge on particle motion in rotary kilns, using cold laboratory scale equipment. We do not attempt to review the literature, except in so far as it relates directly to the work described.

2   Steady State – Rolling Mode

2.1   Saeman's Model

The first published description of granular motion in a rotary kiln was by Sullivan *et al.*[2] The first theoretical model of which we are aware is that of Pickering *et al.*,[3] which treated a bed of uniform depth. This was closely followed by the work of Saeman,[4] Vahl and Kingma,[5] and Kramers and Croockewit,[6] which considered beds of varying depth. The simple picture of granular motion is that an individual granule rotates at the same rotational speed as the kiln wall until it reaches the free surface, and then cascades down the free surface in a path that is inclined to the horizontal at the angle of repose. From the bottom half of the free surface, the granule is absorbed once again into the rigid body rotation, and starts another cycle. This is indicated in Figure 1. The model makes the following assumptions: (*i*) the time spent falling down the free surface is small compared to the time in the rigid body rotation, (*ii*) the bed surface is locally flat, and (*iii*) the angle of inclination of the kiln and the variation of bed depth with axial position are both small. Then using geometry, the distance of axial transport can be related to the distance travelled down the free surface in a single cascade. The model gives the following expression for the dependence of volumetric flow rate, Q, on radius of kiln, R, rotational speed, n (rps), angle of inclination of the kiln, β, angle of repose, γ, bed depth, h, and distance from discharge, x; see Figure 2:

$$Q = \frac{4\pi n}{3}[(2R - h)h]^{3/2}\left(\frac{\tan\beta}{\sin\gamma} + \frac{dh/dx}{\tan\gamma}\right) \tag{1}$$

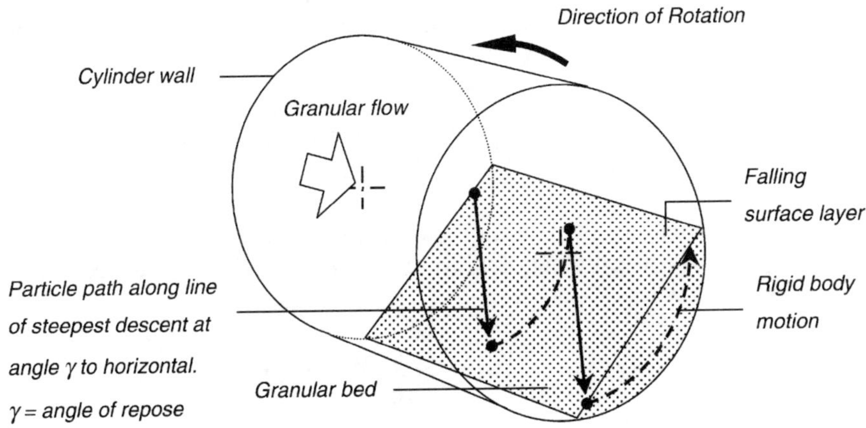

Figure 1   *Schematic diagram of particle motion in the granular bed*

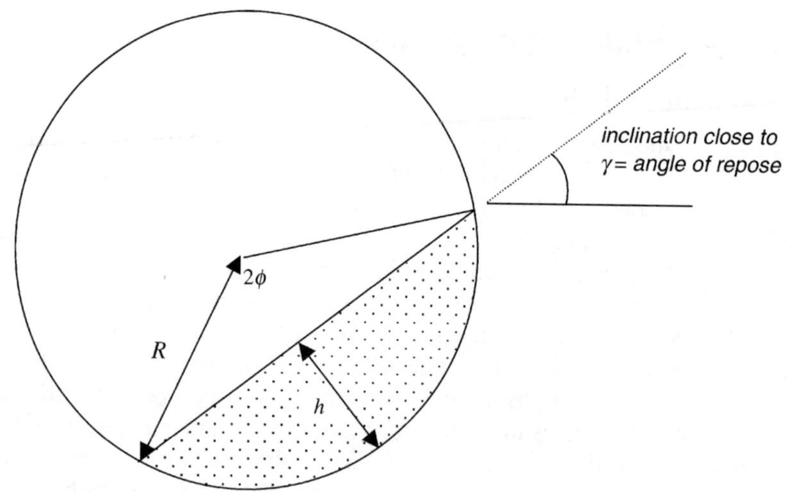

Figure 2   *Cross section of cylinder*

2.2   Residence Time

Equation 1 can be integrated to give the axial profile of bed depth, and thus the axial profile of the cross sectional area of the granular bed, A, which, from the geometry of Figure 2, is given by

$$A = R^2[\cos^{-1}(1 - y) - (1 - y)\sqrt{y(2 - y)}] \quad \text{where } y = h/R \tag{2}$$

Knowing A, the hold-up and thus the mean residence time, τ, can be calculated. For a constant volume chemical reactor, τ is inversely proportional to the feed flow rate. However, this is not true for a rotary kiln, because increased flow rate means increased hold-up. For the special case of uniform depth, which can be achieved by using an annular exit dam of appropriate height, the model gives

$$\tau = \frac{3L\sin\gamma}{4\pi nR\tan\beta}\left[\frac{\cos^{-1}(1 - y) - (1 - y)\sqrt{y(2 - y)}}{[y(2 - y)]^{3/2}}\right] \tag{3}$$

where L is the length of the kiln. Equation 3 is compared to data of McTait *et al.*[7] in Figure 3, in which dimensionless residence time, T_d, is plotted against a dimensionless flow rate, Q_d, where

$$T_d = \frac{\tau nD\tan\beta}{L\sin\gamma}, \quad Q_d = \frac{Q\sin\gamma}{nD^3\tan\beta}, \quad D = 2R \tag{4}$$

Note that the results in Figure 3 are for a uniform bed depth, $dh/dx = 0$; Equation 1 then gives the relation between Q and y. An increase in Q, with other operational parameters fixed, leads to an increase in bed depth, and a small increase in τ; to a first approximation τ is almost independent of flow rate, Q,

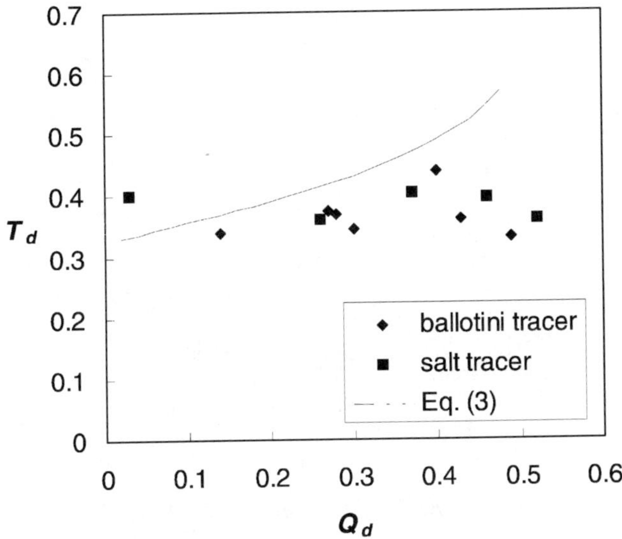

Figure 3   *Dimensionless residence time T_d versus dimensionless flow rate Q_d for uniform bed depth, dh/dx = 0*
(Data from McTait *et al.*[7])

because of the afore-mentioned change of hold-up with flow rate. Note that Equation 3 gives the average residence time for all particles: McTait *et al.*[7] gave the corresponding expression for the residence time of outermost particles.

Experimental measurements of the Residence Time Distribution (RTD) have been reported by a number of authors (e.g. 8–16). The RTD seems to consist of a narrow, approximately symmetrical peak, though McTait[7,15] found the central peak to have a tail for longer times, which may have been caused by segregation of finer tracer particles. The mean residence time is fairly well approximated by that of the model. A working formula for the standard deviation, σ, of residence time can be found by modelling the kiln by well-stirred tanks-in-series, with the number of tanks being given by the number of cascades experienced by an average particle. For a bed of uniform depth this gives, with the assumption that the average cascade length is $0.5D$ (see References 7 & 15),

$$\sigma / \tau = [(R\tan\beta)/(L\sin\gamma)]^{1/2} \tag{5}$$

2.3   Bed Depth

As we mentioned, Equation 1 can be integrated to give the axial profile of bed depth (see refs. 4–6, 11, 16, 17). Figure 4 shows results of experiments carried out by Spurling.[18] There is good agreement between model and experimental data. There is similarly good agreement between theory and experiment for flow over a variety of exit and internal dams, as demonstrated by Spurling[18] and Lim.[19]

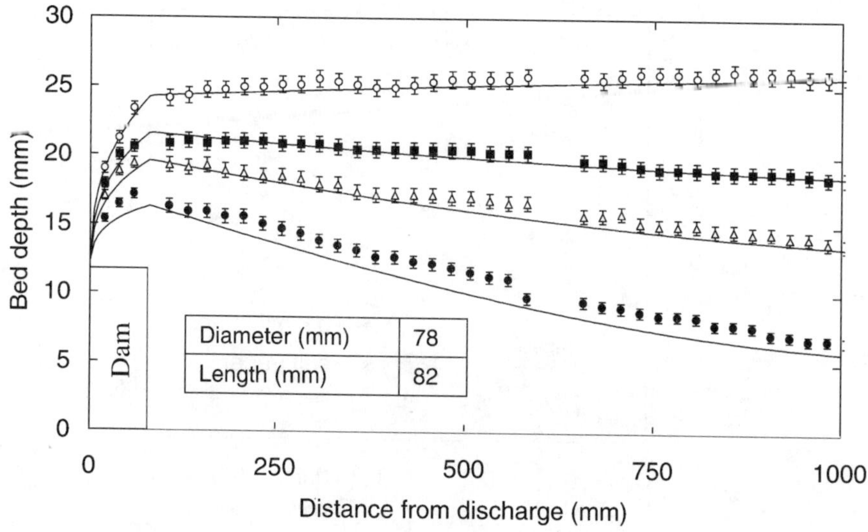

Symbol	Mass flow rate	Model hold-up	Model residence times
	g/s	% of cylinder volume	s
●	0.23	4.9	2894
△	0.72	9.7	1826
■	1.17	13.3	1543
○	1.94	18.8	1312

Figure 4   *Axial bed depth profile over a cylindrical dam. Kiln length 1 m, diameter 0.105 m, speed 0.097 rps. Angle of inclination 1°. Granular material: sand of mean diameter 460 μm. The curves are from integration of equation (1).*
(From Spurling.[18])

2.4   No Flow

In this section we discuss the motion of granular material in a rotary kiln with no net flow of material along the axis of the kiln. Two practical examples of this situation are as follows.

(*i*)   For rotary kilns with continuous throughput, the feed is sometimes put in at a point somewhat downstream of the top end of the kiln; at the top end of the kiln there is always an orifice of diameter less than the diameter of the kiln, to prevent spill-back of the granular material. Between this feed-end orifice and the feed point, there is a *no-flow* region: the material put in at the feed point, e.g. by a screw conveyer, all goes downstream, but there is granular material in the kiln between the orifice and the feed point; this granular material simply rotates with the kiln with no net flow.

(*ii*) The second example is the dish granulator, which is a short cylinder rotating about its axis which is inclined to the horizontal; this angle of inclination is usually not small. The lower end of the cylinder is closed by a flat plate normal to the axis, and the upper end is open, thus forming a cylindrical flat-bottomed dish. There is a small or zero axial flow of granules. Thus a rolling bed of granular material is formed within the dish: the shape of the surface of the bed is predicted by the theory given below.

We have constructed a model for the no-flow situation[20] (see also Chadwick and Bridgwater[21]) based on the hypothesis that, as observed experimentally, the surface of the bed of granular material is flat. It is assumed that the angle of slope of the flat surface is the angle of repose of the granular material. The orientation of the surface is found by postulating that a granule following an outermost track has a closed path: starting from the top of the free surface at a point, P, it cascades down the free surface along the line of steepest descent, which is at the angle of repose; this line is in a plane normal to the kiln axis. The plane of the path of the granule in rigid body rotation contains the granule's line of steepest descent. The granule then rotates with the kiln in rigid body rotation until it reaches the free surface back at point P. Geometry gives the orientation of the plane of the bed surface, which intersects the wall of the kiln in an ellipse. Note that in this analysis the angle of inclination of the kiln, β, need not be small.

The model gives the angle of inclination of the bed surface to the kiln axis, θ, which is

$$\sin \theta = \sin \beta / \cos \gamma \tag{6}$$

The length in the axial direction of the above-mentioned ellipse is

$$L_{ellipse} = 2R\sqrt{\cos^2 \gamma - \sin^2 \beta} / \sin \beta \tag{7}$$

Figure 5 shows a comparison between experiment and theory; details of theory and experiments are given by Spurling *et al.*[20]

2.5 Residence Time in the Free Surface Cascade

Positron Emission Particle Tracking (PEPT) is a non-invasive technique for following the motion of a single radioactively labelled tracer particle. The technique is based on positron emission tomography, which is widely used in medicine, and was developed for particle tracking at the University of Birmingham, UK, by Hawkesworth *et al.*,[22] Parker *et al.*,[23] and Parker *et al.*[24] Spurling[18] carried out a series of experiments using the facility at the University of Birmingham to study the mean velocity field in the transverse cross section of a horizontal cylinder. He found that in the rolling mode, the ratio of the mean particle residence time per circuit of the granular bed measured experimentally

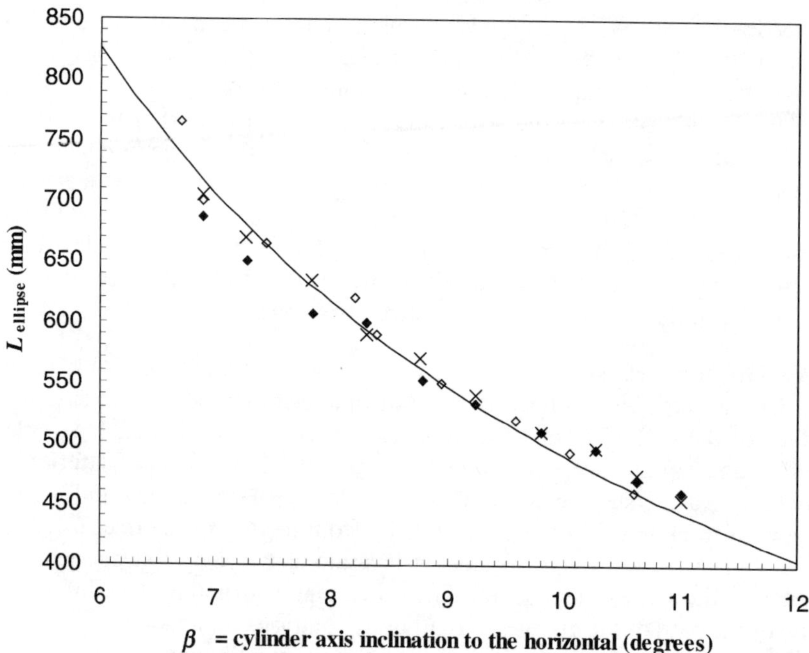

Figure 5 *Measurements of axial length, $L_{ellipse}$, of intersection between bed surface and cylinder for dry sand and titanium dioxide powder. Cylinder diameter = 105 mm. Each point is an experimental mean. Closed diamonds: sand, slumping bed, speed ~1 rpm. Open diamonds: sand, rolling bed, speed ~6 rpm. Crosses: titanium dioxide powder, bed slumping and rolling, speed 3.5–7 rpm. The curve is from equation (7), using the value of the angle of repose $\gamma = 34°$*
  (See Spurling et al.[21])

to that calculated from the model was between 1.15 and 1.75, indicating a significant discrepancy between theory and experiment, which is not apparent in Figure 4. Lim[25] suggested that the discrepancy might be due to the method of measuring bed depth. Lim measured bed depth mechanically, using a pointer, and found the bed surface to be slightly convex. Spurling observed the points of intersection of the bed surface with the containing walls of the cylinder, and calculated the bed depth assuming its surface was flat, which may have led to an underestimate of the bed depth.

3   Transients

Saeman's model provides a good description of the shape of the bed in the steady state, e.g. Figure 4. We[18,26] have extended the model to describe transient behaviour. This is accomplished by first performing a material balance on an element of length δx, as indicated in Figure 6, which gives

$$\frac{\partial A}{\partial t} = \frac{\partial Q}{\partial x}$$

(8)

Here A, the cross sectional area of the granular bed, is given by Equation 2. Then the steady-state model is used to give the relationship locally between Q and h, shown by Equation 1. The result is the following nonlinear diffusion equation, which is solved numerically.

$$2R^2 \sin^2 \phi \frac{\partial \phi}{\partial t} = \frac{\partial}{\partial x}\left\{ \frac{R^3 \sin^3 \phi}{C_A}\left[R\sin\phi \frac{\partial \phi}{\partial x} + C_B \right]\right\} \qquad (9)$$

where 2ϕ is the angle subtended by the bed at the axis of the kiln, see Figure 2, and

$$C_A = \frac{3\tan\gamma}{4\pi n}, \; C_B = \frac{\tan\beta}{\cos\gamma} \qquad (10)$$

The model contains no extra parameters beyond those used for the steady state model.

Some sample results will now be given. Figure 7 (a) shows the response in discharge flow rate and Figure 7 (b) in bed depth due to a large step increase in the feed flow rate, all other operational parameters being kept constant. The model gives a good prediction of the transients. Similarly good results are found[18,26] for step decreases in the feed flow rate, and for step changes in rotational speed, and cylinder inclination, for cylinders of length 1 m and 2 m.

A breakthrough time, t_b, can be defined as the time after a step change when the discharge flow rate has changed by 1% of its initial value. For a diffusion-like process, $t_b \propto L^2$, and for a wave-like process $t_b \propto L$, whereas for the unsteady state behaviour in our experiments, the nonlinear diffusion Equation 9 gives[18,26] $t_b \propto L^{1.6}$.

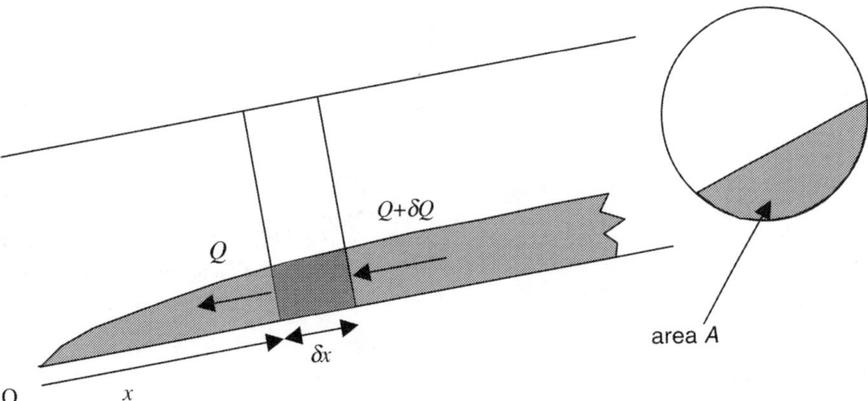

Figure 6　*Element of kiln of length δx. The volumetric flows Q and Q+δQ relate to the unsteady state material balance.*

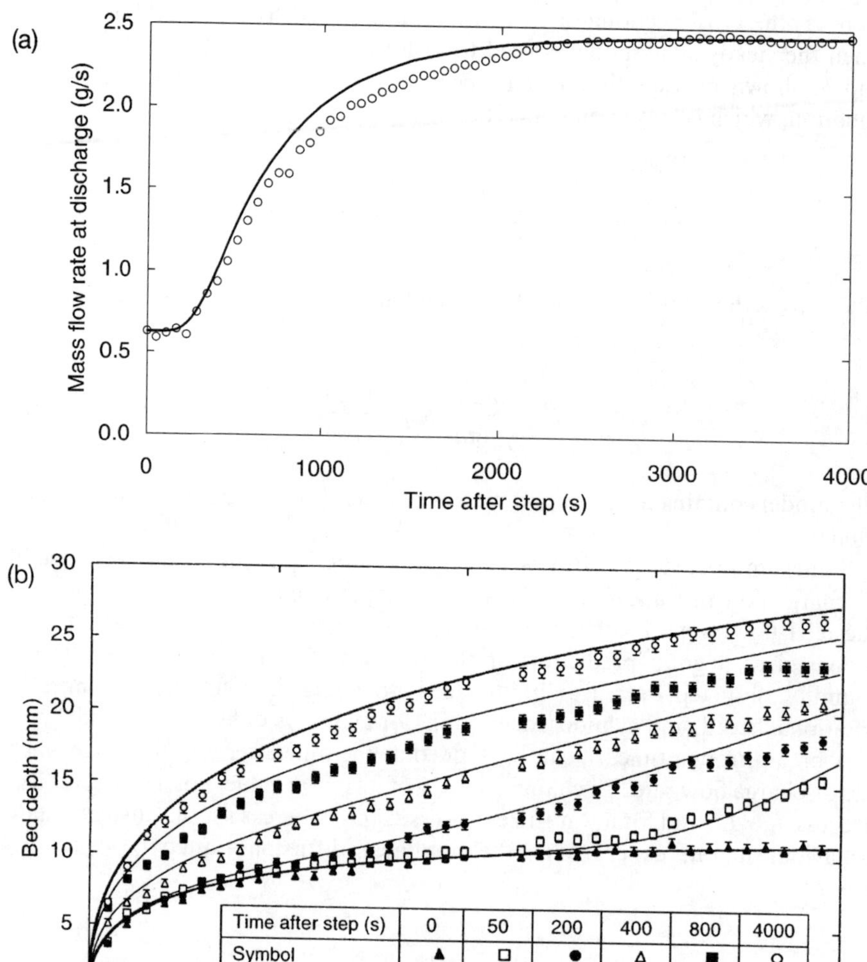

Figure 7   *Development of (a) discharge flow rate and (b) axial bed depth profile with time following a step up in the feed rate. Kiln length 1 m, diameter 0.105 m. Angle of inclination 1°. Initial feed rate 0.625 g/s, final feed rate 2.45 g/s. Speed 0.0855 rps. See Spurling[18] and Spurling et al.[26] The curves are theoretical, from numerical solution of equation (9).*

4   Avalanching

Avalanching is a common mode of granular motion in an industrial rotary kiln. The transition from avalanching to rolling has been addressed, amongst others, by Henein *et al.*,[1] who proposed a Froude number as a suitable parameter to define the transition:

$$Fr = \omega^2 R/g \tag{11}$$

where ω (rad/s) is the angular velocity of the kiln. Their experiments showed that very low values of Fr, less than 10^{-5}, are consistent with avalanching. Then there is a transition region where $10^{-5} < Fr < 10^{-4}$; values of Fr above 10^{-4} are consistent with the rolling mode. The use of Fr is important for scale-up. If the transition from avalanching to rolling were uniquely described by Fr, then large kilns would have the transition at lower speeds than small kilns.

We have studied[27,28] this transition in horizontal drums of diameter 194, 288 and 500 mm using sand in the size range 300–500 μm, and TiO_2 powder as discharged from a rotary kiln. It was found that Fr gives a rough guide to the transition between the flow regimes, but (*i*) the transition values of Fr are markedly different between sand and raw TiO_2, and (*ii*) the transition value Fr tends to increase as the drum diameter decreases. The overall conclusion is that although Fr can give a very rough guide to behaviour, there is certainly not a unique value of Fr for each transition. There must therefore be grave doubts in the use of these transition Froude numbers for industrially-sized kilns.

With this in mind we studied cycle times in the avalanching or slumping mode. Figure 8 shows the sequence observed at low rotational speeds. An avalanche starts at state 1, when the bed surface is at maximum angle γ_s, the static angle of repose. Then an avalanche brings the granular bed into state 2, when the inclination of the bed surface is γ_d. Between states 2 and 3 the granular bed rotates with the drum in rigid body rotation. It is assumed that as ω varies the avalanche time between states 1 and 2, t_{12}, is approximately constant. During the 2–3 sequence, there is rigid body rotation, and so the time for the whole cycle, t_{13}, is given by

$$t_{13} = t_{12} + (\gamma_s - \gamma_d)\pi/(180\omega) \tag{12}$$

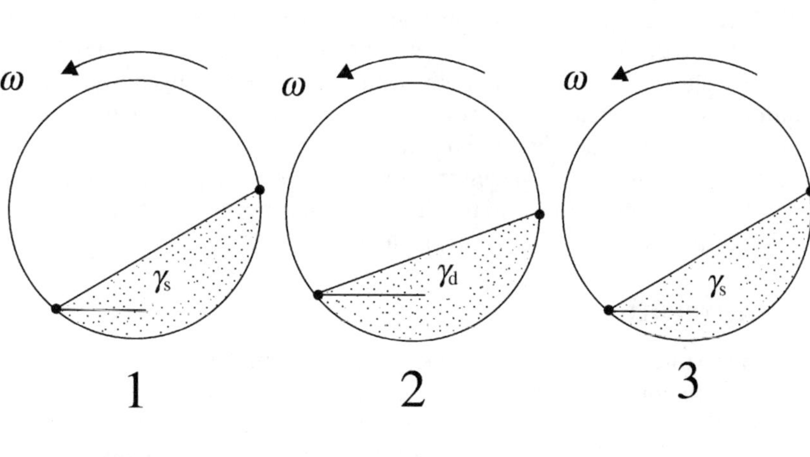

Figure 8 *Cyclical slumping or avalanching. (1) start of slump at angle γ_s. (2) end of slump at angle γ_d. (3) start of next slump at angle γ_s.*

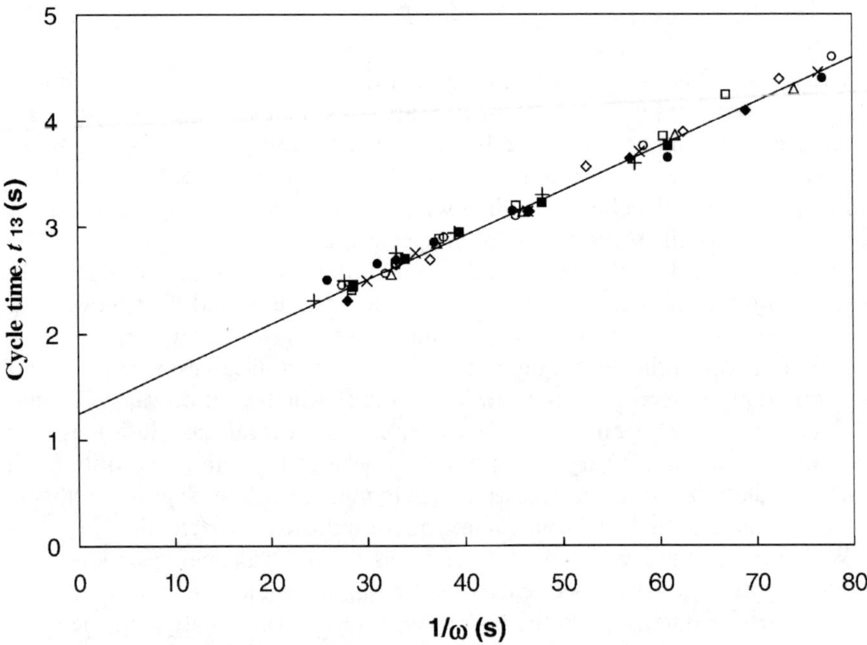

Figure 9 *Slump cycle time against 1/(angular speed) for sand in 500 mm diameter drum. The volumetric fill varies from 3.6% to 23.8%. See Bird and Herbert[27] and Davidson et al.[28]*

This linear dependence of t_{13} on $1/\omega$ is borne out by experiment; an example is shown in Figure 9.

The path of an individual granule in the avalanching mode has been followed by Lim[19,29] using PEPT. A particle rotates with the cylinder in rigid body rotation until it reaches the 'active layer', the region adjacent to the free surface, and then moves down the active layer in a number of discrete avalanches until it is absorbed again into the rigid body rotation. Experiment shows that as the rotational speed increases, and consequently *Fr* increases, the mean number of avalanches experienced by a particle in its traverse down the active layer decreases; see Figure 10. When this mean number of avalanches approaches unity, the transition from avalanching to rolling occurs: from Figure 10, it is clear that the transition is not sharply defined.

5 Segregation

When a mixture of two distinct species of granular material is rotated slowly in a cylinder with horizontal axis, the granules forming a rolling bed filling about one third of the cylinder volume, it is possible for the mixture to separate into distinct bands, each band rich in one of the components of the mixture. The first reported observation of such axial banding of which we are aware is by Oyama,[30] and a great deal of work has been done since then. A recent review is given by Levine.[31]

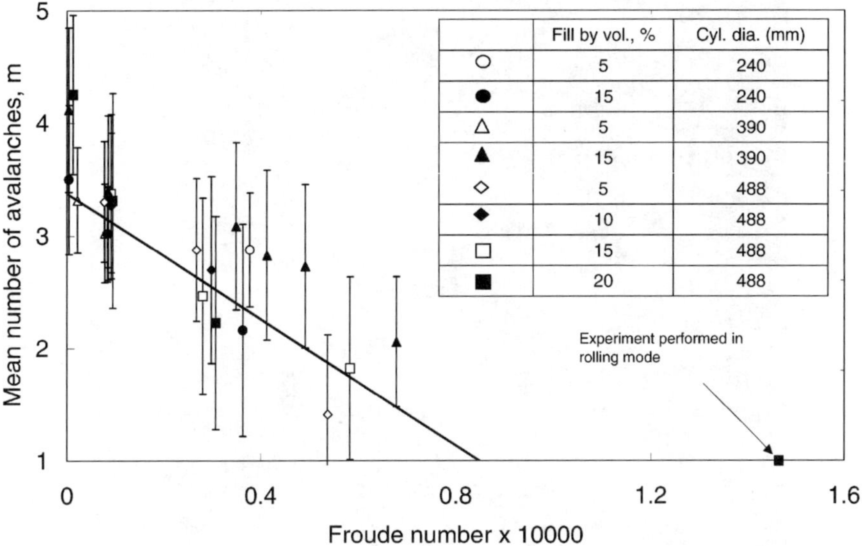

Figure 10 *The mean number of avalanches needed for a tracer particle to traverse the active layer versus Froude number. See Lim et al.*[29]

One condition under which banding occurs is when the two species are of different sizes, and the smaller species has the larger angle of repose; see Donald and Roseman.[32] A physical mechanism which could then produce the banding has been proposed by Das Gupta *et al.*[33] and Zik *et al.*,[34] as follows. The process starts with a spontaneous fluctuation in the composition, initially uniform. An axial region with a depletion in the large particles will have a free surface with a larger inclination to the horizontal than the rest of the bed of particles: thus the bed will be locally elevated in comparison with the axially adjacent sections. The larger particles, being more mobile, will tend to roll axially away from the elevated region, thus making it even more depleted in the larger particles.

Published work so far has been concerned with batch processes, with the cylinder axis horizontal. However in many practical applications of rotating cylinders, there is continuous flow of the granular material, with feed at one end and discharge at the other; the axis of the cylinder is usually inclined at a small angle to the horizontal, with the inlet higher than the outlet. We[35–37] have investigated continuous segregation in the rolling mode using the apparatus described earlier. The discharge end of the cylinder was fitted with a suitably chosen annular dam to control the height of the bed to be approximately uniform along the cylinder. Granular material was continuously fed to the cylinder from hoppers using vibratory feeders. The granular material used was a mixture of 300 μm white sand, with an angle of repose of 36°, and 1 mm black glass ballotini, with an angle of repose of 26°. Thus the smaller species had the larger angle of repose. Visually, the two species were easily distinguishable. Observations were made for the existence of bands, that is adjacent regions with visible difference in composition; an example is shown in Figure 11. Near the feed point

Figure 11 *Stripes formed in a cylinder (length 2.00 m, diameter 0.105 m) fed with a mixture of 49wt% sand (300–600 µm) and black ballotini (1.00–1.25 mm). Angle of inclination 1.4°. Feed flow rate 4.11 g/s. Speed 0.125 rps. Bed depth 21 mm.*

and for some distance downstream, the bed surface was black, because of segregation causing the black ballotini to be on top of the rolling bed and the white sand within it: this exhibited the well-known segregation effect in each plane normal to the cylinder axis; the small particles segregate into the middle of the bed because the large particles roll more readily down the sloping surface of the bed. Band formation is a gradual process and it is difficult to identify accurately the point at which band formation begins. Attempts were made to measure the distance from the feed end to the place where the bands were well-formed, indicated by smooth, sharp edges, and Figure 12 shows this distance as a function of the angle of inclination of the cylinder for three bed depths. The data suggest that the distance from the feed to the place where the bands become well-formed is approximately proportional to the cylinder inclination, β, for fixed bed depth, h, all other parameters remaining (approximately) unchanged, and that the constant of proportionality increases with increasing h.

The composition of the discharge flow was measured by taking samples of the discharge in polystyrene cups, and sieving. A typical result is shown in Figure 13, where it can be seen that there is a considerable degree of segregation.

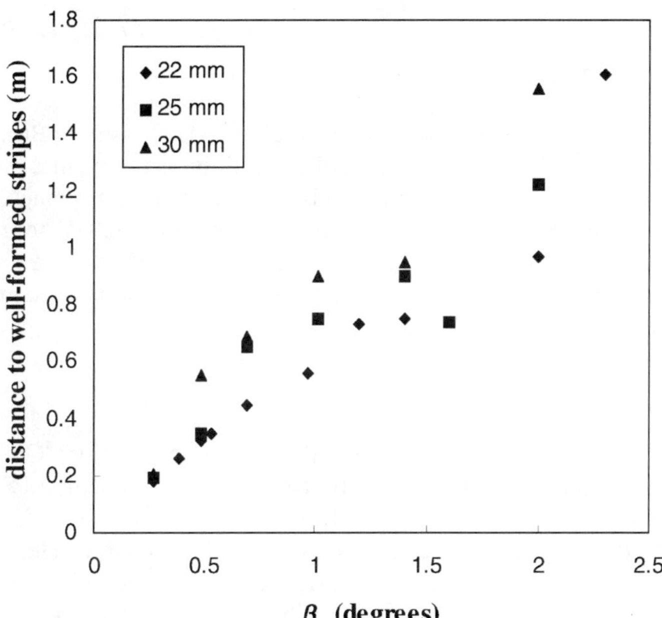

Figure 12 *Distance from feed end to location of formation of well-formed stripes in a cylinder (length 2.00 m, diameter 0.105 m) fed with a mixture of approximately 30 wt% sand (300–600 μm) and black ballotini (1.00–1.25 mm), as a function of angle of inclination. Feed flow rate adjusted to give uniform bed depth indicated on figure). Speed 0.125 rps.*

(Data from Cumming and Hibbs,[35] Guillard and Ladislaus,[36] and Chai and Lam.[37])

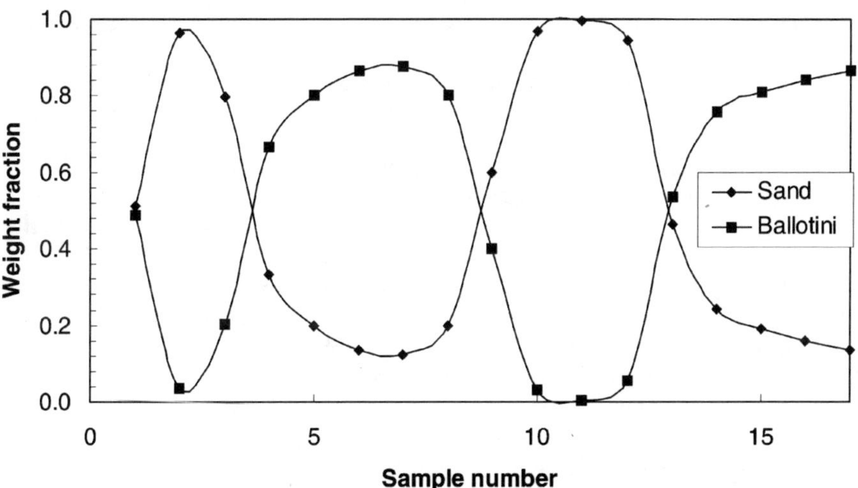

Figure 13 *Composition of discharge from a cylinder (length 2.00 m, diameter 0.105 m) fed with a mixture of approximately 49 wt% sand (300–600 μm) and black ballotini (1.00–1.25 mm). Angle of inclination 1.4°. Feed flow rate 4.11 g/s. Speed 0.125 rps. The discharge was sampled and analysed approximately every 5 s.*

(Data from Guillard and Ladislaus.[36])

6   Related Work

Related work, partly collaborative, is ongoing at the University of Birmingham, with Professor Seville's group, using PEPT for the experimental work. The objectives include (*i*) to characterise solids motion, (*ii*) to determine the effects of process variables on residence time and on solids motion, (*iii*) to investigate the avalanching to rolling transition, (*iv*) to determine scale-up rules, and (*v*) to formulate and test a comprehensive model for solids motion. Some results are given in References 38, 39 and 40.

Acknowledgments

We thank the Cambridge Commonwealth Trust, Her Majesty's Government, Foreign and Commonwealth Office, Huntsman Tioxide, and the EPSRC, UK, for financial support. We are grateful to our Ph.D. students G. McTait, R. J. Spurling, and S.-Y. Lim, and our M.Eng. students P. A. Bird, O. Herbert, A. A. Powell, H. V. M. Ramsay, D. J. Cumming, P. J. Hibbs, N. Guillard, P. J. Ladislaus, T. L. Chai and B. Lam for their contributions to this work. We are grateful to R. Horne and M. Short of Huntsman Tioxide for helpful advice, and to J. P. K. Seville, D. J. Parker, R. Forster, A. Ingram, Y. Ding (now at Leeds) and other members of the PEPT group at the University of Birmingham for all the help they have given us.

References

1. H. Henein, A. P. Watkinson, and J. K Brimacombe, *Metall. Trans. B*, 1983, **14B**, 191.
2. J. D. Sullivan, C. G. Maier, and O. C. Ralston, U.S. Bureau of Mines, Technical Paper No. 384, 1927.
3. R. W. Pickering, F. Feakes, and M. L. Fitzgerald, *J. Appl. Chem.*, 1951, **1**, 13.
4. W. C. Saeman, *Chem. Eng. Prog.*, 1951, **47**, 508.
5. L. Vahl and W. G. Kingma, *Chem. Eng. Sci.*, 1952, **1**, 253
6. H. Kramers and P. Croockewit, *Chem. Eng. Sci.*, 1952, **1**, 259.
7. G. E. McTait, D. M. Scott, and J. F. Davidson, 'Residence Time Distribution of Particles in Rotary Kilns', p. 397 in *Fluidization*, Vol. IX, 1998, eds. L.-S. Fan and T. M. Knowlton, New York: Engineering Foundation.
8. H. M. Ang, M. W. Sze, and M. O. Tade, *The AusIMM Proceedings*, 1998, **1**, 11.
9. J. Des Boscs, 'Granular Motion in Rotating Drums', M. Phil. Thesis, Faculty of Engineering, University of Birmingham, UK, 1998.
10. M. Hehl, K, Schugerl, H, Helmrich, and H. Kroger, *Powder Tech.*, 1978, **20**, 29.
11. V. K. Karra and D. W. Fuerstenau, *Powder Tech.*, 1978, **19**, 265.
12. J. Mu and D. D. Perlmutter, *AIChEJ*, 1980, **26**, 928.
13. R. S. C. Rogers and R. P. Gardner, *Powder Tech.*, 1979, **23**, 159.
14. P. S. T. Sai, K. Sankaran, Z. G. Philip, V. Suresh, A. D. Damodaran, and G. D. Surender, *Metall. Trans. B, 1990*, **21B**, 1005.
15. G. E. McTait, 'Residence Times and Solid Flows in Rotary Kilns', Ph.D. Thesis, University of Cambridge, UK, 1998.
16. R. Hogg, L. G. Austin, and K. Shoji, *Powder Tech.*, 1974, **9**, 99.
17. A. Z. M. Abouzeid and D. W. Fuerstenau, *Powder Tech.*, 1980, **25**, 21.

18. R. J. Spurling, 'Granular Flow in an Inclined Rotating Cylinder: Steady State and Transients', Ph.D. Thesis, University of Cambridge, UK, 2000.

19. S.-Y. Lim, 'Granular Flow in an Inclined Rotating Cylinder', Ph.D. Thesis, University of Cambridge, UK, 2002.

20. R. J. Spurling, J. F. Davidson, and D. M. Scott, *Chem. Eng. Sci.*, 2000, **55**, 2303.

21. P. C. Chadwick and J. Bridgwater, *Chem. Eng. Sci.*, 1997, **52**, 2497.

22. M. R. Hawkesworth, G. Jonkers, N. L. Jeffries, J. F. Crilly, P. Fowles, and D. J. Parker, *Nucl. Instrum. Meth. A*, 1991, **236**, 592.

23. D. J. Parker, P. A. McNeil, M. R. Hawkesworth, P. Fowles, and C. J. Broadbent, *Nucl. Instrum. Meth. A*, 1991, **310**, 423.

24. D. J. Parker, P. A. McNeil, T. D. Fryer, P. Fowles, C. J. Broadbent, and M. R. Hawkesworth, *Nucl. Instrum. Meth. A*, 1994, **348**, 583.

25. S.-Y. Lim, 'Granular Flow in an Inclined Rotating Cylinder', Certificate of Postgraduate Studies Thesis, University of Cambridge, UK, 2000.

26. R. J. Spurling, J. F. Davidson, and D. M. Scott, *Trans. IChemE.*, 2001, **79** Part A, 51.

27. P. A. Bird and O. Herbert, 'Particle Dynamics in a Rotary Cylinder', MEng Reports, Department of Chemical Engineering, University of Cambridge, UK, 1996.

28. J. F. Davidson, D. M. Scott, P. A. Bird, O. Herbert, A. A. Powell, and H. V. M. Ramsay, *KONA Powder and Particle*, 2000, **18**, 149.

29. S.-Y. Lim, J. F. Davidson, R. N. Forster, D. J. Parker, D. M. Scott, and J. P. K. Seville, 'Avalanching of Granular Material in a Slowly Rotating Horizontal Cylinder: PEPT Studies', 4th World Conference on Particle Technology, Sydney, Australia, July 2002.

30. Y. Oyama, *Bull. Inst. Phys. Chem. Res. Japan Rep.*, 1939, **18**, 600 (in Japanese).

31. D. Levine, *Chaos*, 1999, **9**, 573.

32. M. B. Donald and B. Roseman, *Brit. Chem. Eng.*, 1962, **7**, 749.

33. S. Das Gupta, D. V. Khakhar, and S. K. Bhatia, *Chem. Eng. Sci.*, 1991, **46**, 1513.

34. O. Zik, D. Levine, S. G. Lipson, S. Shtrikman, and J. Stavans, *Phys. Rev. Lett.*, 1994, **73**, 644.

35. D. J. Cumming and P. J. Hibbs, 'Axial Segregation of Binary Mixtures in a Rotating Cylinder', MEng Reports, Department of Chemical Engineering, University of Cambridge, UK, 2000.

36. N. Guillard and P. J. Ladislaus, 'Axial Segregation of Granular Materials in a Rotating Cylinder', MEng Reports, Department of Chemical Engineering, University of Cambridge, UK, 2001.

37. T.-L. Chai and B. Lam, 'Flow of Granular Materials in a Rotating Cylinder', MEng Reports, Department of Chemical Engineering, University of Cambridge, UK, 2002.

38. Y. L. Ding, R. N. G. Forster, J. P. K. Seville, and D. J. Parker, *Chem. Eng. Sci.*, 2001, **56**, 1769.

39. Y. L. Ding, R. N. G. Forster, J. P. K. Seville, and D. J. Parker, *Powder Tech.*, 2002, **124**, 18.

40. Y. L. Ding, R. N. G. Forster, J. P. K. Seville, and D. J. Parker, *Int. J. Multiphase Flow*, 2002, **28**, 635.

CHAPTER 13

Granular Motion in the Transverse Plane of Rotating Drums

YULONG DING[1], S. JOSEPH ANTONY[1] and JONATHAN SEVILLE[2]

[1]Institute of Particle Science and Engineering, University of Leeds,
Leeds LS2 9JT, United Kingdom
[2]Department of Chemical Engineering, University of Birmingham,
Birmingham B15 2TT, United Kingdom

1 Introduction

A granular material is a collection of a large number of discrete solid particles and there is normally a fluid such as air or water in the interstices. Depending on the prevailing stress and strain conditions, it shows both discrete and continuous characteristics. Such fascinating features determine solids behaviour in rotating drums, one of the most popular devices for processing granular materials.

A rotating drum is usually a cylinder rotating about its central axis, which is either horizontally positioned or inclined at a few degrees to the horizontal. Devices based on such a configuration play an important role in the chemical, metallurgical, food, detergent and pharmaceutical industries, in which they are used to perform mixing, drying, heating and chemical reactions. Although the concept is simple, solids motion in rotating drums is very complicated. Six modes of solids motion have been observed in the transverse plane (Figure 1). With increasing rotational speed, they are slipping, slumping, rolling, cascading, cataracting, and centrifuging modes.[1] This chapter is concerned with both experimental and theoretical aspects of solids motion in horizontally positioned drums loaded with less than 50% by volume solids, and operated in the rolling mode. Section 2 presents experimental observations of free-flowing materials in rolling drums. Section 3 describes experimental results with cohesive powders. Theoretical analyses using both discrete and continuous approaches are performed in Section 4. Section 5 concludes this chapter. Experimental results

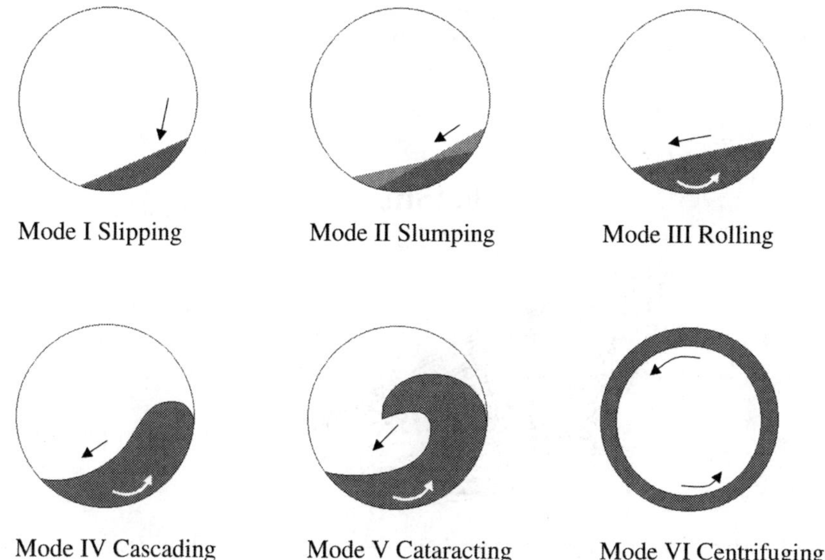

Figure 1　　*Models of solids motion in the transverse plane of a rotating drum*

presented in the chapter are obtained exclusively by using the Positron Emission Particle Tracking (PEPT) technique. The principle of such a technique can be found elsewhere.[2]

2　Experimental Observations of Free-Flowing Materials in Rolling Drums

2.1　Surface Shape, Dynamic Repose Angle and Bed Expansion

Surface shape of particle bed depends on both physical properties of particles and drum operating conditions. For both mono-sized particles and binary mixtures, bed surface is approximately flat in the rolling mode except for positions close to the drum wall.[3–5] This can be seen from a typical velocity vector map for 1.5mm glass beads in a 240mm drum (Figure 2). The velocity vector map can be used to calculate the dynamic repose angle, which is found in a range of 24–27° for glass beads.[4,5] The velocity vector map can also give an indication of bed expansion. Experimental results show that bed expansion occurs mainly in the surface region and increases with increasing rotational speed. At low to medium rotational speeds, bed expansion is less than 10%.[5]

2.2　Bed Structure and Velocity Profiles in Rolling Drums

In the rolling mode, bed material can be divided into two distinct regions, namely, a 'passive' region where particles are carried upward by the drum wall and a relatively thin 'active' region where particles flow down the sloping upper

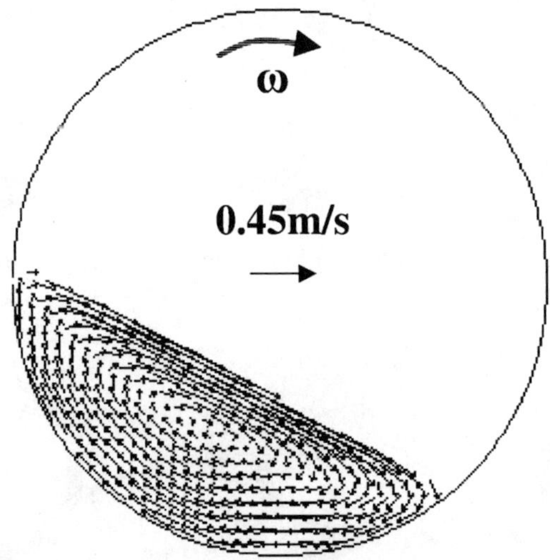

Figure 2   *Map of the velocity vectors for 1.5mm glass beads in a 240mm drum operated at 9.0rpm, 25% fill by volume (PEPT measurements)*

bed surface (Figure 3). Such a structure has been observed for both mono-sized particles[5] and binary mixtures.[4] Figure 4(a) shows x-direction velocity (u) profiles for mono-sized particles at two sections normal to the bed surface, where R is drum radius and L is the half-chord length; see Figure 3. At a given section, the maximum x-wise velocity in the active region occurs at the bed surface, similarly to chute flows. However, the maximum bed surface velocity occurs at a position close to the mid-chord, i.e. the bed material accelerates until reaching a position close to the mid-chord where deceleration starts. This is significantly different from chute flows, in which continuous acceleration is often observed.

Velocity profiles for binary mixtures are similar to those for mono-sized particles at relatively low rotational speeds but difference occurs at high rotational speeds due to velocity-slip between the small and large particles; see Figures 4(b) and 4(c). PEPT measurements also indicate a rather thick active region for binary mixtures,[4] which cannot be revealed by visual observations.

Comparison of Figures 4(a), 4(b) and 4(c) also shows that the so-called free surface (zero shear) boundary condition does not always apply to granular flows in rotating drums. At low to medium rotational speeds, $(du/dy)|_{y=0}$ approaches zero (Figures 4a and 4b).[5] However, $(du/dy)|_{y=0}$ is not zero at high rotational speeds (Figure 4c).[3,4]

2.3   Segregation of Binary Mixture of Free-flowing Materials in Rolling Drums

A mixture of granular materials with different physical properties such as size, density, shape, roughness and resilience, tends to segregate.[6] Segregation occurs in both the radial and axial directions of rotating drums. Segregation in the

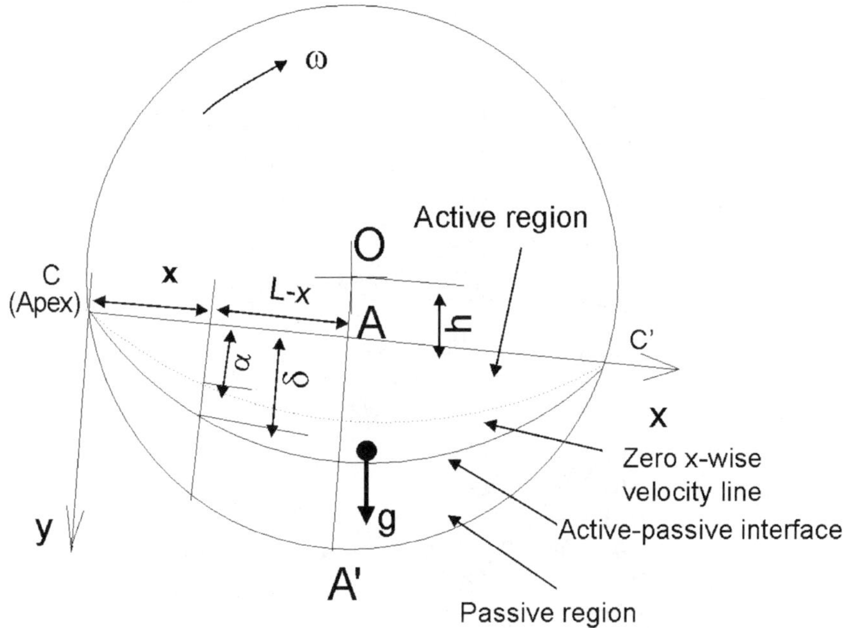

Figure 3   *Bed structure of the transverse plane of rolling drums*

transverse plane occurs mainly through percolation, random sieving and expulsion, and trajectory mechanisms,[6,7] while the axial segregation may be attributed to difference in the repose angles of different materials. Axial segregation proceeds slowly (usually hundreds or even thousands of revolutions),[8] while radial segregation takes place rapidly (often within several drum revolutions).[9] In the transverse plane, fine, rough and dense particles are found to concentrate in the core region by most workers.[10] Coarse particles are also observed to concentrate in the core region under certain conditions.[11] For axial segregation, alternative bands of coarse and fine particles have been observed.[12]

One way to characterise segregation is the use of the so-called occupancy plots for each component of the mixture. Typical examples are shown in Figure 5 for a mixture of small and large particles with the same density, which are obtained from the PEPT measurements. Here the occupancy of a component is defined as the ratio of the time that the component spends at a given position to the total observation time. Such a parameter can be regarded as an indication of concentration for highly agitated systems. From Figures 5(a) and 5(b), a clear core-shell structure can be seen, which suggests small particles tend to occupy the core region of the bed, while large particles spend most of their time in the shell region. Figures 5(a) and 5(b) also reveal a gradual change in the occupancies of both small and large particles from the core to shell regions, an indication of diffusive-like mixing (continuous behaviour).[4] Figures 5(c) and 5(d) show the axial occupancy plots measured over a period of about two hours, which suggest that the axial mixing is very poor. The results shown in

(a)

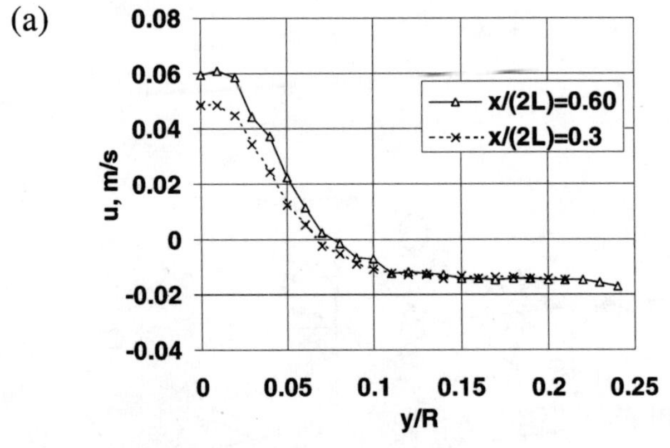

(b)

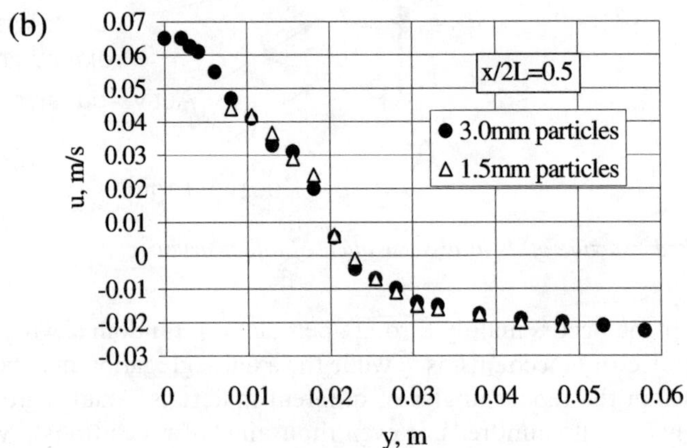

(c)

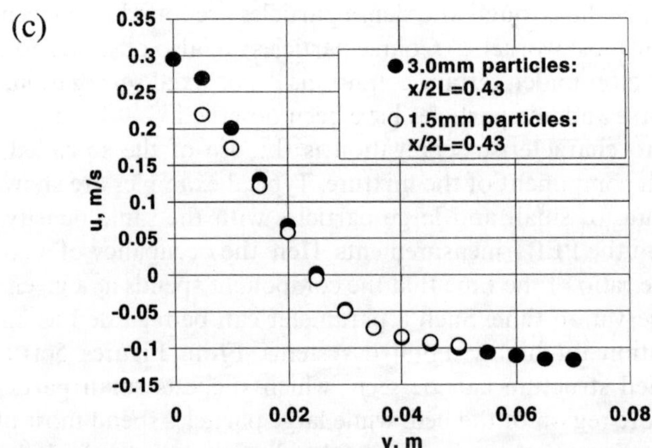

Figure 4   *Velocity distributions in x-direction*

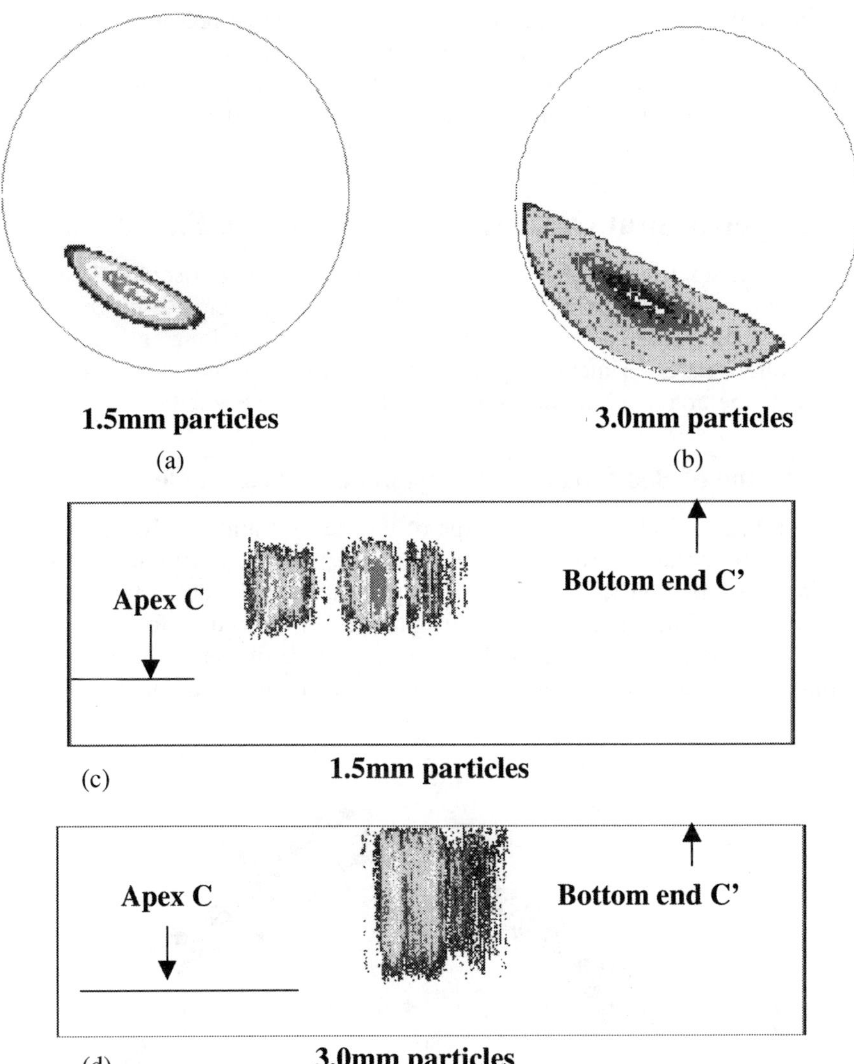

Figure 5 *Occupancy plots in the transverse and axial directions: 30% 1.5mm / 70% 3.0mm glass beads, 240mm drum, 25% fill, 9.55 rpm*

Figures 5(c) and 5(d) also indicate that small particles (in the core) have a higher axial mobility.

2.4 Surface Velocity Profiles

It is interesting to compare the surface velocity (u_s) under different operating conditions. Figure 6 shows the results for glass beads and sand.[4] The results have been normalised in terms of the velocity at the mid-chord position (x=L). It can be seen that experimental data in the second half agree reasonably well, but deviation occurs in the first half. The exact reasons for such difference are unknown. It seems to be associated with difference in physical and mechanical properties and hence different dissipative characteristics of the two types of materials.

3 Experimental Observations of Cohesive Particles

Cohesion arises from strong interparticle forces in comparison with gravity. Cohesive particles thus exhibit different behaviour from their free-flowing counterparts, which creates considerable difficulty for their handling and processing. In the following, experimental work on cohesive titanium dioxide particles with 1.2mm diameter and a bulk density of ~760kg/m^3 are presented.

3.1 Shape of Bed Surface and Dynamic Repose Angle

As described in Section 2.1, the shape of bed surface and the dynamic repose angle can be obtained from velocity vector maps. Experiments with titanium dioxide powders show that bed surface is approximately flat but differs from that of glass beads. There are small local avalanches on the bed surface. These avalanches may be associated with the cohesive nature of the particles since the avalanches do not start from the apex of the bed surface as observed in the

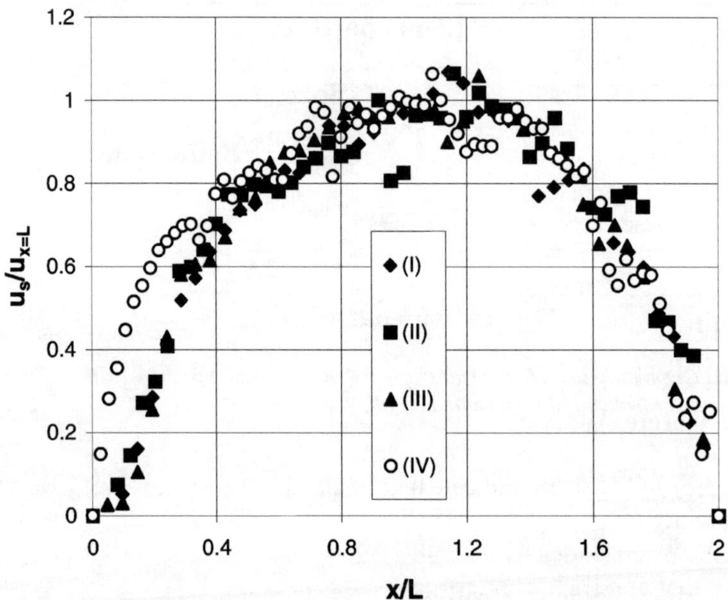

Figure 6 *Surface velocity profiles: (I) 240mm drum, 25% fill of binary mixture of 30% 1.5mm and 70% 3.0mm glass beads, 9.55rpm. (II) 400mm drum, 7.5% fill of 1.5mm glass beads, 0.72rpm. (III) 240mm drum, 15% fill of 3.0mm glass beads, 7.55rpm. (IV) 400mm drum, 30% fill of 0.5mm sand particles, 2rpm*

slumping mode. PEPT data show that the dynamic repose angle is ~32°, about 6° larger than that of glass beads.

3.2 Surface Velocity of Titanium Dioxide Powder

Typical surface velocity profiles for 1.2mm titanium dioxide in a 240mm drum are shown in Figure 7. In contrast to the single peak for glass beads (Figure 6), two peaks occur at positions of x/(2L) ~ 0.3 and 0.7. The first peak velocity is about 10% lower than the second one. The trough between the two peaks occurs at x/(2L) ~ 0.45, and the velocity at the trough is approximately 75% of the higher peak velocity. Interestingly, the dual peak velocity profiles can be fitted to two 2nd order polynomials (parabolic curves) with quite high regression coefficients (>0.94). It is also interesting to note that, for a given drum fill (14.5%), the maximum surface velocity of TiO_2 particles is higher than that of glass beads even though the rotational speed of the former is lower. This may be due to the higher dynamic repose angle of TiO_2, which makes the particles at the apex have more potential energy.

The dual-peak velocity distributions have also been observed for long rice and limestone particles by using the fibre-optic technique,[3] and instabilities have been proposed to be responsible which develop as a result of granular energy dissipation – a phenomenon depending strongly on the solid fraction and particle restitution coefficient.[13] The instabilities may also be responsible for the behaviour of titanium dioxide particles. Another possible reason is that titanium dioxide particles are cohesive and may aggregate to form relatively large agglomerates. The velocities of these large granules would be higher than that of primary particles and hence formation of the first peak. However, the high velocity may lead to breakage of the aggregates during rolling down and results in a decrease in the particle velocity and hence formation of the trough. After a short distance, particles would pick up speed so that the second small peak is observed. These reasons are supported by the observation of local avalanches made during the experiments though the first reason is more convincible as the dependence of the surface velocity on particle diameter seems weak.

4 Analysis of Free-Flowing Particles in Rolling Drums

4.1 Discrete Analysis

4.1.1 Motion of a Single Particle on the Bed Surface – a Kinematic Analysis

Consider a drum loaded with mono-sized particles and rotates at a relatively low speed so that an ideal rolling mode is established. Assume that particle motion in the active layer is not energetic enough to generate a highly dilated granular flow. In this case, a particle rolling down the bed surface from the apex has little chance to find a void to percolate through the surface layer. With reference to Figure 3, an x-wise direction force balance on the particle gives:

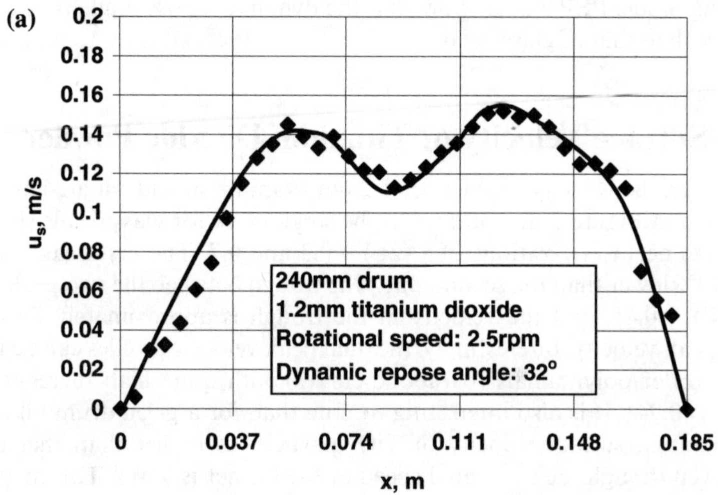

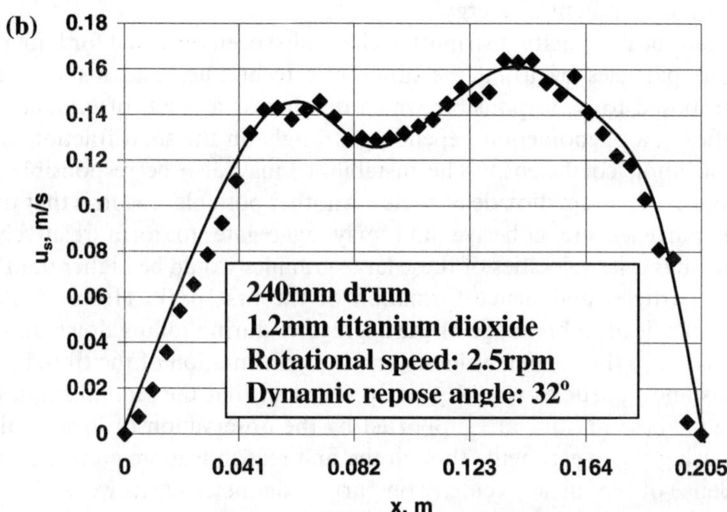

Figure 7 *Surface velocity profiles: 1.2mm TiO₂ particles, 240mm drum*

$$m_p g \sin \beta - F_{pp} = m_p \frac{du_p}{dt} = m_p u_p \frac{du_p}{dx} = \frac{1}{2} m_p \frac{du_p^2}{dx} \tag{1}$$

where m_p is the particle mass, g is the acceleration due to gravity, β is the dynamic repose angle, u_p is discrete particle velocity and F_{pp} is particle-particle interaction force. In Equation 1, the forces due to gas drag and the centrifugal action are neglected. If the particle-particle interaction force is ignored then solution to Equation 1 gives

$$u_p = \sqrt{(2g \sin \beta)x} \tag{2}$$

On the other hand, if particle-particle interaction force is independent of particle velocity and is approximated by the usual frictional force $F_{pp} = \mu_p m_p g . \cos\beta$, then solution to Equation 1 gives:

$$u_p = \sqrt{2g(\sin \beta - \mu_p \cos \beta)x} \tag{2a}$$

where μ_p is the particle-bed surface friction coefficient. Note that both Equations 2 and 2a are obtained by using $u_p = 0$ at $x = 0$ as the boundary condition. Equations 2 and 2a have been used by some workers in the derivation of particle residence time.[14–16] However, these two expressions show a constant increase in particle velocity with increasing distance from the apex, which is obviously in contradiction to the experiments as shown in Figure 6 where maximum velocities occur at positions close to the middle chord. An obvious reason for this is that the particle-particle interaction force is not properly accounted for. At positions close to the bottom end of the chord (C' in Figure 3), particles collide with drum wall and nearby particles, which consumes most of their kinetic energy gained from the conversion of the potential energy. Such particle-wall and particle-particle interactions at the bottom end may travel upstream and affect the motion of upstream particles. In other words, upstream particles may feel the interactions occurring at the bottom end. Based on this argument, it is hypothesised that F_{pp} has the following form:

$$F_{pp} = \mu_p m_p g \cos \beta + K_1 \cdot \frac{1}{2} m_p u_p^2 / (2L - x)^n \tag{3}$$

where K_1 and n are constants. The first term accounts for the local particle-particle interactions (approximated by the usual frictional force). The second term accounts for the effect of particle-particle and particle-wall interactions at the bottom end, which is assumed to be proportional to particle kinetic energy (more kinetic energy implies more intensive interactions), but in inverse proportion to the distance from the bottom end of the chord. Substitution of Equation 3 into Equation 1 and rearrangement gives:

$$\frac{du_p^2}{dx} + K_1 \frac{u_p^2}{(2L - x)^n} = 2g(\sin \beta - \mu_p \cos \beta) \tag{4}$$

Equation 4 is a first order ordinary differential equation with an analytical solution depending on *n*. Considering the simplest case, *n* = 1, the solution can be easily obtained by using $(2L - x)^{-K_1}$ as the integrating factor and $u_p = 0$ at $x = 0$ as the boundary condition:

$$u_p = \sqrt{\frac{2g(\sin \beta - \mu_p \cos \beta)(2L)^{1-K_1}}{1 - K_1} \left[(2L - x)^{K_1} - \frac{2L - x}{(2L)^{1-K_1}} \right]} \tag{5}$$

Taking $K_1 = 2$, one has:

$$u_p = \sqrt{\frac{g(\sin \beta - \mu_p \cos \beta)}{L}(2Lx - x^2)} \qquad (5a)$$

Equation 5a suggests that the maximum velocity occurrs at the mid-chord position, in agreement with the experiments using glass beads and sand (Figure 6). It should be noted that the model proposed here lumps the effects of drum operating conditions and particle properties in two parameters K_1 and n. It is only intended to serve as a qualitative model.

4.1.2   *Maximum Surface Velocity – an Energy Analysis*

From the viewpoint of energy transfer, particles are carried up toward the apex by the drum wall so that they gain a certain amount of potential energy before rolling down. However, particles at the apex have little kinetic energy as can be seen from the surface velocity profiles (Figures 4 and 6). Considering particles in contact with the drum wall, they must also carry some kinetic energy because they move up at the same velocity as the drum wall if there is no particle-wall slippage. It seems that the kinetic energy is dissipated as a result of inelastic collisions occurring in the region close to the apex where material build-up is often observed in experiments (material build-up also suggests that compression of granular material may take place at the apex). During rolling down, besides overcoming dissipation due to local inelastic particle-particle collisions and non-local particle-particle and particle-wall interactions mentioned above, the potential energy of these particles gradually converts to their kinetic energy. Most of the kinetic energy is owned by the mean flow, i.e. macroscopic kinetic energy. The rest is in the form of random energy (granular temperature). The partition of the two energy forms depends on the shear rate in the active layer. For drums operated at low to medium rotational speeds, solids motion is usually in the quasi-static regime and the random kinetic energy is small.[5] On reaching the bottom end of the chord, due to strong particle-drum wall interaction, compression may also occur hence loss of the kinetic energy.

The above analysis suggests that the maximum macroscopic kinetic energy of a particle rolling down the bed surface from the apex, $m_p u_m^2/2$, should be proportional to $m_p gL\sin\beta$, the potential energy difference between the apex and the mid-chord position, where u_m is the maximum particle velocity. For a given drum and bed material, this implies:

$$Fr_m = \frac{u_m}{\sqrt{2gL\sin\beta}} = \text{constant} \qquad (6)$$

where Fr_m is the modified Froude number based on u_m. Plots of Fr_m against drum fill and rotational Froude number Fr_r ($= \omega^2 R/g$) are shown in Figures 8(a) and 8(b) for 3mm glass beads in a 240mm drum, 0.5mm sand in a 400mm drum,

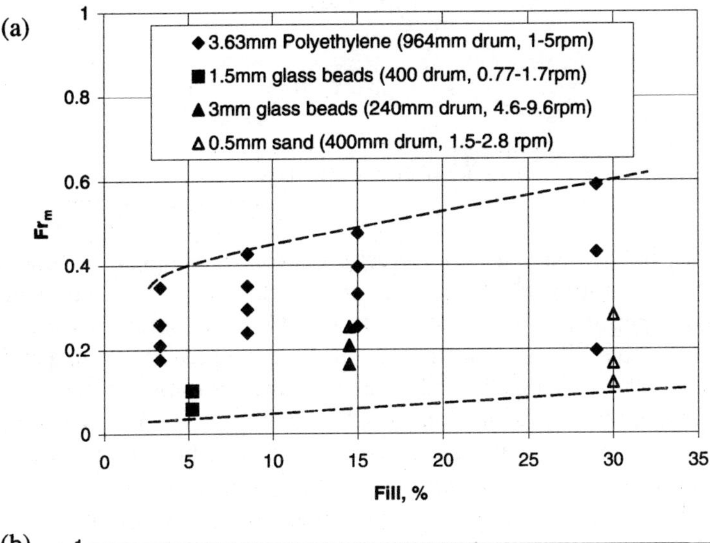

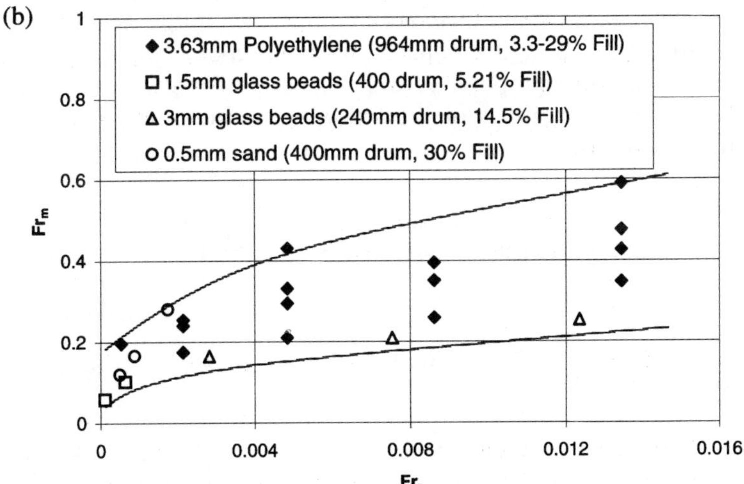

Figure 8 *Effects of (a) drum fill and (b) rotational speed on the Fr_m*

1.5mm glass beads in a 400mm drum, and 3.63mm polyethylene in a 964mm drum. Although scattering, Fr_m is seen to change only slightly with increasing drum fill level and rotational speed, in particular when $Fr_r > 0.004$.

4.2 Continuous Analysis of Solids Motion for Mono-sized Particles

Assume that the granular material is a continuous medium; the flow field can be obtained from the following conservation equations of mass, momentum, and pseudo-energy:

$$\frac{d\rho}{dt} + \rho\nabla\cdot\vec{u} = 0 \tag{7}$$

$$\rho\frac{d\vec{u}}{dt} = \rho\vec{g} - \nabla\cdot\vec{p} \tag{8}$$

$$\frac{3}{2}\rho\frac{dT}{dt} = -\vec{p}:\nabla\vec{u} - \nabla\cdot\vec{q} - \gamma \tag{9}$$

where ρ is the bulk density, $\vec{u}$ is the velocity vector, $\vec{p}$ is the pressure tensor, T is the granular temperature, q is the pseudo-energy flux due to conduction, and γ is the energy dissipation due to inelastic collisions. However, solution to Equations 7–9 is often too complicated as available constitutive equations for $\vec{p}$, q and γ are very complicated in most cases and involves many assumptions which are not always applicable to solids motion in rotating drums.[17,18] As a consequence, development of relatively simple models is required, which should be able to catch the main features of solids motion in rotating drums. In the rolling mode, bed material has been shown to consist of an active layer and a passive region. The particle bed in the passive region can be described by the rigid body model. As the active region is thin compared with the chord length, a thin layer approximation has been proposed for this region; see for example References 4 and 5. Such a model consists of the following integro-differential mass and momentum equations:

$$\frac{d}{dx}\int_0^\delta u\,dy = \omega\left[(L-x) - (h+\delta)\frac{d\delta}{dx}\right] \tag{10}$$

$$\frac{d}{dx}\int_0^\delta u^2\,dy = \omega^2(h+\delta)\left[(h+\delta)\frac{d\delta}{dx} - (L-x)\right] + g\cos\beta(\tan\beta - \tan\xi)\delta \tag{11}$$

where u is the continuous solids velocity in x-direction, h is the shortest distance between drum centre and bed surface, ω is the drum angular velocity, δ is the active layer depth, and ξ is the internal frictional angle; see Figure 3. The derivation of Equations 10 and 11 involves the following additional assumptions: (a) flat bed surface; b) non-cohesive particles; and c) negligible centrifugal force compared with gravitational action. Equations 10 and 11 only consist of x-wise velocity (u); y-direction velocity (v) can be obtained from the differential continuity equation (Equation 7).

The solution of Equations 10 and 11 involves:

(a) Select a suitable profile for u, usually, a polynomial or other simple functions deduced from experiments,
(b) Apply the boundary conditions (and even the differential governing equations) to obtain the coefficients of the polynomial, which are functions of δ,

(c) Apply Equations 10 and 11 to obtain δ and therefore u, and
(d) Apply Equation 7 to obtain v.

Depending on the rheological properties of bed material and shear rate, the flow in the active layer may mimic pseudoplastic (indicated by slightly convex velocity profiles), Newtonian (linear velocity profiles), or dilatant (concave velocity profiles). As dilatant flows are observed in most reported work on dry granular flows in rotating drums, only the concave velocity profiles are considered here. For dilatant flows, experiments have shown two cases depending on the value of $(du/dy)|_{y=0}$. For drums loaded with spherical particles of relatively high restitution coefficients (e.g. glass beads) and operated at low to medium rotational speeds, bed dilation is small, $(du/dy)|_{y=0}$ is approximately 0 and the velocity u can be approximated by:[5]

$$u = \frac{\omega(h+\delta)}{1-\Lambda^2}\left[\Lambda^2 - \left(\frac{y}{\delta}\right)^2\right]$$
(12)

where $\Lambda\ (\equiv a/\delta)$ is a parameter characterising the rheological behaviour of granular material, and lies within 0.75–0.90 at $x/(2L) = 0.10$–0.90 for mono-sized particles,[3,5] and a is the zero x-wise velocity position; see Figure 3. Substitution of Equation 12 into Equation 10, and considering $\delta = 0$ at $x = 0$, an analytical solution for δ can be obtained:[5]

$$\delta = \frac{1}{3\Lambda^2 + 1}\left[\sqrt{6(1-\Lambda^2)(3\Lambda^2+1)\left(Lx - \frac{x^2}{2}\right) + 4h^2} - 2h\right]$$
(12a)

Note that Equation 12a holds even when the parameter Λ is a function of x. For drums operated at relatively high rotational speeds, a slightly curved bed surface occurs, and the surface cannot be regarded as a free surface if it is still assumed to be flat, i.e. $(du/dy)|_{y=0} \neq 0$. In this case, the following expression has been shown to give good approximation:[4]

$$u = u_s - \left[\frac{1+\Lambda}{\Lambda}u_s - \frac{\Lambda}{1-\Lambda}\omega(h+\delta)\right]\left(\frac{y}{\delta}\right) + \left[\frac{u_s}{\Lambda} - \frac{\omega(h+\delta)}{1-\Lambda}\right]\left(\frac{y}{\delta}\right)^2$$
(13)

where u_s is the surface velocity. Substitution of Equation 13 into Equation 10 and application of the boundary condition $\delta = 0$ at $x = 0$ gives:

$$\delta = \frac{-\left[\frac{3\Lambda-1}{\Lambda}u_s + \frac{4-3\Lambda}{1-\Lambda}\omega h\right] + \sqrt{\left[\frac{3\Lambda-1}{\Lambda}u_s + \frac{4-3\Lambda}{1-\Lambda}\omega h\right]^2 + \frac{12\omega^2(2Lx - x^2)}{1-\Lambda}}}{2\left(\frac{\omega}{1-\Lambda}\right)}$$
(13a)

The boundary conditions used to derive Equation 13a are $u = u_s$ at $y = 0$, $u = -\omega(h+\delta)$ at $y = \delta$, and $u = 0$ at $y = a$. Equations 13 and 13a can be simplified to give Equations 12 and 12a if $(du/dy)|_{y=0}$ is taken as zero. Note again that Equation 13a holds even when the parameter Λ and surface velocity u_s are functions of x. Equations 12 and 13 are compared with experiments in Figure 9, which shows reasonably good agreement.

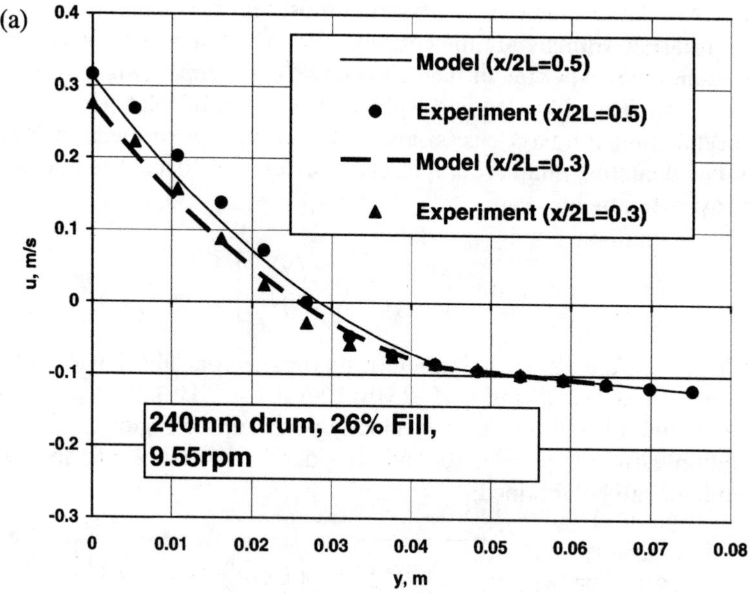

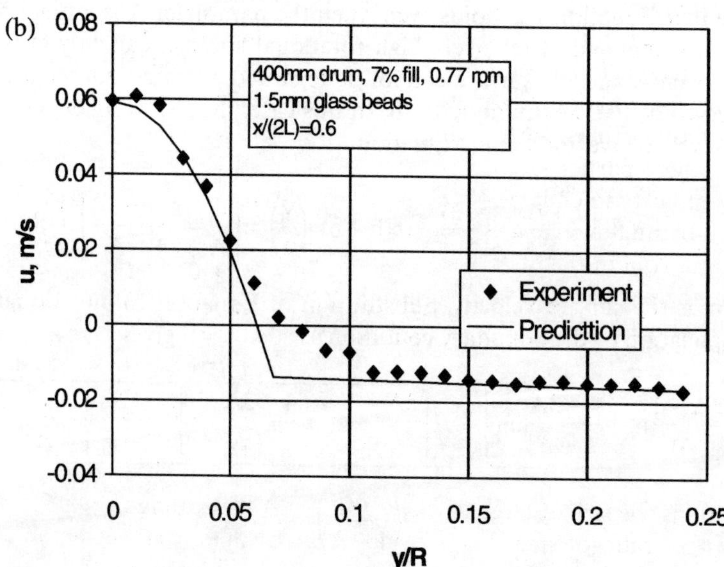

Figure 9   *Comparison between model predictions and PEPT measurements of x-wise velocity*

4.3 Continuum Analysis of Particle Segregation for Binary Mixtures

Consider a horizontal drum filled with a mixture of small and large particles of identical density and operated in a steady-state batch mode. Assume that:

(a) convection and diffusion in the axial direction are negligible,
(b) the mixture is incompressible with the small and large particles interpenetrateable,
(c) x-direction diffusion and percolation are negligible compared with the convective flux, and
(d) the bulk velocity does not depend on concentration, then a mass balance on the small particles gives:[5]

$$u\frac{\partial C_f}{\partial x} + v\frac{\partial C_f}{\partial y} + \frac{\partial[v_p C_f(1 - C_f)]}{\partial y} = \frac{\partial}{\partial y}\left(D_y \frac{\partial C_f}{\partial y}\right) \tag{14}$$

where C_f is the concentration of small particles, u and v are respectively x- and y-wise components of the bulk velocity, v_p is the percolation velocity of small particles in y direction, and D_y is the diffusivity of fines in y direction due to random motion. The 3rd term in the left-hand side of Equation 14 is due to percolation.[19] Equation 14 applies to both the active and passive regions. Its solution requires the bulk velocities u and v, the percolation velocity v_p, and the diffusivity D_y.

As shown in Section 2.2, velocity difference between small and large particles at relatively low rotational speeds is negligible. This suggests that the flow model for the active region described in Section 4.2 can be applied to binary mixtures at relatively low rotational speeds. The passive region follows the rigid body model.

In rotating drums, particle mixing and segregation mainly occur in the active region where particles cascade down the slope of the bed. This process bears some similarity to chute flows. As a consequence, the segregation model for the inclined chute flows can be adopted.[19] This model considers the probability for forming a void in the underlying layer of a particle with a size large enough to capture the particle. The net percolation velocity v_p as a result of this model has the following form:[19]

$$v_p = v_{p0} d_{pl}\left(\frac{du}{dy}\right) \tag{14a}$$

where d_{pl} is the size of large particles, and v_{p0} is the scaling factor that depends on number ratio of small to large particles, mean void diameter, and particle packing; see References 4 and 19 for details.

Due to shearing, vigorous particle collisions may occur in the active region. The instantaneous velocity of a particle can be considered to consist of a mean velocity plus a fluctuation component. It is this fluctuation component that leads to the diffusive-like mass transfer. According to the kinetic theory, D_y can be given as:

$$D_y = \frac{1}{2}\lambda \cdot v' \tag{14b}$$

where v' is the fluctuation velocity in y-direction and λ is the mean free path of particles. The mean free path of particles can be approximated by the mean distance separating adjacent particles (s) which can be in turn linked to the solid volumetric fraction (v_s):[20]

$$\lambda \approx s = \left(\sqrt[3]{\frac{\pi}{3v_s}} - 1 \right)d_{av} \tag{14c}$$

where d_{av} is the average particle size.

Due to the thin nature of the active layer, the y-direction fluctuation velocity is approximated by the following expression developed for uniform rectilinear shear flows:[20]

$$v' = l_p \frac{du}{dy} \tag{14d}$$

with l_p given by:

$$l_p = \left[\frac{(1+e_p)(0.05+0.08\mu_p)d_{av}^2(1+s/d_{av})}{\dfrac{3}{2}\dfrac{C_D\rho_f}{(d_{av}/s)\rho_p} + \dfrac{1}{8}(1-e_p^2) + \dfrac{1}{2\pi}\mu_p(1+e_p) - \dfrac{1}{8}\mu_p^2(1+e_p)^2} \right]^{1/2} \tag{14e}$$

where C_D is the drag coefficient due to interstitial fluid, e_p is the restitution coefficient of particles, ρ_f is the density of interstitial fluid, ρ_p is the material density of particles, and μ_p is the kinetic friction coefficient.

Equations 14a and 14b apply to the active region. In the passive region, D_y and v_p can be taken as zero if the ratio of small to large particles is larger than ~0.12.

Solution to Equations 12–14 gives the concentration distribution of small particles. Figure 10 compares the modeling and experimental results, where C and O denote concentration and occupancy, respectively, subscripts f and L stand for small and large particles, and m denotes the maximum in a given x position. It can be seen that reasonably good agreement has been achieved.[4]

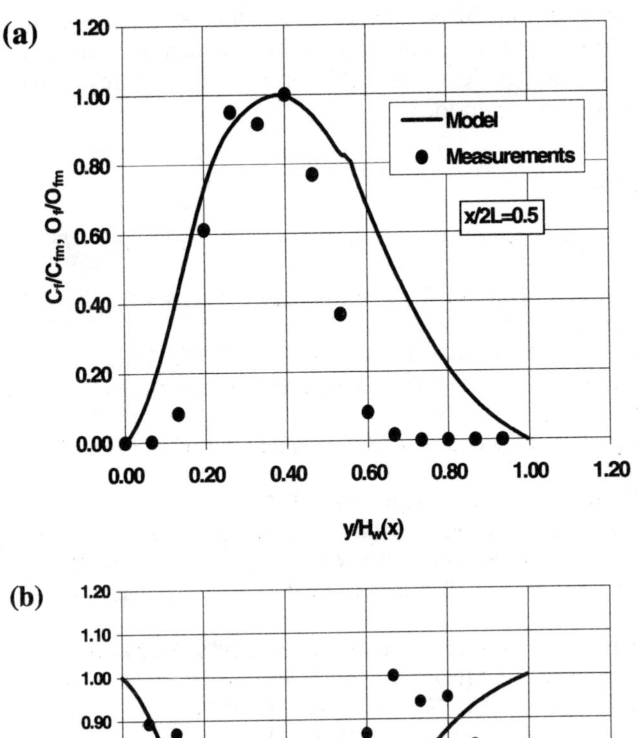

Figure 10   *Comparison between modelling and experiments – 240mm drum, 25% fill by volume, 30% 1.5mm/70% 3mm glass beads, 9.55 rpm: (a) 1.5mm particles; (b) 3mm particles*

5   Concluding Remarks

Solids motion in rotating drums exhibits both discrete and continuous characteristics. As a result, many fascinating features have been observed. However, the underlying physics of these features has not been fully understood despite the considerable effort that has been made by both engineering and physics communities in the past century. Examples include transition between different modes and segregation. Most reported work has been focused on free-flowing particles in drums operated in or close to the rolling mode. More work is clearly needed on cohesive powders and multi-component mixtures in drums operated in other than the rolling modes.

References

1. H. Henein, J. K. Brimacombe and A. P. Watkinson, 1983, *Met. Trans. B*, **14B**, 191–205.
2. D. J. Parker, A. E. Dijkstra, T. W. Martin and J. P. K. Seville, *Chem. Eng. Sci.*, 1997, **52**, 2011–2022.
3. A. A. Boateng and P. V. Barr, *J. Fluid Mech.*, 1997, **330**, 233–249.
4. Y. L. Ding, R. N. Forster, J. P. K. Seville and D. J. Parker, *Int. J. Multiphase Flow*, 2002, **28**, 635–663.
5. Y. L. Ding, R. N. Forster, J. P. K. Seville and D. J. Parker, *Chem. Eng. Sci.*, 2001, **56**, 1769–1780.
6. J. C. Williams, *Powder Technology*, 1976, **15**, 245–251.
7. J. Bridgwater, W. S. Foo and D. J. Stephens, *Powder Technology*, 1985, **41**, 147–158.
8. M. B. Donald and B. Roseman, *Brit. Chem. Eng.*, 1962, **7**, 749–753.
9. H. Henein, *Rotary Kiln Technology*, World Cement Publication, London, 34–42, 1987.
10. B. L. Pollard and H. Henein, *Can. Met. Quart.*, 1989, **28**, 29–40.
11. C. Wightman and F. J. Muzzio, *Powder Technology*, 1998, **98**, 125–134.
12. M. Nakagawa, *Chem. Eng. Sci.*, 1994, **49**, 2540–2544.
13. S. B. Savage, *J. Fluid. Mech.*, 1992, **241**, 109–123.
14. W. C. Saeman, *Chem. Eng Prog.*, 1951, **47**, 508–514.
15. G. E. McTait, PhD Thesis, University of Cambridge, 1998.
16. T. Kohav, J. T. Richardson and D. Luss, *AIChE J.*, 1995, **41**, 2465–2475.
17. S. B. Savage and D. J. Jeffrey, *J. Fluid Mech.*, 1981, **110**, 255–272.
18. C. K. K. Lun, S. B. Savage, D. J. Jeffrey and N. Chepurniy, *J. Fluid Mech.*, 1984, **140**, 223–256.
19. S. B. Savage and C. K. K. Lun, *J. Fluid Mech.*, 1988, **189**, 311–335.
20. H. Shen and N. L. Ackermann, *J. Eng. Mech.*, 1982, **108**, 748–763.

Subject Index